T0350902

Multiwavelength Optical LANs

Multiwavelength Optical LANs

G.I. Papadimitriou
Aristotle University, Greece

P.A. Tsimoulas
Aristotle University, Greece

M.S. Obaidat
Monmouth University, USA

A.S. Pomportsis
Aristotle University, Greece

WILEY

Telephone (+44) 1243 779777

Email (for orders and customer service enquiries): cs-books@wiley.co.uk
Visit our Home Page on www.wileyeurope.com or www.wiley.com

Other Wiley Editorial Offices

John Wiley & Sons Inc., 111 River Street, Hoboken, NJ 07030, USA

Jossey-Bass, 989 Market Street, San Francisco, CA 94103-1741, USA

Wiley-VCH Verlag GmbH, Boschstr. 12, D-69469 Weinheim, Germany

John Wiley & Sons Australia Ltd, 33 Park Road, Milton, Queensland 4064, Australia

John Wiley & Sons (Asia) Pte Ltd, 2 Clementi Loop #02-01, Jin Xing Distripark, Singapore 129809

John Wiley & Sons Canada Ltd, 22 Worcester Road, Etobicoke, Ontario, Canada M9W 1L1

Wiley also publishes its books in a variety of electronic formats. Some content that appears in print may not be available in electronic books.

Library of Congress Cataloging-in-Publication Data

Multiwavelength optical LANs / G.I. Papadimitriou ... [et al.].
p. cm.
Includes bibliographical references and index.
ISBN 0-470-85108-2 (alk. paper)
1. Local area networks (Computer networks) 2. Optical communications. 3. Wavelength divison multiplying. I. Papadimitriou, G. I. (Georgios I.), 1996–
TK5105.7.M85 2003
004.6′8–dc22 2003057191

British Library Cataloguing in Publication Data

A catalogue record for this book is available from the British Library

ISBN 0-470-85108-2

Typeset in 10/12pt. Times by TechBooks, New Delhi, India
Printed and bound in Great Britain by TJ International Ltd, Padstow, Cornwall, UK
This book is printed on acid-free paper responsibly manufactured from sustainable forestry in which at least two trees are planted for each one used for paper production.

To my children Zoi and Ilias
(Georgios I. Papadimitriou)

To my parents Antonios and Ioanna and my brother John
(Paraskevas A. Tsimoulas)

To Balqies: for her unwavering love, support and encouragement
(Mohammad S. Obaidat)

To my parents, to my wife Zoi and to our children, Sergios and George
(Andreas S. Pomportsis)

Contents

Preface

The existing network infrastructure in a world-wide scale is at present largely based on copper cable and optical fibre carrying a single wavelength. Single-wavelength (or *first generation*) optical networking has allowed transmissions at higher bit rates over longer distances in comparison to copper cable-based networking. Even so, the existing infrastructure of today overall cannot satisfy the ever-increasing demand for communication bandwidth and cannot support new Internet services and applications with increased bandwidth requirements. This is because electronic devices can only handle bit rates up to a few Gb/s and therefore exploit only a small fraction of the available bandwidth in an optical fibre. On the other hand, laying new fibre is generally not a practical solution.

Wavelength-Division Multiplexing (*WDM*) is probably the most powerful technique to unlock the enormous bandwidth in optical fibre and thus overcome the electronic bottleneck without laying new fibre. WDM implies dividing the entire optical bandwidth of a fibre into multiple non-interfering wavelengths serving as channels that can be independently and simultaneously accessed by electronic devices at reasonable—for electronic circuitry—speeds. Accordingly, *multiwavelength optical networking* is the most promising technology for meeting both our present and future information networking needs; the corresponding WDM-based networks belong to the so-called *second generation of optical networks*.

Multiwavelength Optical LANs, G.I. Papadimitriou, P.A. Tsimoulas, M.S. Obaidat and A.S. Pomportsis.
 ISBN 0-470-85108-2.

After many years of research, conferences, workshops, a large amount of published work in relevant journals and a number of experimental prototypes and testbeds, we have recently reached a point where multiwavelength optical networks have eventually broken through (out of the laboratories) into commercial reality. A constantly growing number of long-haul network operators have adopted WDM solutions to their networks in an effort to increase their capacity and be able to respond to the high demand for broadband services. It is noteworthy for example that in the last couple of years the North American WDM transport market for optical backbone networks has grown by at about 50% (in billion dollars).

This book, however, focuses on the application of multiwavelength optical networking in the *local area*. Even though most of the early research studies on optical networks were centred on the local area too, it is a fact that the (hitherto) high cost of applying optical WDM networking techniques in local area networks (LANs) has been the greatest obstacle to their commercial viability. The necessary optical components for such implementations account for a significantly higher cost as compared to their electronic counterparts. Consequently, thus far, it is hard for a corporation to justify the cost of using a multiwavelength optical LAN when it can cover its needs quite well and much more economically with more established options, like Gigabit Ethernet, for example. The problem of cost-effectiveness remains to some extent for large lightwave networks as well.

On the other hand though, the underlying optical component technology is still experiencing a phase of rapid progress with the objective of achieving both further technical developments and cost levels comparable to traditional electronic technology. It would not be an overstatement to say that we are still at the beginning of the learning curve for much of the enabling optical technology, and many of the limitations it imposes, both technical and economical, are expected to be overcome in the near future. After all, as far as the main obstacle for multiwavelength optical LANs is concerned, i.e. cost, and according to the observation made by Paul E. Green, Jr. in reference [12] of Chapter 1, there is nothing in the law of physics that says optical components such as lasers, multiplexers, switches and so forth, have to be so expensive. Taking all these into account, it would be normally expected that optical WDM will develop into a viable alternative for implementing high-performance LANs sooner or later.

Thus, the main purpose of this book is to investigate the major architectural, topological and protocol issues regarding multiwavelength optical LANs today. Considering the previous discussion about constant advances on optical component technology that make WDM optical LANs all the more feasible for wide commercial deployment, the book investigates thoroughly the crucial latter topic, i.e. the *Media-Access Control* (*MAC*) *protocols* that should be used. A noteworthy part of the vast literature on such protocols is overviewed after some general distinguishing key protocol characteristics are provided. Furthermore, a recent significant class of promising protocols whose operation is based on network feedback information comprises a focal point; in this way, these *adaptive protocols* achieve an overall higher performance in comparison to many other non-adaptive schemes.

The main idea, i.e. the essence of each protocol, is described neither too briefly (making further study necessary as in most overviews found in other books and journal articles) nor including all the details that may be tiresome for the reader. Moreover, the survey of protocols provided in this book is more complete than most other reports, since sample schemes that are representative of all the key protocol classes are presented.

Hence, we hope the book serves as a valuable, up-to-date and concise reference item for students, researchers, network designers and network operators, who are interested in multiwavelength optical LANs. The book intends to shortly introduce these people to the spirit of most protocols and sharpen their mind to easily understand others and possibly design new ones. An additional objective of the book is to provide an aid-back, yet complete and comprehensive, introduction to more general aspects of optical networking, including first generation optical networks, other classes of second generation optical networks (besides multiwavelength optical LANs) and the underlying enabling device technology. Thus, in addition to introducing people who wish to become familiar with optical networks to the field, this may also serve as an all-embracing overview helping people who already have some knowledge of optical networking to reorganize things in their mind.

OVERVIEW OF THE BOOK

Chapter 1 provides a general introduction to optical networks. This comprises solid background knowledge on optical networking in any case and also for the rest of the book. Slightly more emphasis is placed on the local area, as this is the main subject of the book. The main distinctive characteristics of multiwavelength optical LANs are discussed (except for MAC protocols which are investigated in-depth in the last two chapters). These include the physical topology, the logical topology and the structure of network nodes. Despite the slight emphasis on LANs, however, the introduction of Chapter 1 remains general and covers many topics on optical networks, such as first generation optical networks and other significant classes of second generation optical networks, as mentioned before.

The background is complemented well with a comprehensive description of the enabling device technology in Chapter 2, since, unlike other technologies, optical networking is closely related to the physical layer. Thus, optical network component technology should definitely be the focus of our attention as well.

The rest of the book focuses on the area of MAC protocols which is of major importance for multiwavelength optical LANs. Chapter 3 discusses some general distinguishing key protocol characteristics. After that, it presents a noteworthy part of the vast literature on MAC protocols as described before.

Finally, Chapter 4 introduces a recent significant class of promising protocols for multiwavelength optical LANs, whose operation is based on network feedback information. As mentioned before, these adaptive schemes achieve an overall higher performance in comparison to many other non-adaptive protocols.

We really hope you will enjoy reading this book and find it helpful.

Acknowledgements

We would like to thank Chrisula Papazoglou for willingly undertaking the task of drawing quite a lot of figures for Chapter 1 and Chapter 4 and, of course, we would also like to acknowledge the understanding and priceless support given to us by our families throughout the process of writing this book.

1

Introduction

We live in a society that has already started to experience the daily effects of the information revolution. The associated change in the way we communicate started to escalate in the last decade of the twentieth century with the proliferation of the Internet which, from a technical point of view, can be seen as a collection of Local Area Networks (LANs) with often different architectures and protocol stacks, glued together with a couple of simple yet highly flexible protocols: the connectionless Internet Protocol (IP) and the connection-oriented Transport Control Protocol (TCP) running on top of IP. As the human appetite for all kind of things grows stronger, it was normally expected that the demand for network capacity (or bandwidth) would follow the same rule. What was not expected, however, was the intensity of this phenomenon which has been unprecedented indeed.

The explosive and continuous spread of the Internet, even when it is just seen as an increase in the number of users, is one of the major causes of higher and higher demand for network capacity. However, the bandwidth required by each individual user has been increasing dramatically too. It is worth noting for example that in the middle 1990s there was an annual factor of eight-fold growth in bit rate required by each user, a really daunting growth rate for any kind of user demand [11]. Thus, it seems that users are constantly growing more in number and, at the same time, becoming more difficult to 'satisfy' in terms of provided capacity.

Multiwavelength Optical LANs, G.I. Papadimitriou, P.A. Tsimoulas, M.S. Obaidat and A.S. Pomportsis.
 ISBN 0-470-85108-2.

In an effort to find the reasoning behind this persistent increase of network capacity demand, we could examine the phenomenon from an economics perspective, i.e. considering the simple rule of the cost-demand relation that is valid for all markets (including the networking one). The high demand imperatively calls for ways of reducing the cost of bandwidth. The latter has been achieved throughout the years, first of all, by means of various kinds of advances in technology. For example considering the access part of the network, the more sophisticated way of exploiting common copper wires introduced by the Digital Subscriber Line (DSL) technology resulted in an overall reduced cost per unit of bandwidth provided to the residential customer. Bandwidth of the order of 1 Mb/s became a reality for such so-called 'broadband' access technologies; another example is cable access, providing similar bandwidth capabilities in places where cable television (CATV) infrastructure was already installed. Several other improvements in electronic technology have allowed in an analogous way business customers to start leasing faster lines than the usual T1 (= 1.54 Mb/s) and this has been especially true for large customers.

What has also been driving the cost of bandwidth to lower and lower levels, apart from the progress in technology, is the high competition that is present in such profitable markets such as the telecommunications services one. This competition has been particularly noticeable after the deregulation of the telecommunications market, which has become a reality in many places all over the world for quite some time now.

This reduction in the cost of bandwidth in turn encourages the development of new more demanding applications that not only take advantage of the offered bandwidth but often ask for even more. Since applications are the easy part (compared e.g. to achieving advances in technology), a state where the applications are one step ahead of the network capabilities in terms of capacity has been a commonplace throughout all the stages of networks evolution. Consequently, it could be said that the situation forms a kind of a vicious circle; high demand calls for ways of reducing bandwidth cost; if and when reduction in cost is achieved, new applications, and thus higher network capacity demand, come in. Although according to this discussion the demand for more capacity may seem interminable and uniformly high across all stages of networks evolution, practically the pressure for more capacity tends to get higher during certain periods and lower in others. For example, right after some major technological breakthrough fundamentally changes things on a wide scale, it is normal to expect that demand will be satisfied and thus significantly subdued at least for a while.

It could be observed that currently we find ourselves in a period where such a technological breakthrough along with its global scale effects is almost desperately expected. Indeed, many new applications that require high bandwidths have recently emerged and the classical implementations of our global mesh of communication networks, e.g. today's Internet and Asynchronous Transfer Mode (ATM) networks, do not have the potential (in capacity) to support them satisfactorily. A few samples just from the predictable part of a future set of applications requiring overall network capacities of the order of Tb/s are the following:

- telemedicine applications, such as high-resolution medical image archiving and retrieval;
- video on demand;
- video phone;
- video teleconferencing;
- Internet browsing (offering enhanced capabilities to the user as compared to now);
- multimedia document distribution;
- remote supercomputer visualization, i.e. connections of workstations to supercomputers giving users ability to manipulate full-motion colour graphics;
- interconnection of mainframe computers along with their peripherals.

Of course, various applications which cannot be predicted right now, as has happened many times in the history of data and computer communications, and which will have at least proportionate bandwidth demands, are very likely to turn up in the near future as well.

There has been strong evidence during the past 30 years or so, and especially over the last decade, that the classical electronic time-division multiplexing[1] (TDM) approaches realized in electronic circuitry of constantly more speed and complexity are not a long-term-solution; in fact, sometimes they are not much of a solution at all, e.g. for some of the forenamed applications that entail masses of visual information and very fast response time requirements. It is obvious for instance that the 1.5 per year increase of available electronic TDM technology in the mid-1990s could not manage to handle well the corresponding eight-fold annual growth in bit rate required by each user, that was mentioned before [11].

On top of all these, there are two other important subjects that have come up quite recently and have to be considered carefully by carriers in the way they build their networks. First, a significant change in the type of traffic that dominates the networks has been observed lately. The growth of data and voice traffic throughout the years is illustrated in Figure 1.1. Voice traffic does not dominate the networks any more though still growing each year, while data, on the other hand, has been growing at impressive rates especially during the past three to four years, overtaking voice by far. For example [24] states that two large carriers, AT&T and MCI

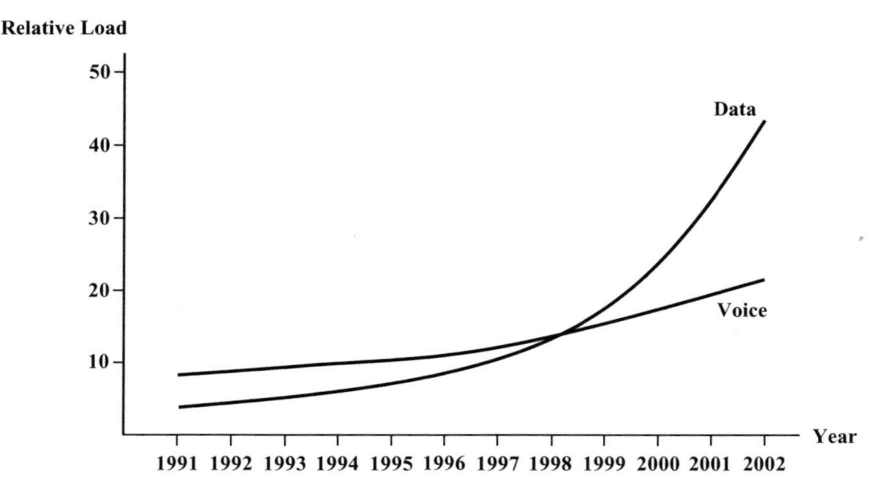

Figure 1.1 Growth of data and voice traffic throughout the years.

Worldcom, are experiencing annual data traffic growths up to 70% (in some of their markets), while their voice traffic is growing at around 15% each year. A straightforward implication of the described change in the traffic mix is the reduced efficiency of current networks, since these were initially designed to carry voice traffic. Therefore it is natural to expect that carriers will seek new ways of building their networks.

In addition, we could observe a recent significant change in the nature of services demanded by customers, which was further encouraged by the increased competition of the last years. To be exact, customers (including big ones that lease high-capacity lines) have recently started demanding connections that can be delivered very quickly, in hours or even minutes. The duration of these connections is sometimes required to be much shorter than usual, a couple of days for instance. Furthermore, these connections must be able to carry traffic in various formats, i.e. be as *transparent* as possible, and may originate in various different locations that can hardly be predicted by the carrier. All in all, it seems that we have moved to a point where networks must be able to provide us with the ability to access information *where* we need it, *when* we need it, and *in whatever format* we need it [24].

The majority of researchers and network engineers believe that *optical networking* is the most promising technology for coping with the aforementioned problems and meeting both our present and future information networking needs, mainly on account of the potentially limitless capabilities of optical fibres.

1.1 ADVANTAGES OF OPTICAL FIBRE AS A TRANSMISSION MEDIUM

Optical networks utilize optical signal transmissions for information exchange. Like all communication systems, an optical communication system comprises a source, a destination and a communications medium. According to the selection of the optical communications medium, optical networks can generally be divided in two categories, *guided* and *unguided* systems [27]. In unguided systems the optical beam that is transmitted from the source widens as it propagates into space resembling microwave transmission. The use of free space as the communications medium introduces interference problems in transmissions inside the Earth's atmosphere.[2] Guided systems use *optical fibre* as the communications medium and are thus also known as *fibre optics communication systems*. Since the majority of today's optical communication systems are fibre-based, the term 'optical networks' (or 'systems') will be used as a synonym of guided systems throughout this book, the same as it is almost always considered in the literature.

We have already stated that optical fibre is probably the key transmission medium in the direction of coping with the current high demands for capacity and deployment of new services, because it has several exceptional capabilities. Here we will review the main advantages of optical fibre (implying guided systems) over other media, when used as a transmission medium. (Advantages of optical fibre are also presented in detail in Chapter 2, since fibre is the basic element of enabling technologies discussed there). In particular, optical fibres offer the following:

- *Huge bandwidth.* Optical fibres offer radically higher bandwidths than alternative transmission media. The available bandwidth of a single optical fibre is up to several tens of THz (around 50 THz theoretically) and it is exploited by using mainly its two low-attenuation areas of about 200 nm centred at 1310 nm and 1550 nm respectively. These areas are illustrated in Figure 1.2. To gain an understanding of the potential of fibres for offering bandwidth, recall that in just the 1.5 micron (1 μm is equal to 10^{-6} m) band of each *single-mode fibre*, the available bandwidth is three orders of magnitude more than the entire usable radio-frequency (RF) bandwidth on Earth, which is about 25 GHz! It is also quite staggering to note that the channel capacity of a single fibre is more than the typical aggregate telephone traffic during a peak period in the United States [32]! A measure of the enormous offered bandwidth can also be realized if we consider the achievable bit rates over a single optical fibre. There is the potential of several tens of Tb/s while, for example, optical transmission systems combining WDM[3] and TDM techniques and arriving at aggregate bit rates of 1 Tb/s were commercially available a couple of years ago.

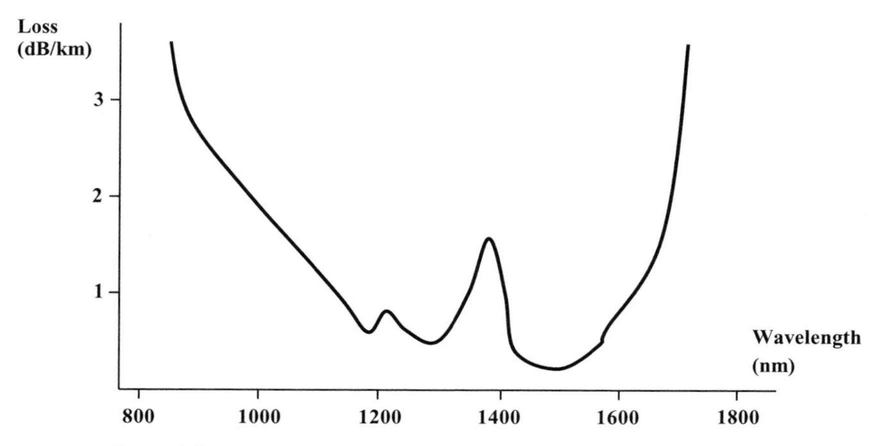

Figure 1.2 The two basic low-attenuation regions of an optical fibre.

- *Better signal qualities.* Since optical transmission is not affected by electromagnetic fields, optical fibres exhibit superior performance to other alternatives, such as copper wires for example. At a given distance the Bit Error Rate (BER) of a fibre-based transmission is significantly better than the BER of a copper or wireless-based transmission. Specifically, for optical telecommunication and computer data links the BER can typically be about 10^{-9} and 10^{-15} respectively, while copper-based links would not achieve bit error rates better than around 10^{-5} only. In other words, the *noise immunity* offered by optical fibre is better than the other transmission media, which suffer from considerable electromagnetic interference.
- *Low signal attenuation (and thus power requirements).* When optical signals are transmitted through optical fibres, their attenuation can be as low as 0.25 dB/km. This is also illustrated in Figure 1.2 where we can see that the loss (or attenuation), which is depicted on the vertical axis and measured in dB/km, can drop down to 0.25 dB/km for the 1550 nm wavelength band. Practically, such levels of loss would imply that an optical signal in this wavelength band could travel a distance of about 120 km before it needs amplification or regeneration. A total of 30 dB loss would correspond to this distance (since 120 km $\times$ 0.25 dB/km $=$ 30 dB) or, in simpler terms, the signal would suffer a loss by a factor of 1000. In the same figure the 1310 nm wavelength band with typical loss around 0.4 dB/km can also be distinguished; this has also been used in optical communications systems. The third low-loss band at 800 nm is not shown in the figure; this exhibits higher losses than the previously mentioned two (around 2.5 dB/km) and was used

especially in the first optical systems. Note that Figure 1.2 and the points made in this paragraph hold for *conventional silica-based fibre*. There are other types of optical fibres like the new *allwave fibre* [24] for example; this provides a more usable optical spectrum by eliminating the 'water-peak window' of conventional fibre located in the neighbourhood of 1385 nm (the region where loss becomes higher for a while, see Figure 1.2).

- *Easy deployment and maintenance.* A good quality optical fibre is sometimes less fragile than a copper-based link. Furthermore, optical fibres are not subject to corrosion which makes them less vulnerable to various environmental hazards. After all, optical fibres are very flexible, weigh less than copper wires and have less space requirements, i.e. in the space required by a single copper wire we can install more than just one fibre.
- *Better security.* Optical fibre provides a secure transmission medium since it is not possible to read or change optical signals without physical disruption. It is possible to break a fibre cable and insert a tap, but this would involve a temporary disruption. For many critical applications including military and e-commerce applications where security is of the utmost importance, optical fibres are preferred over e.g. copper transmission media which can be tapped from their electromagnetic fields.

1.2 BASIC MULTIPLEXING TECHNIQUES

We have already referred to two multiplexing techniques in previous sections, namely *Time-Division Multiplexing* (*TDM*) and *Wavelength-Division Multiplexing* (*WDM*). Here we will explain in some detail the notion and need of multiplexing. We will also discuss the aforementioned multiplexing techniques that can be used in optical networks and thus are of practical interest to us.

In order to understand the need for multiplexing, consider two communicating stations and the necessary data link connecting them. It is quite typical for the capacity of the link to be much higher than the one utilized by just the two stations. Such an inefficient use of the link would definitely be undesirable. Consequently, if we wanted to use the data link in a cost-effective way, it would be wise to *share* all of its capacity among several communicating stations. A common term for this sharing is *multiplexing*. Obviously, recalling the extraordinary bandwidth capabilities of optical fibre, as well as the fact that installing more fibres is quite costly, we can gain an understanding of how helpful multiplexing can be in the case of optical networking, i.e. when the above mentioned 'data link' is physically an optical fibre.

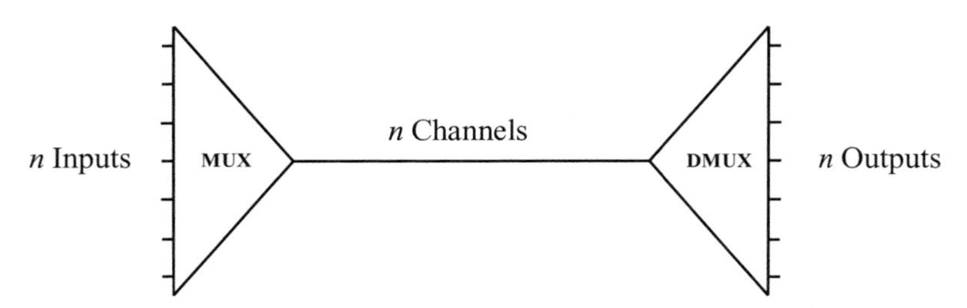

Figure 1.3 Generic form of multiplexing and demultiplexing.

Multiplexing is depicted in a simple and generic form in Figure 1.3. There are n inputs to the multiplexer (MUX) and, accordingly, n outputs from the demultiplexer (DMUX). The MUX combines the n input data streams into a single signal that is transmitted over the high-capacity data link. The single signal essentially 'contains' *n separate channels*, the exact form of which determines the specific type of multiplexing used. The DMUX demultiplexes (extracts) the data streams out of the signal and delivers them to the appropriate outputs.

Let us now examine two specific forms of multiplexing. These generally differ on whether the frequency or time dimension is 'sliced' in separate channels carrying the data from the various input streams.

In *Frequency-Division Multiplexing* (*FDM*) each data stream is modulated onto a different carrier frequency. The frequency spectrum is thus divided or 'sliced' in channels, each one intended to serve an individual input station. The channels must be appropriately separated by guard bands, i.e. unused portions of the spectrum, so that interference between them is avoided during the simultaneous transmission of their data over the single physical medium. FDM is shown in Figure 1.4 (a), where C_1, C_2, etc. stand for channel 1, channel 2 and so on. Observe that transmissions in channels are not interrupted in the time dimension, i.e. are simultaneous. FDM has been widely used in radio systems for many years and also has other applications such as in broadcast and cable television where the bandwidth of the coaxial cable is shared among different customers.

Wavelength-Division Multiplexing (*WDM*) is in essence exactly the same as FDM, but the term WDM has prevailed in cases where we consider division of the optical frequency spectrum in channels (or wavelengths). It could very well be called Optical FDM. These wavelengths must be kept sufficiently apart from each other to avoid interference in this case too. The great thing about WDM networking which explains the huge research and commercial interest in this technology, is its compliance with the limited (compared to the capabilities of a fibre) speed

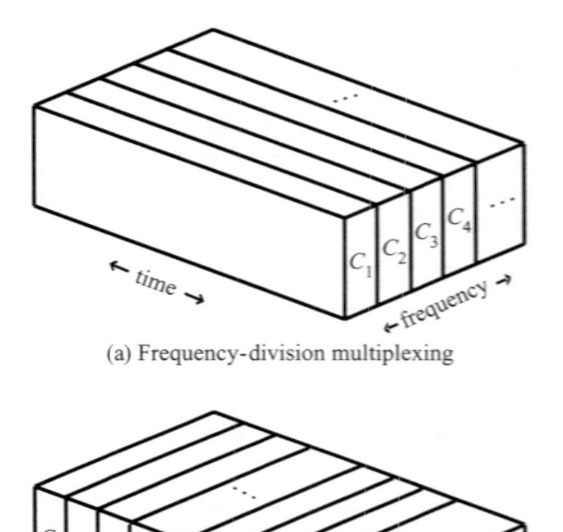

(a) Frequency-division multiplexing

(b) Time-division multiplexing

Figure 1.4 Generic illustration of (a) frequency-division multiplexing and (b) time-division multiplexing. C_1, C_2, C_3, etc. stand for channel 1, channel 2, channel 3, etc. respectively.

of the stations' electronic circuits. Practically, considering that the maximum rate at which an end-user[4] can access the network is limited by electronics to a few Gb/s, WDM offers an excellent way of exploiting the huge bandwidth of optical fibres by introducing concurrency among multiple users transmitting at ('electronically') feasible rates. Optical networks adopting the WDM technique are termed *multiwavelength (or multi-channel) optical networks*. They achieve a much more cost-effective way of using fibres than *single-channel optical networks*, since it is definitely wiser to use multiple '*virtual fibres*' inside a single fibre instead of wasting multiple different fibres. Several classes of WDM networks will be studied in another section. The transmission inside an optical fibre can be either unidirectional or bidirectional as depicted in Figure 1.5.

In *Time-Division Multiplexing (TDM)* the time dimension is 'sliced' instead, to form usable channels for different data streams. This is depicted in Figure 1.4(b). At a certain point in time the whole of the transmission medium capacity is occupied

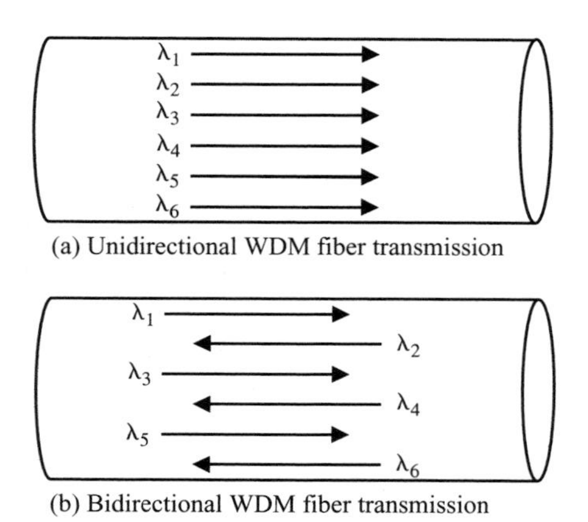

Figure 1.5 (a) Unidirectional and (b) Bidirectional WDM fibre transmission.

with serving a specific input stream; we can say that essentially it is a round robin use of all the available frequency. Practically, *time slots* are assigned *periodically* to inputs for transmitting a small portion of their signals. The small portions of the various signals are interleaved in time, i.e. combined in so-called TDM frames, and transmitted as a single high-speed signal over the data link. For example, 48 52 Mb/s input data streams may be time-division multiplexed into a single 2.5 Gb/s stream.

There are basically two different types of TDM. In *synchronous TDM* a frame consists of one complete cycle of time slots. Thus the number of slots in frame is equal to the number of inputs. Exactly the same time slot is allocated to each input data stream at all times, whether or not it is active, i.e. the corresponding input device has something to transmit. Time slot 1, for example, is assigned to input 1 alone and cannot be used by any other device. An example of how synchronous TDM works is shown in Figure 1.6(a).

In Figure 1.6(b) the same example for the second type of TDM is illustrated, namely for *asynchronous or statistical TDM*. Each slot in a frame is not dedicated to a fixed input device. Each slot contains an actual portion of an input data stream along with an index of the input it corresponds to. Thus, the number of slots in a frame does not have to be equal to the number of input devices and is usually smaller. More than one slot in a frame can be allocated for an input device. Statistical TDM maximizes utilization of the link, since in this case time slots are not allocated wastefully to input devices when they have nothing to send. It is also called *intelligent TDM* after this intelligent allocation of time slots to inputs.

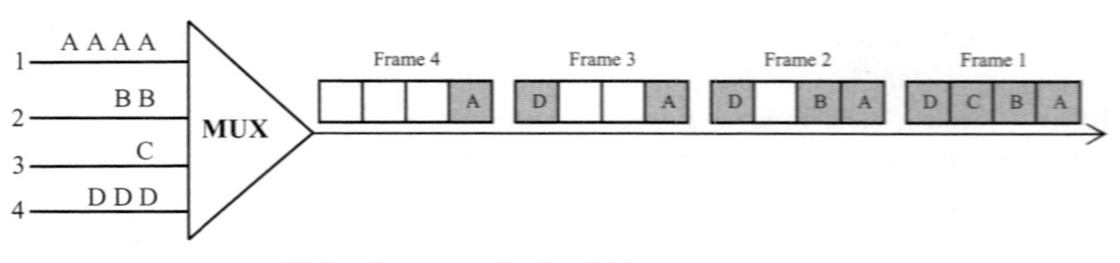

(a) Synchronous or fixed multiplexing

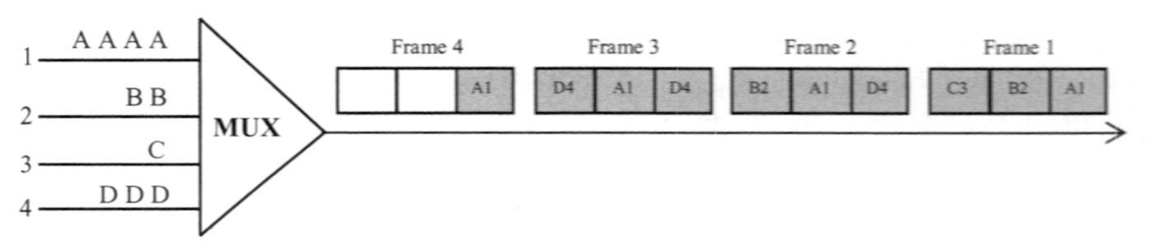

(b) Asynchronous or statistical multiplexing

Figure 1.6 Examples of (a) synchronous and (b) asynchronous or statistical time-division multiplexing.

Notice that statistical time-division multiplexing is very suitable for bursty traffic, such as Internet traffic for example. It is used on the Internet for multiplexing IP packets ('datagrams') from various data streams (users). It is likely that at any given time only some streams are active, while the remaining ones are idle. Thus, focusing on a single link between users, a smaller link bandwidth could serve more users than in the case of assuming all users are active simultaneously, such as synchronous TDM does. However, there may be times when more users are active simultaneously (transmitting IP packets) than the data link can handle. Some of the packets will therefore have to be buffered or even rejected if congestion is so high that buffers overflow. TCP runs on top of IP and one of its tasks is to ensure that these packets are retransmitted. Note also that IP packets may take different routes across the network; TCP also has to put them in the right sequence at the destination points. Thus, this *packet-switched* service offered by today's Internet is well known as *best-effort* service; the name implies that the network does its best every time to serve a specific packet, but without any guarantees concerning time of arrival at the destination or even the success of the transmission itself. Statistical TDM plays a central role in the implementation of this service by the network.

In a previous paragraph we referred to an example of time-division multiplexing several input data streams (48 52 Mb/s to be exact) into a single signal of aggregate 2.5 Gb/s bit rate. It should be noted that the highest transmission rate in similar commercially available electronic TDM systems has recently risen up to 40 Gb/s for shorter distances than the previous milestone of 10 Gb/s, yet it seems quite difficult to push electronics technology beyond this rate. As a result, electronic TDM turns out to be insufficient when higher speed networking requirements are taken into consideration.

Far more interesting for the optical transmission systems, though still needing time to mature (at least commercially), is *Optical Time Division Multiplexing* (*OTDM*), whereby there is an effort to perform the (de)multiplexing functions optically. The bit rates of the individual data streams are at such high levels (e.g. 10 Gb/s) that the multiplexing and demultiplexing operations are best performed in the optical domain. In a quite analogous way as in electronic TDM, OTDM can be fixed (resembling synchronous TDM) or statistical (similar to statistical TDM). OTDM networks appear fairly futuristic for today's technology in that there are still some serious problems that have to be overcome before their commercial deployment. Networks employing OTDM will be studied in some greater detail in a following section examining the various classes of optical networks.

It should be pointed out that the TDM and WDM techniques are complementary to each other. That is to say the application of one does not preclude the use of the other as well. In fact, when high-performance optical networking is considered, the best approach is to combine both techniques so as to maximize utilization and exploit the potential of optical fibres bandwidth in the most advantageous way. As has already been mentioned in a previous section, optical transmission systems combining WDM and TDM and arriving at aggregate bit rates of 1 Tb/s, were commercially available a couple of years ago; even higher bit rates are naturally expected in the near future.

Last of all, another interesting multiplexing technique that can be combined with WDM is so-called *Subcarrier Multiplexing* (*SCM*). According to this technique, within a certain wavelength there is another level of electronic FDM subdividing the channel bandwidth into many radio frequency channels, each at a different microwave frequency. A certain MAC protocol using this technique is described in Chapter 3 as an example.

1.3 EVOLUTION OF OPTICAL NETWORKING—MAJOR TECHNOLOGICAL MILESTONES

Now that the notions of TDM (electronic, optical) and WDM have been explained, and before proceeding to the analysis of the most important optical networks categories, we shall make a brief reference to the evolution of optical networking through some key technological advancements that were actually the major milestones in its progress. Note that certain devices mentioned here are parts of the enabling technology and will be examined more thoroughly in Chapter 2.

The idea of using a high-speed optical transmission system can be traced back to the time when laser was invented, i.e. around the late 1950s. Following that, in the 1960s, several experiments showed that *waveguides* are capable of transporting information encoded in light signals. However, optical fibre transmission (or *guided* transmission) really became practically possible when the first *low-loss optical fibre* was invented around the early 1970s. From about the late 1970s until the mid-1990s fibre transmission capacity roughly doubled each year due to the constant advances in optical fibre transmission systems technology. The advances witnessed primarily concerned the improvement of *optical transceivers* and the reduction of fibre loss.

The first type of fibre used was *multimode fibre*, in which light propagated in multiple *modes*, each travelling over a different path and at a different velocity. The transmitters were basically *Light-Emitting Diodes* (*LEDs*) or *Multi-Longitudinal Mode* (*MLM*) *Fabry-Perot lasers* in two of the three low-loss wavelength bands of silica-based optical fibre, 0.8 μm and 1.31 μm (recall that the third one exhibits the lowest loss and is located at 1.55 μm, see Figure 1.2).

In the early 1980s *single-mode fibre* provided a great improvement by eliminating *modal dispersion*, the main deficiency of its predecessor multimode fibre. As its name suggests, within single mode fibre light is travelling in just a single mode and thus its core diameter is not required to be more than 10 μm (actually it can drop down to just 8 μm). The transmitters, theretofore MLM lasers, were transmitting in the 1.31 μm waveband, which demonstrates a higher loss than the 1.55 μm band, as has already been mentioned above.

Attempts to take advantage of the low-loss 1.55 μm band started in the late 1980s and eventually led to the substitution of the broadband transmitters used until then by the narrowband *distributed feedback* (*DFB*) *laser*. The latter managed to eliminate the main drawback of transmitting in the 1.55 μm region, namely *chromatic dispersion*. Chromatic dispersion is shown schematically in Figure 1.7

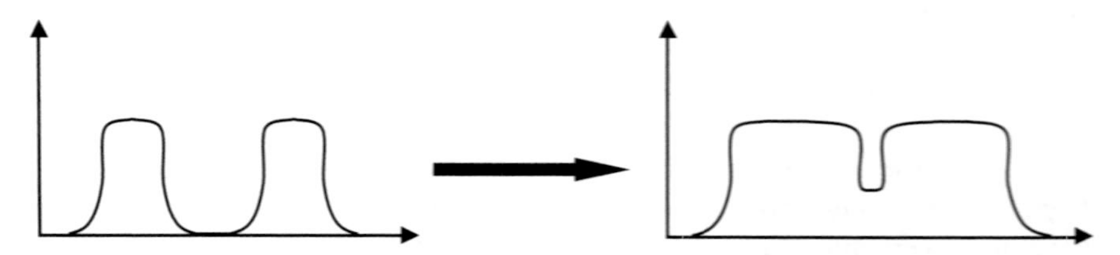

Figure 1.7 The effect of chromatic dispersion on a transmitted optical signal.

and comprises a weathering of the optical signal after travelling a certain distance due to the different velocities of its different spectral (frequency) components. As optical-fibre deployment increased, no standards existed to control how network elements should format the optical signal. Therefore, there was a big push for optical standards, in order to make it easier for network providers to interconnect equipment from different vendors and avoid bit errors of asynchronous (thus far) transmission. Thus, during the same period and in parallel with the introduction of DFB lasers operating at the 1.55 μm low-loss window, there were some standardization efforts concerning optical fibre transmission systems, which finally evolved into the *Synchronous Digital Hierarchy* (*SDH*) standard in Europe and Asia, and the quite analogous *Synchronous Optical Network* (*SONET*) standard in North America. The wide deployment of SDH and SONET networks which started from then on resulted in the situation today where the majority of the telecommunications infrastructure core in a worldwide range is of that kind.

As an aside, we just mention here that SONET and SDH networks belong to what we usually refer to as *first generation optical networks*; these will be examined in the following section. The main characteristic of networks of that kind is that optical fibre is used purely as a transmission medium just replacing the copper cable. Note that they could not even be termed 'optical' networks (at least in a strict sense), since electronics take over all the necessary switching, routing and bit processing procedures at the ends of the physical fibre link connections.

Moreover in the late 1980s, we witnessed the deployment of some enterprise data communication metropolitan-area networks (MANs) clearly belonging to the same first generation of optical networks, namely the 100 Mb/s *Fibre Distributed Data Interface* (*FDDI*) and the 200 Mb/s IBM's *Enterprise Serial Connection* (*ESCON*). The *Distributed-Queuing Double Bus* (*DQDB*) also belongs to the same category.

It is worth noting that until the late 1980s light signals transmission in long distances required the use of *electronic regenerators* after several kilometres. The main task of these expensive devices was to convert the attenuated light signal into electrical form and then to regenerate it back and transmit it as a new optical signal, a copy of the one that arrived in the first place. The first optical trans-Atlantic cable using such regenerators was laid in 1988.

A major progress of great significance, realized in the late 1980s and early 1990s, concerned the replacement of these electronic regenerators by the *Erbium-Doped Fibre Amplifiers* (*EDFAs*) for coping with the light signal attenuation after travelling a certain distance inside the optical fibre. EDFAs are called thus, because amplification in the optical domain is achieved by doping a small strand of fibre with the rare earth metal Erbium. Except for being cheaper, EDFAs have the remarkable virtue of being sufficiently broadband to be able to amplify signals at different wavelengths simultaneously. On the basis of this property, a natural and straightforward consequence of EDFAs' development was a great boost to Wavelength Division Multiplexing (WDM) systems, so far primarily regarding point-to-point links where full-duplex transmission was achieved by using two wavelengths, one in each of the 1.31 μm and 1.55 μm low-loss windows. Note that optical amplification performed by EDFAs provides total *data transparency* and is known as *1R (or optical) regeneration*. The electronic regeneration mentioned in the previous paragraph could be *3R regeneration*, meaning that regeneration, reshaping and reclocking of the signal are carried out (transparency is lost in this approach used in networks like SDH) or it could be of another type coming somewhere in between in terms of transparency, involving regeneration and reshaping only (without retiming), and thus called *2R regeneration*. 1R regeneration, that is to say optical amplifier technology, is continuously improving in terms of both lower noise and flatter gains; this is essential for achieving practical (so-called) *dense wavelength-division multiplexed systems (DWDM)*.[5]

OTDM and WDM networks offering more than just optical fibre point-to-point links, namely performing several routing and switching functions in the optical domain, are usually referred to as *second generation optical networks*. There are several classes of such networks which will be explored in a following section. From the early 1990s on, there has been an increasing interest in such networks, with WDM ones being the most popular due to their compliance with electronics speeds. Indeed, apart from a noteworthy activity on WDM testbeds that had begun as early as the mid-1980s, continued during the 1990s and has not stopped yet, what we most remarkably observe (just in the last couple of years) is a sudden growing interest in the commercial deployment of WDM networking. In this context, many new ventures are ready to supply multi-wavelength transmission and switching products that allow real optical networking to be implemented.

1.4 FIRST GENERATION OPTICAL NETWORKS

In this section we shall refer to the main characteristics of first generation optical networks, and comment on the utilization of optical fibre potential in the context of such networks. Thus, we can gain an understanding of how optical fibres have been used for many years in the existing infrastructure whereupon next generation optical network technologies are to be applied gradually. A detailed analysis of first generation optical networks (e.g. SDH or FDDI) is beyond the scope of this introduction.

Enormous quantities of optical fibre have been going into the ground and also under the sea worldwide for something like 25 years now and especially during the past decade. It is quite staggering that in the United States alone, as of the end of 1998, around 19.6 million fibre miles had been deployed; 16 million fibre miles had been installed by local-exchange carriers and the rest by inter-exchange ones![6] [29]. The exclusive task and main reason of installing for all these huge amounts of fibre was to serve as a high capacity transmission medium substituting for copper wires in the existing network infrastructures (wherever possible legally, economically, and so on). Thus, we have reached a point where a great part of the public telecommunication networks worldwide has followed this trend to a large extent. We should observe, however, that these networks in fact remain purely electronic besides the optical transmission medium. That is to say, at the ends of the physical fibre link connections, electronics take over all the necessary switching, routing and bit processing procedures. End users are transmitting and receiving electrical signals which are only converted to and from optical signals for the sake of transmission between point-to-point links. Additionally, all the intermediate nodes can similarly process only electrical signals and, thus, such conversions from the electronic to the optical domain and back, are carried out unavoidably after every hop from source to destination. Networks of this type are called *first generation optical networks* (though the term 'optical' is quite questionable considering the above observations).

1.4.1 SDH/SONET NETWORKS

Probably the most representative example of first generation optical networks are *SDH* networks in Europe and Japan, and the analogous *SONET* networks in North America. SDH (Synchronous Digital Hierarchy) and SONET (Synchronous Optical Network) are standards developed by ITU[7] (Europe, Japan) and ANSI (North America), respectively, around the mid to late 1980s and defining, first of all, a hierarchy of digital data rates along with an efficient multiplexing scheme to combine multiple lower speed signals into higher speed (in the hierarchy) ones.

The lower speed signals could conform to the SDH/SONET standards or could very well agree with the *plesiochronous* (*asynchronous*) *digital hierarchy* (*PDH*) which was the basis of the existing (before SDH) infrastructure, e.g. the ITU's E1 = 2.048 Mb/s and E3 = 34.368 Mb/s or the AT&T's T1 = 1.54 Mb/s and T3 = 44.736 Mb/s could be some of these PDH lower speed signals. Observe that the PDH rates proposed by ITU (Europe) were different from the AT&T ones (USA, Canada and Japan), and, consequently, some of this disorder was inevitably

Table 1.1 SDH/SONET transmission rates.

SDH Name	SONET Name	Optical Equivalent	Line Rate (Mb/s)
None	STS-1	OC-1	51.84
STM-1	STS-3	OC-3	155.52
STM-4	STS-12	OC-12	622.08
STM-16	STS-48	OC-48	2 488.32
STM-64	STS-192	OC-192	9 953.28
STM-256	STS-768	OC-768	39 813.12

spread to SDH and SONET as well. The lower speed signals could also be ATM traffic streams at various bit rates.

Furthermore, the SDH and SONET standards specified the types of network elements required, network architectures that vendors could implement, and the function that each node must perform. All these ensured at least a basic level of interoperability between different vendors' optical equipment. From the very beginning and for the reason described in the previous paragraph there was some conflict between ANSI and ITU, but eventually a compromise was reached, according to which SONET bit rates are a subset of SDH rates. The hierarchy of bit-rates for both standards is presented in Table 1.1.

For SDH the lowest rate is 155.52 Mb/s, which is designated as STM-1 (Synchronous Transport Module—Level 1). This comprises the basis for the various STM-N signals that are defined as multiples of the basic 155.52 Mb/s rate (N times 155.52 Mb/s for STM-N where Table 1.1 shows the most typical values for N). SONET defines a lower rate as a basis, namely STS-1 (Synchronous Transport Signal—Level 1) which corresponds to 51.84 Mb/s. Multiple STS-1 signals can be combined by byte interleaving to form an STS-N signal in the same way as STM-N is formed in SDH. The compromise mentioned before between SDH and SONET is in fact the relation between the basic signals, i.e. STM-1 being exactly equal to three times STS-1. Higher speed SONET signals are multiples of three of the basic STS-1 signal and thus have an equivalent STM-N signal each, so compatibility between SONET and SDH is achieved. Note that since end users are transmitting and receiving electrical signals, by 'STS-N' for example we refer to the electrical signal. The optical equivalent of an STS-N signal, that is to say the one coming up from the necessary conversion to optical form for transmission through the optical fibre, is represented by OC-N (Optical Carrier—Level N).

The basic network elements for SONET (analogous equipment exists for SDH), besides regenerators that are used after proper distances, are:

- *Terminal Multiplexers* (*TMs*) or *Line Terminals* (*LTs*);
- *Add/Drop Multiplexers* (*ADMs*);
- *Digital Crossconnect Systems* (*DCSs*).

Terminal Multiplexers (*TMs*) perform the multiplexing of low speed streams (PDH like T1 or T3, or OC-M low speed optical signals) into a single high speed optical signal OC-N. Apparently, they include electrical-optical signal conversion when necessary. They also perform the reciprocal function of demultiplexing at the other end. *Add/Drop Multiplexers* (*ADMs*) are used to add one or more low speed streams to a high speed stream (Add function) or select one or more low speed streams out of a high speed stream (Drop function); the remaining (not dropped) part of the traffic is allowed to pass through to other (possibly ADM) nodes of the network. *Digital Crossconnect Systems* (*DCSs*) can cross-connect (switch) a large number of individual streams, both PDH and SONET. They can also perform the ADM functions.

Figure 1.8(a)–(c) shows the basic SONET network configurations, namely a *point-to-point*, a *ring* and a *linear* network respectively. Ring is the most common topology for SDH/SONET networks both in the access and the backbone parts, due to its high restoration capability in the case of failures and its simplicity. ADMs are interconnected in ring topologies with two-fibre links; one is the *working fibre* and the other is the *protection fibre* offering a high degree of availability in case some kind of failure occurs in the working fibre. Backbone rings may involve, for example, a 2.5 Gb/s (OC-48 see Table 1.1) signal, while an example rate for access rings could be 622 MB/s (OC-12). In Figure 1.8(d) we can see an example of using DCS as a hub interconnecting two simple ring networks and being part of a linear network at the same time. Note that network elements performing similar operations, and thus named quite analogously, are used in some (second generation) WDM networks, as we will see. There are many other aspects of SDH/SONET networks that are considered beyond the scope of this introduction (for details please refer e.g. to [13]).

1.4.2 EXAMPLES OF OTHER FIRST GENERATION OPTICAL NETWORKS

At the same time as SDH/SONET networks prevail in public telecommunication carriers, there are some enterprise local or metropolitan area networks that clearly belong to the first generation of optical networks as well. Specifically, they use optical fibre as a substitute of copper wires too, so that higher speeds can be achieved, while conversion of the transmitted signals to electrical form takes place at every intermediate or end node just as in SDH/SONET.

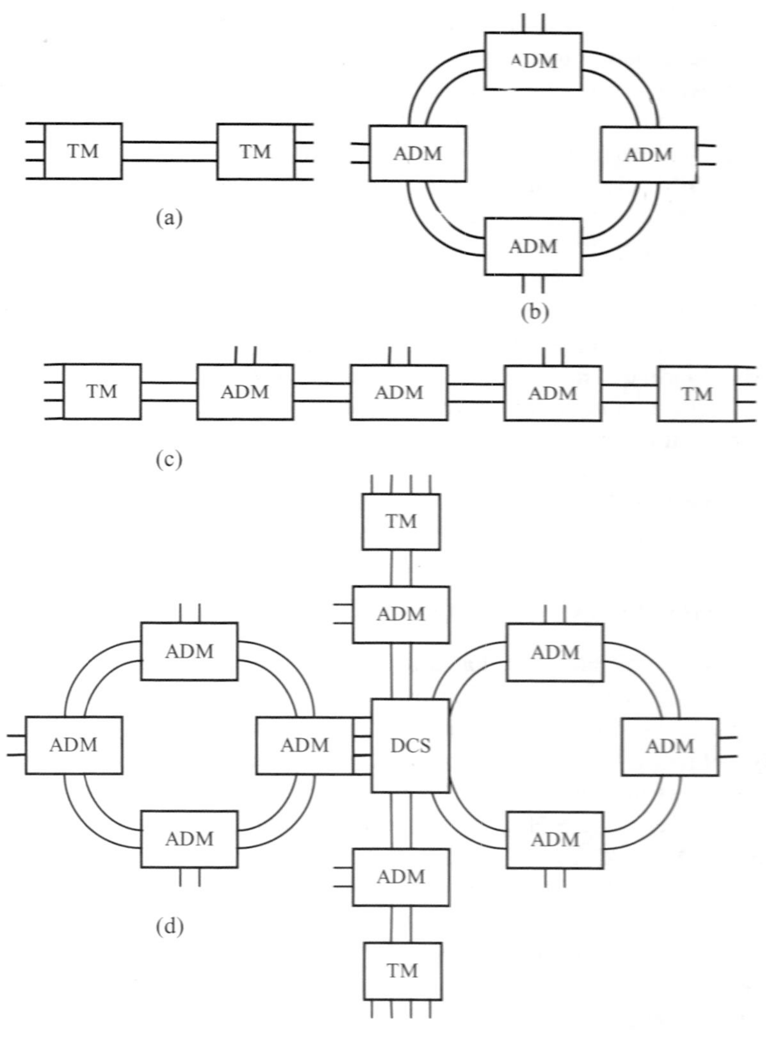

Figure 1.8 (a) Point-to-point (b) Ring (c) Linear SONET network configurations and (d) an example of using a DCS as a hub interconnecting simple networks.

Some examples have been already mentioned in the section describing the evolution and major technological milestones of optical networking and include, first of all, the *FDDI* (*Fibre Distributed Data Interface*) networks. These networks are used in metropolitan areas improving the performance of the well-known token ring networks by allowing a higher bit rate equal to 100 Mb/s, as a result of the superior transmission medium used (i.e. fibre). The dual-fibre ring topology of FDDI networks provides high and instant restoration capability in the event of some failure occurring in a certain part of the ring.

In order to interconnect efficiently mainframe computers along with their peripherals, IBM developed *ESCON* (*Enterprise Serial Connection*) networks. Each mainframe can connect to hundreds of other mainframes and/or peripherals through channels operating at 200 Mb/s each, while the overall network equipment may spread up to a metropolitan range.

Other networks of this type include the *Fibre Channel* network appropriate for the same application like ESCON, but reaching higher speeds per channel (up to 800 Mb/s) and the *Distributed-Queuing Double Bus* (*DQDB*). The latter metropolitan or local area fibre-based network operates in a broadcast mode by using two buses, each of which transmits fixed size cells in opposite directions. Each bus can operate at hundreds of megabits per second. In a DQDB network nodes may also be interconnected in a 'looped bus' (essentially a ring) configuration, apart from the typical 'open bus' configuration whereby the two edge nodes stay unconnected.

1.4.3 REMARKS ON THE UTILIZATION OF FIBRE

Comparing the bit rates mentioned in the first generation optical networks described above with the theoretical bandwidth potentiality of fibre explained in Section 1.1, one can without doubt conclude that optical fibres have been used really inefficiently for many years. Actually, fibres have been underused by about four orders of magnitude [12]. This roughly means that you get, for example, 2.5 Gb/s bit rate (in an SDH network), while the actual potentiality of the medium could provide e.g. 25 Tb/s. This is still the case to a large extent today, even though second generation WDM networks have started steadily penetrating the backbone for a couple of years now and an outburst is expected for the years to come. For one thing, it comprises a great challenge to mine all this bandwidth resource, if we just think that optical fibres are widely deployed in all kinds of telecommunication networks all over the world.

But why hasn't all this fibre been used in the most productive way? This would be a most logical question. First and foremost, fibres, as a rule, have not been interconnected in a proper architecture that would make the most of their exciting characteristics. Moreover and from an economics point of view, it is at least as important that investors in new optical technologies supporting new types of services[8] have rather been too cautious, since for many years they could not see in reality any market for these services (as these services didn't really exist in the first place yet). It is reasonable for any carrier to wait for tangible evidence of revenue, before spending money on large investments, however promising the new technology might sound in theory. In addition to the above two principal reasons, there have been various other legal, political and technological issues that have hindered the efficient exploitation of the bandwidth resources inside installed fibres. Concerning technology, for example, the residential access part of the network (the so-called 'last mile') has been rather primitive as compared to a prospective ultra high-speed fibre backbone.

In Section 1.1 we presented several benefits of optical fibres. All these are at the same time advantages of optical networks over other types of networks inherited in a straightforward way *just by using optical fibre as the communication medium*. If fibres were used in some more 'efficient way', even more benefits would be gained for optical networks, such as *reduced cost*, the ability to offer *wavelength (instead of fibre) services* and *improved restoration capabilities*. This is the main task of second generation optical networks which, along with their classes, are examined next.

1.5 SECOND GENERATION OPTICAL NETWORKS—MAIN CLASSES

When we talk about optical networks we basically mean second generation optical networks. These networks are something more than a set of fibres terminated by electronic switches, in that some of the routing, switching and intelligence is moving into the optical domain, giving rise to the utilization of optical fibres.

It should be noted, however, that the term 'optical networks' does not necessarily imply purely optical networks. Several operations are better performed by electronics, at least in the current state-of-the-art. These are called *nonlinear* operations and include, for example, certain functions requiring special intelligence, like packet headers processing, network control and management. It seems that electronics are slow but clever, while optics are fast yet a bit dumb. So, proper cooperation is needed for practical implementation of optical networks. Electronics are also necessary at the end-nodes for the conversion of the signal to optical form so that it can enter the optical network. Furthermore, electronic regeneration may be needed in some networks especially along lengthy optical routes.

Nevertheless, in second generation optical networks, electronics are relieved from some of the serious amount of load as far as *linear* operations are concerned, in contrast to first generation ones. In a network like the ring SONET of Figure 1.8(b), every node must handle electronically not only the data destined for it, but also all the data intended for other nodes and trying just to pass through. A key idea

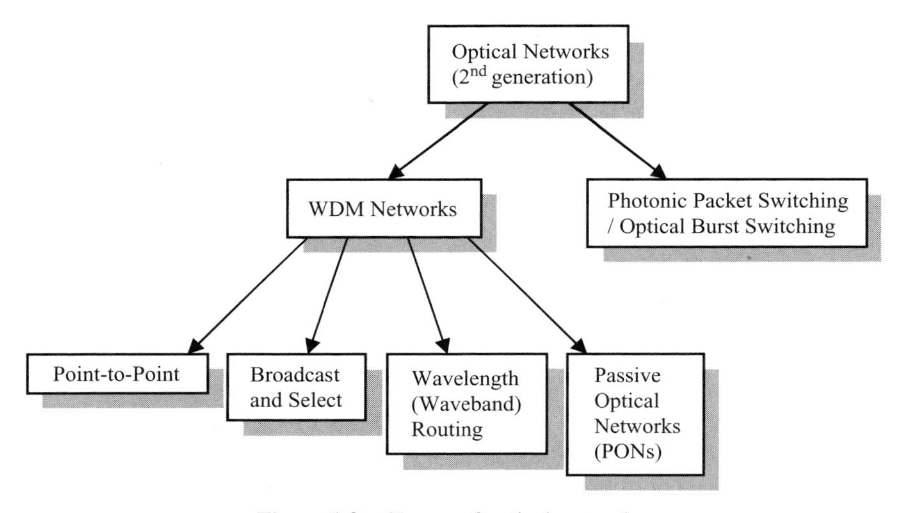

Figure 1.9 Classes of optical networks.

leading to the development of real optical networking[9] is to route data intended for other nodes in the optical domain.

There are various kinds of optical networks appropriate for all different ranges (access part, local, metropolitan and wide area). They also differ in the extent to which they can be practically implemented today. As we shall see, the ones based on the wavelength-division multiplexing (WDM) technique are generally more feasible and have already been penetrating our networks (especially the backbone), while optical time-division multiplexing (OTDM) ones still need some time to mature before commercial deployment.

The main classes of optical networks[10] are shown in Figure 1.9 and will be explored in the following subsections.

1.5.1 WDM POINT-TO-POINT LINK NETWORKS

The first step towards penetration of WDM optical technology in the existing first generation infrastructure is naturally the upgrade of individual links to *WDM point-to-point links*. However, in this class of optical networks, these WDM links are interconnected via non-optical (i.e. electronic) equipment. On account of WDM the two communicating nodes for every WDM link are capable of exchanging data by using many different channels (wavelengths) simultaneously. This allows each node to have many input/output ports. The communication over the fibre could be in both directions (bidirectional) or alternatively we could have a dual fibre WDM point-to-point link where each fibre would be used exclusively for data flow in one direction (unidirectional case, see Figure 1.5).

Most of the leading service providers in North America, Europe and Japan have started investing money to follow the trend of upgrading to WDM links for some years now. Such upgrades carry on at a constantly increasing rate for these and for other new providers as time goes by. The main drivers for doing so are the constantly rising bandwidth demand from their customers and the fact that WDM point-to-point links often seem to be the most cost-effective solution as compared to other alternatives for increasing capacity. For example, a study in [21] compared three possible solutions to upgrading the capacity of a point-to-point transmission link from 2.5 Gb/s (OC-48) to 10 Gb/s (OC-192):

1. Installation of additional fibres.
2. Using higher speed (OC-192) electronic TDM.
3. Upgrading to a WDM point-to-point link with four channels (wavelengths).

It turned out that for distances longer than 50 km the WDM link solution was the most cost-effective. For instance, as compared to the first alternative, it offered four 'virtual fibres' on account of the WDM technique used, and thus avoided the costly installation of new fibres into the ground. Note that in this example the fibre connecting the two end points of the WDM link was used for unidirectional transmission.

An example of a WDM point-to-point link occupying a single fibre for unidirectional transmission is depicted in Figure 1.10.

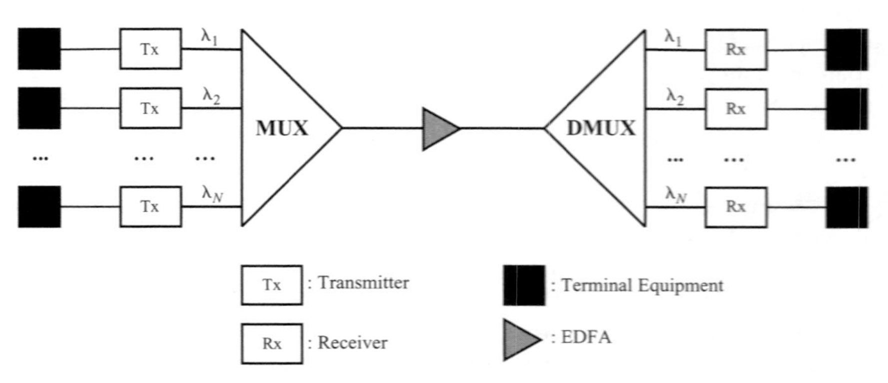

Figure 1.10 Example of a WDM point-to-point link.

Generally, a WDM point-to-point link comprises the following elements, most of which appear also in the example of Figure 1.10:

- two nodes (at the two edges of the link) having many input/output ports;
- different interfaces per port allowing communication via various protocols over the link;
- electro-optical converters which include lasers for transmission on different channels (wavelengths) at the two endpoints of the link;
- the appropriate receivers for each wavelength converting the signal back to electronic form at the two endpoints of the link;
- WDM Multiplexers/Demultiplexers located right after the transmitters (for WDM multiplexing) and before the optical signal reaches the receivers (for WDM demultiplexing);
- amplifiers to provide adequate signal power gain when this drops below certain levels.

It should be noted that devices such as lasers, receivers and WDM multiplexers/demultiplexers relate to the enabling technology and thus will be described in Chapter 2.

1.5.2 WDM BROADCAST-AND-SELECT NETWORKS

As has been already mentioned, WDM Broadcast-and-Select networks will be the centre of attention in this book. An example of such a network is shown in Figure 1.11. It consists of one *optical $N \times N$ passive star coupler* and N nodes each equipped with one *fixed-tuned transmitter* and one *tunable receiver*. Transmitters and receivers are connected via separate fibres to the optical star coupler; thus each node actually uses a pair of fibres to communicate with the star coupler. Star couplers, transmitters and receivers are obviously part of the enabling technology and will be described in Chapter 2. However, here we will point out as much as necessary of their characteristics, just enough to understand this significant class of optical networks.

A fixed-tuned transmitter, as the name suggests, is capable of transmitting on a specific wavelength only, in which it is fixed-tuned *a priori*. Tunable receivers are capable of receiving on various wavelengths by reconfiguration, i.e. by retuning when needed on the desired wavelength. The optical $N \times N$ star coupler is generally just a passive, unpowered (therefore exceptionally reliable and easy to manage) device. Nonetheless, note that there are architectures and protocols where the central hub takes on a more active role contributing usually to an overall higher

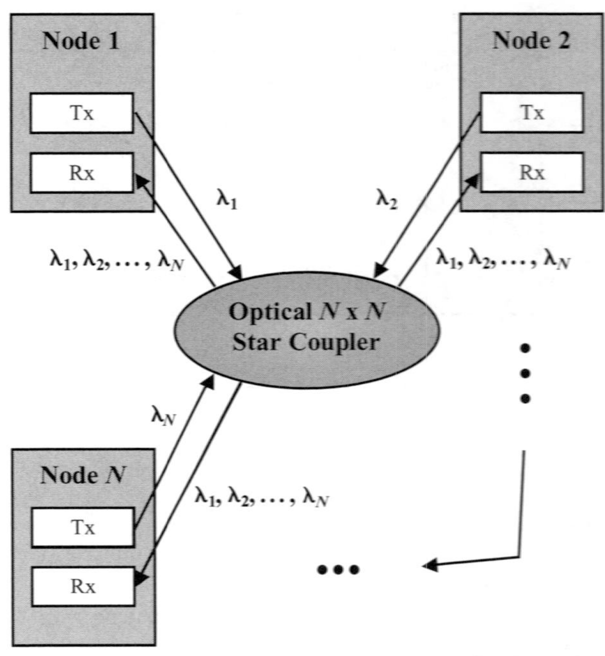

Figure 1.11 An example of a WDM broadcast-and-select star network with N nodes. Each node is equipped with one transmitter fixed-tuned to one of the N available wavelengths and one tunable receiver.

performance at the cost of losing the advantages of passiveness. The operation of a 4×4 passive star is shown in Figure 1.12 [24]. A signal inserted on a given wavelength from an input fibre port will have its power equally divided among all output ports, in which it will appear exactly on the same wavelength as it entered. For example, Figure 1.12 shows how a signal on wavelength λ_1 from input fibre 1 and another on wavelength λ_4 from input fibre 4 are broadcast to all output fibre ports. The remainder of broadcasts through input fibres 2 and 3 are only shown in dashed lines to keep some clarity in the figure.

Now, turning to the specific example of Figure 1.11, which is quite representative by the way, each node i uses its transmitter, denoted by Tx in the figure, to transmit a signal on a unique wavelength λ_i pre-assigned to it (since Tx is fixed-tuned). The optical $N \times N$ star coupler, which is located at the centre of the network in the

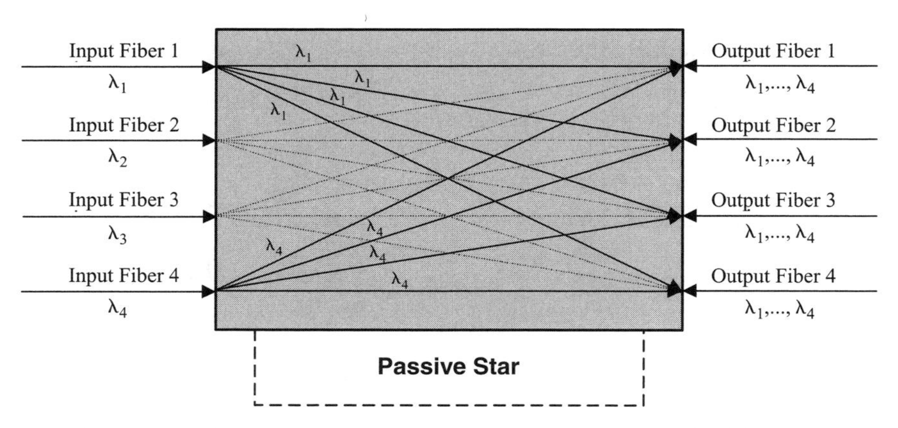

Figure 1.12 The operation of a 4 × 4 passive star coupler [24]. (© 2000 IEEE)

figure, undertakes the task to combine the signals received from each node and send the result (mix) to all nodes, making use of the fibres that are intended for node reception. This explains the *broadcast* part of the characterization of this WDM network architecture, as this is exactly what happens in any network of a broadcast architecture: the network simply sends signals received from each node to all the nodes. Subsequently, each node has to use its receiver, denoted by Rx in the figure and assumed to be a tunable filter, in order to *select* the desired wavelength. Thereby the term 'broadcast-and-select' is evident. A nice attribute of this class of networks is that they can easily provide *multicast* services apart from the apparent unicast transmissions (Figure 1.13). This is achieved in a straightforward way just by having all desirable destination nodes of a multicast transmission on some wavelength tune their receivers to this wavelength.

It should be noted that in broadcast-and-select networks two transmitters sending optical signals at the same time must choose distinct wavelengths for doing so. This is called the *distinct channel assignment constraint* and characterizes the broadcast-and-select architecture. Otherwise we say a *collision* occurs, meaning that the signals interfere with each other and information arrives corrupted, thus useless, to the receivers. Obviously such collisions are undesirable and should be avoided by any means. In broadcast-and-select networks the available channels (wavelengths) constitute a resource that has to be *shared* among the various nodes, i.e. users. Any kind of a shared resource requires a set of rules, a *protocol* in this case, that would be able to coordinate the prospective users of the resource and determine how it should be used. Such protocols in broadcast-and-select networks

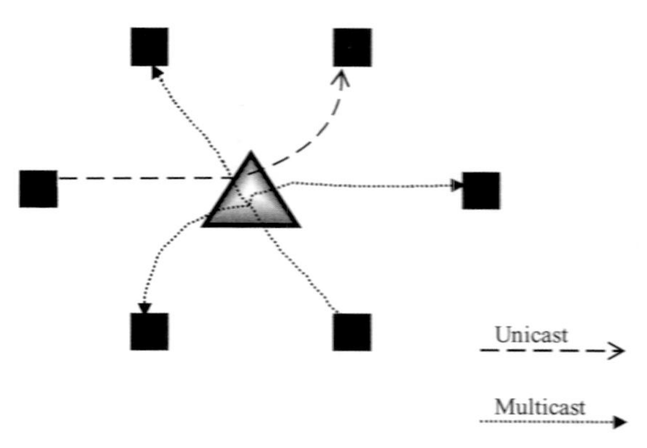

Figure 1.13 Unicast and multicast services offered by a WDM broadcast-and-select network.

are thus necessary and typically try to avoid the above mentioned collisions so as to achieve a better utilization of the shared medium (channels).

Apart from the distinct channel assignment, there is another constraint associated with this class of optical networks. As a result of using an optical $N \times N$ passive star coupler, only a fraction of the signals arriving at the coupler's input ports is delivered to its output ports. In fact, if we assume in our specific case that all N nodes transmit using the wavelengths assigned to them at a certain point in time, the coupler will deliver approximately $1/N$ of the power of each arriving signal to its output ports. Broadcast-and-select networks are therefore proper for local and metropolitan area networks (LANs and MANs) where N is not too large. Apparently, it is not be possible to have one single star coupler and thousands or millions of users connected to it, as would be required in the case of a wide area network (WAN).

Finally, we observe that the total number of wavelengths (N) used in the network of Figure 1.11 is equal to the number of nodes and that each node has exactly one transmitter and one receiver, the types of which were described above. It should be noted that in the general case of a broadcast-and-select network, however, the total number of nodes can also be greater than the number of wavelengths and each node can have multiple transmitters and/or receivers (that may be fixed-tuned or tunable). Moreover the topology should not necessarily be a star. Anyway, we defer the discussion about topologies, transceivers and other interesting aspects of WDM broadcast-and-select networking to the end of this introductory chapter, where we shall examine this interesting class of optical networks in more depth.

1.5.3 WAVELENGTH (AND WAVEBAND) ROUTING NETWORKS

In the beginning of this section about optical networks, we remarked that the key idea for progressing to real optical networking is to have some of the routing, switching and intelligence transferred to the optical domain. This is what happens first and foremost in *wavelength routing* networks. WDM point-to-point links are in fact just one first step supporting the gradual penetration of the wavelength routing architecture into the network, while broadcast-and-select networks are rather static in nature and not scalable to large sizes. The WDM wavelength-routing architecture comes as a more general and less static proposal aiming, from a physical point of view, to be deployed mainly as a backbone network for large regions, e.g. for nationwide and global coverage. Thus, a likely form that the overall network might take in the future is a combination of broadcast-and-select LANs interconnected by a wavelength-routing network, while other well-established LAN technologies will not be precluded of course.

One main advantage of wavelength routing networks over broadcast-and-select ones is that they support *reusing wavelengths* in different parts of the network. Recall that this cannot be the case for broadcast-and-select networks. Suppose for example that node 2 in Figure 1.11 is transmitting data on wavelength λ_2 to node 1 and node N wishes to communicate with another node. Node N is constrained not to use wavelength λ_2 of the initial transmission for this, even though such a communication concerns another part of the network (nodes 1 and 2 seem to have nothing to do with it), on account of the broadcast nature of this architecture. In particular, if node N did transmit on λ_2, its data would be broadcast to all other nodes including node 1, which at that time would be still receiving data on the same wavelength from node 2. Contrary to this, *wavelength reuse* is provided by wavelength routing networks as we will see. On the basis of this property scalability is increased allowing a large number of nodes to be served by relatively few channels (today close to around 160 per single fibre).

Before proceeding to the details of the components constituting a wavelength routing network, let us first spot the kind of services offered by a wavelength routing network and the way it operates. For this purpose consider Figure 1.14(a) which shows an example of a small wavelength routing mesh network. Five *intermediate (wavelength routing) network nodes* are represented schematically by circles in the figure, from 1 to 5. These are interconnected between them forming a simple mesh physical topology and, also, each of them is connected with five *end-nodes* represented as rectangles, from A to E, where intermediate node 1 connects to end-node A and so on. We must point out here that some of the intermediate wavelength routing nodes may not connect to any end-node in the general case of a wavelength

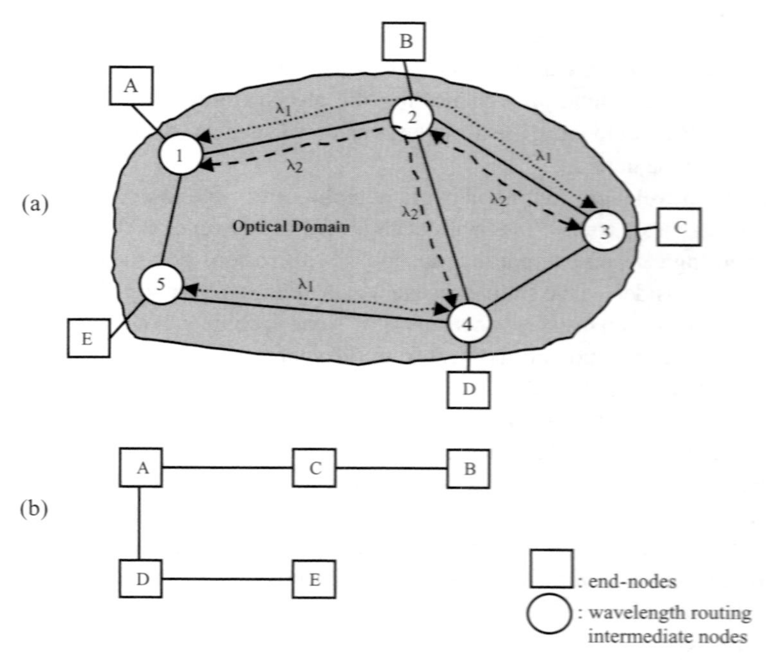

Figure 1.14 (a) Example of a small wavelength routing mesh network (b) Virtual (logical) topology of the network in (a).

routing network. The end-nodes are the sources and sinks of traffic moving around in the network. The intermediate nodes themselves form an *all-optical network* in that traffic stays in optical form throughout the network and just gets converted to electronic only at the edges (end-nodes). Thus, it is imperative for each end-node to be equipped with optical transceivers, enabling it to convert the electronic data it generates into optical form before transmitting to the desired destination (through the intermediate node they attach to) and the optical data received from the neighbouring intermediate node back to electronic form before it can actually process it.

It should be remembered here that such a purely all-optical network (excluding the end-nodes) is not practical when large spans are considered. In fact some kind of optoelectronic-optical (OEO) conversions might take place e.g. in electronic regenerators after great distances. The use of electronics in so-called all-optical networks was discussed in the introduction to this section. Thus, it is more common in practice to encounter several 'islands' of all-optical subnetworks interconnected to form the overall network. The latter is sometimes called an 'opaque' network, on account of the loss of data transparency introduced by the necessary use of electronics in the links interconnecting the aforementioned all-optical islands. Figure 1.14(a) could more precisely be considered such an island (subnetwork) of a larger opaque network.

What a wavelength routing network actually offers to end-nodes is the ability to make (logical) connections with each other. At any time, each end-node may maintain logical point-to-point connections with various other end-nodes in the network. Considering the simple case in which intermediate nodes do not have any ability to perform *wavelength conversions*,[11] one specific wavelength is assigned to carry this connection all the way from the source end-node to the destination. Provided that the paths taken by any two connections do not have any common fibre links, i.e. do not overlap at any point in the network, they can be carried by the same wavelength; in simple terms, same wavelengths can be *reused* in various parts of the network. If we imagine a large network consisting of e.g. thousands of nodes with quite a few physical (fibre) links between them, we can gain an understanding of how much wavelength reuse will take place, resulting in a dramatically decreased number of wavelengths required to support numerous logical connections between pairs of end-nodes.

Apart from wavelength reuse it is also noteworthy that in wavelength routing networks optical signals on certain wavelengths are not broadcast to every other end-node (as in broadcast-and-select systems), but are routed along the appropriate logical connection only to the desired destination. Thus, wavelength routing networks eliminate the other weakness of broadcast systems, namely the waste of signal power due to the splitting and broadcasting of the signal to many nodes, most of which are typically uninterested.

In such wavelength routing networks, as in the case of Figure 1.14(a), an all-optical path provided to a pair of end-nodes in order to serve as a high-speed logical connection between them is termed a *lightpath*. The bit rates supported along a certain lightpath are as high as a few Gb/s, e.g. 10 Gb/s (or even 40Gb/s but for much shorter distances before electronic regeneration is needed, ceasing the all-optical characteristic of transmission and thus constraining lightpaths to shorter spans). Recall that such bit rates offered by lightpaths, each served by one wavelength, comply well with the potential of electronic circuits at the end-nodes; this is exactly how WDM offers a great collaboration of optics and electronics as mentioned in a previous section.

In Figure 1.14(a) we can distinguish four established lightpaths, which are considered to be bidirectional, i.e. comprise pairs of unidirectional fibre links in opposite directions; this is the case for both the links in the access part (between end-nodes and the all-optical network) and the ones between intermediate routing nodes. The first lightpath is between end-nodes A and C on wavelength λ_1 passing through the intermediate wavelength routing nodes 1, 2 and 3; the second lightpath is between A and D on λ_2 through intermediate nodes 1, 2 and 4; the third lightpath is between D and E on λ_1 through nodes 4 and 5; and the fourth lightpath is between B and C on λ_2 through nodes 2 and 3. Note that the DE lightpath reuses wavelength λ_1 which is already used by lightpath AC without any problem of interference, since the two lightpaths do not share any common fibre links. This is the case with wavelength λ_2 and lightpaths AD and BC as well.

Let us assume that the only available wavelengths in our network of Figure 1.14(a) are λ_1 and λ_2. Suppose that we wanted to set up a logical connection between end-nodes E and B. Notice that assuming *no wavelength conversion* capability for the intermediate node 4, the network would have to block such a lightpath request, i.e. it would be unable to provide this lightpath or else the same wavelength would have to be used for two different lightpaths over the same fibre link. This would be undesirable because of the interference between the lightpaths as we have already described.

Figure 1.14(b) illustrates the *virtual topology* of the network with the physical topology shown in Figure 1.14(a). The virtual topology of a wavelength routing network is the topology that we get if we use a link to connect every end-node with all end-nodes with which it has established lightpaths. In simple terms, in order to get the virtual topology we just have to replace lightpaths with direct links. As we will see in the end of this chapter when we look closer at the broadcast-and-select WDM networks, the notion of virtual topologies is applicable to those networks as well. Designing proper virtual topologies to be *embedded* over the underlying physical topologies is quite an important subject matter and has been studied by several researchers.

So far we have not referred more specifically to what those rectangles of Figure 1.14, i.e. end-nodes, might be. End-nodes are the users that will benefit from the services (including above all the establishment of high-speed lightpaths between them) offered by the wavelength routing network. First it would be helpful to view a wavelength routing network (this can be extended to second generation networks in general) as a separate *optical layer* falling for the most part within the physical layer in the well-known network layer hierarchy proposed by the ISO (International Standards Organization). This optical layer acts as a *server layer* supporting a variety of client layers above it, which may be first generation optical network layers such as SDH/SONET and ESCON or other widely used network layers like IP and ATM. The overall network may comprise various combinations e.g. IP over ATM over SDH/SONET over optical, (straightforwardly) IP over

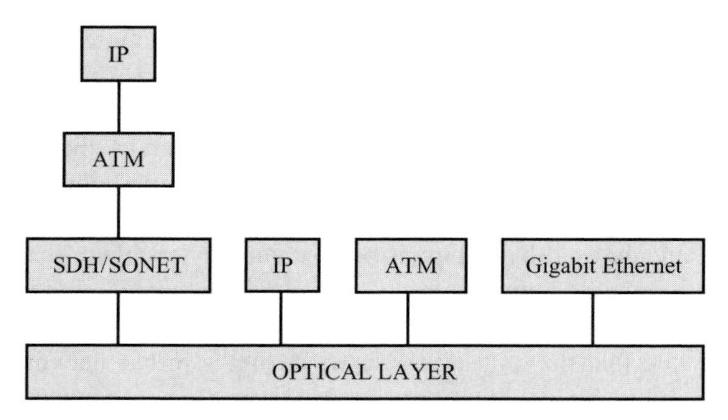

Figure 1.15 Various client layers supported by the optical layer.

optical, ATM over optical and Gigabit Ethernet over optical. These combinations are shown in Figure 1.15 and are just samples of the possible combinations.

The need for such layered architectures for the overall network is explained if we think that the optical layer, due to the state-of-the-art and nature of optics, can handle data at very high bit rates in a much more cost-effective way than the layers above it. On the other hand, the layers above the optical are more cost-effective at processing in more detail relatively lower bit-rate data.

The users (rectangles of Figure 1.14) of a wavelength routing network are in effect SDH/SONET terminals, IP routers, ATM switches and so on. The first case especially is very common. That is mainly why in the first stage of penetration, wavelength routing networks have been configured in the same way as the existing SDH/SONET networks, i.e. as rings. As they become more mature, mesh topologies have been in use constantly more. The SDH/SONET terminals just cannot know that they are being served by lightpaths (logical connections) provided by an underlying optical layer instead of the archetypical dedicated fibre connections between them. Thus the optical layer can offer a dedicated communications backbone for a logical network, e.g. SDH or ATM. Moreover, the optical layer does not just offer permanent lightpaths serving as physical links, but also switched lightpaths, naturally corresponding to connections provided by the familiar circuit-switched services today (as in traditional networks for telephony). Remote access to supercomputers is an example of a temporary activity that would require assignment of logical connections (switched lightpaths) on demand.

1.5.3.1 The Problem of Routing and Wavelength Assignment

The problem of *Routing and Wavelength Assignment* (*RWA*) is crucial in wavelength routing networks. It can be formally stated as follows. Given a set of lightpaths that need to be established on a particular network topology, determine the routes over which these lightpaths should be established and the wavelengths that should be assigned to them, using the minimum possible number of wavelengths.

We have already described two constraints that have to be kept in mind by the approaches trying to solve RWA. In fact the constraints concern the second part of the problem, in which routes of lightpaths are assumed to be known in advance and what remains to be done is the *wavelength assignment* (*WA*). The constraints are summarized below:

1. *Distinct wavelength assignment constraint*. All lightpaths sharing a common fibre must be assigned distinct wavelengths to avoid interference. This applies not only within the all-optical network but in access links as well.
2. *Wavelength continuity constraint*. The wavelength assigned to each lightpath remains the same on all links it traverses from source end-node to destination.

The first constraint holds for solving the WA problem in any wavelength routing network. Recall that a similar constraint was applicable in broadcast-and-select networks too. The second constraint applies *only* to the simple case of wavelength routing networks that have no wavelength conversion capability inside their nodes.

In a wavelength routing network the traffic can be considered either *static* or *dynamic*. That is to say the lightpath requests can be either *offline* or *online* respectively. In a static traffic pattern (offline lightpath requests), a set of lightpaths are set up all in advance and remain in the network for a long period of time; they can be considered *permanent* in a sense. The RWA problem is known as *offline RWA* in this case. In a dynamic traffic pattern (online lightpath requests), a lightpath is set up for each connection request as it arrives, and the lightpath is released after some finite amount of time. The lightpaths are *switched* in this case and provided in a circuit-switched fashion. The RWA problem here is called *online RWA*. There is a vast literature in this research area; a review of RWA algorithms can be found in [37]. Most RWA algorithms have the following general form:

- Arrange all admissible fibre paths for a given source-destination pair (i.e. lightpath request) in some prescribed order in a path list.
- Arrange all wavelengths in some prescribed order in a wavelength list.
- Attempt to find a feasible route and wavelength for the requested lightpath starting with the path and wavelength at the top of the lists.
- Repeat for all lightpath requests.

In order to set up a lightpath a *signalling and reservation protocol* is required to exchange control information among nodes and reserve resources along the way. In many cases the signalling and reservation protocols are closely integrated with the RWA protocols to form a so-called *control and management mechanism (or protocol)*. Upon the reception of a lightpath request, this mechanism will select an appropriate route, assign the appropriate wavelength(s) to the lightpath and configure the network's switches accordingly. The control and management mechanism is also responsible for keeping track of which wavelengths are currently used in each optical link in order to enable correct routing decisions. Note that the mechanism can be either *centralized* or *distributed*, with distributed ones being preferable due to their increased robustness [27]. The research in this area has four main goals [27], which generally may conflict:

1. Minimize the blocking probability of lightpath requests, i.e. establish as many lightpaths as possible.
2. Minimize lightpath set-up delays.
3. Use the minimum possible amount of bandwidth for control messages.
4. Maximize system scalability.

Control and management protocols dealing with the *dynamic lightpath establishment* (*DLE*) problem (which relates to the online RWA) can be found in [38]. For the *static lightpath establishment* (*SLE*) problem (relating to the offline RWA) [37] is a good starting point.

1.5.3.2 Building Blocks of a Wavelength Routing Network

In this subsection we look closer at the basic building blocks of a wavelength routing network, besides optical fibres and regenerators (including optical amplifiers) which we have already discussed.

The principal building block of a wavelength routing network is a node that was referred before as the *intermediate wavelength routing node* and was shown to be the main component of the all-optical network depicted in Figure 1.14(a). To keep the analogy with SDH/SONET networks, where we had digital cross-connect systems (DCSs), these nodes are often called *wavelength cross-connects* (*WXCs*) or *optical cross-connects* (*OXCs*). We shall use the former term from here on. Just as a DCS cross-connects (switches) many PDH and SDH/SONET data streams, a WXC has generally many *input/output ports* and switches the light signals entering on its input ports on to its output ports. The routing decision for each incoming signal is based on its input port and on its wavelength (that is why these networks are called wavelength routing). A DCS also has some local ports (we saw it can perform the actions of an ADM), where digital signals can be added or dropped locally. This is the case as well with WXCs, which, in addition to their input/output ports, have various *local ports* where lightpaths start and end. These can be electrical or optical and as in the example of Figure 1.14(a) before, they are used to connect WXCs with end-nodes.

Figure 1.16 shows WXC_2 of Figure 1.14(a) in some more detail. This node has three input and three output ports and one bidirectional local port connecting it with the end-node B. In point of fact, when we refer to an illustration of a network like the one in Figure 1.14(a), it would be more appropriate to talk about pairs of input/output ports, since we consider the bidirectional (via fibre pairs) connections of this WXC with others. Thus, for example the input port 1 and output port 1 form the first pair of input/output ports of WXC_2 (Figure 1.16), which corresponds to the pair of unidirectional fibres connecting WXC_2 to WXC_1 and is shown as a single (bidirectional obviously) link in Figure 1.14(a). The necessary *wavelength multiplexers/demultiplexers* (*WMUX/DMUX*) for the outgoing/incoming light signals that leave/enter (respectively) the WXC node are also shown in the figure. We assume multiplexing and demultiplexing of only two wavelengths in accordance with the assumption made for the network of Figure 1.14(a). The figure additionally depicts how the switching operation is performed for lightpath AD on wavelength λ_2. Based on the facts that the signal enters the node from input port 1 and wavelength λ_2 is used for the lightpath of interest, the WXC node decides to route the λ_2 part of the signal to output port 3 towards WXC_4. It is obvious that some kind of control and management of the WXC must exist in order to make all the necessary routing decisions. Even in all-optical WXC nodes this should be done *electronically* according to the current state-of-the-art (recall that electronics have the intelligence). The module for management and control is not shown in Figure 1.16.

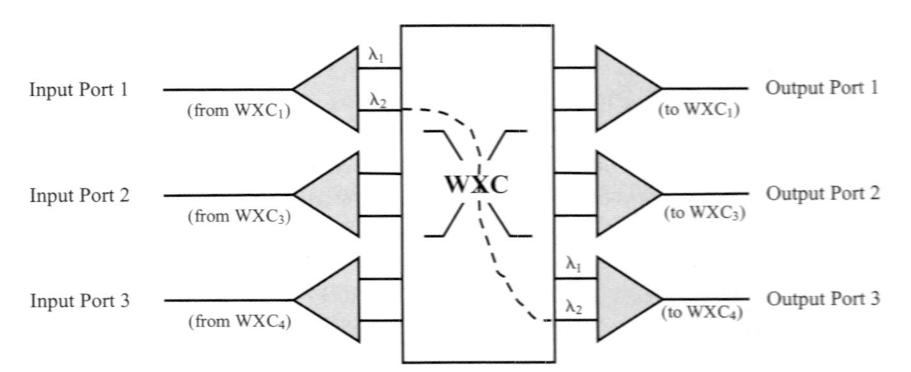

Figure 1.16 The WXC node 2 of Figure 1.14(a) in more detail. The necessary wavelength multiplexers/demultiplexers and the switching for lightpath AD on $\lambda_2$are also shown.

When a WXC has only two pairs of input/output ports (or equivalently can be connected only with two other WXCs through two bidirectional links) without taking into account any additional ports probably used for protection, it is called a *wavelength add/drop multiplexer* (*WADM*) or *optical add/drop multiplexer* (*OADM*). (We shall mainly use the former term.) These wavelength routing network elements are obviously analogous to ADMs, encountered in the case of SDH/SONET networks. For example, in the simple network of Figure 1.14(a), intermediate nodes 1, 3 and 5 are WADMs, while 2 and 4 (having three pairs of input/output ports) are WXCs, according to the way we defined these network elements. Like normal WXCs, WADMs can have a number of local ports too.

As we saw, wavelength multiplexers/demultiplexers are also included in the building blocks of wavelength routing networks. We have already referred to these devices that were shown in Figure 1.16 to separate wavelengths in input ports (demultiplexers) or combine wavelengths in output ports (multiplexers). A WMUX/DMUX is also called *optical line terminal* (*OLT*) to keep the analogy with line terminals (LTs) or terminal multiplexers used in SDH/SONET networks.

1.5.3.3 Inside WXCs—Static vs. Reconfigurable WXCs and Networks

There are various types of wavelength cross-connects (WXCs) and wavelength add/drop multiplexers (WADMs) that determine the operation of a wavelength routing network. We shall confine the discussion to WXCs since they are more general and in a sense include WADMs as was explained in the previous subsection.

A WXC node always comprises a number of wavelength multiplexers/demultiplexers (WMUXs/DMUXs), which can be considered a part of it (even if in the previous subsection it helped to look at WMUXs/DMUXs as separate network elements).

Except for this certain fact, there is quite a large variety of WXC implementations. First of all, a WXC may have the ability to perform wavelength conversions, adding to the overall flexibility of the network in establishing lightpaths between end-nodes. *Wavelength converters* will be examined in more detail in Chapter 2, but we shall briefly refer here those of their characteristics that are of some interest in implementing WXCs. Looking back at Figure 1.14(a) it is easy to see for example that if WXC_4 was capable of wavelength conversions, it would be possible to establish a lightpath connecting end-nodes E and B through WXC_5, WXC_4 and WXC_2. More than one wavelength is assigned to a single lightpath in the case of wavelength conversion-capable intermediate WXC nodes. Thus, in this example, in order to establish lightpath EB, wavelength λ_2 should be used along the link from WXC_5 to WXC_4 and wavelength λ_1 along the link from WXC_4 to WXC_2, thereby interference with previously set up lightpaths would be avoided. Given a number of available wavelengths, wavelength conversion capability at some WXCs allows establishment of more lightpaths than in the case of no wavelength conversion, i.e. decreases the *blocking probability* of connection requests.

However, generally, not only the existence of wavelength conversion capability should concern us but also its kind. The various kinds of wavelength conversions in a decreasing order of offered flexibility are:

- *Full* wavelength conversion, implying the capability of conversions from (every) one wavelength to any other within the set of available wavelengths.
- *Limited* wavelength conversion, meaning that a wavelength can be converted only to some wavelength included on a certain subset of the available wavelengths, but not to any one (as in the previous case). This offers less flexibility than full wavelength conversion.
- *Fixed* wavelength conversion, implying that (every) one wavelength is *always* converted to another specific and fixed wavelength. This offers no flexibility at all.
- *No* wavelength conversion, offering no flexibility (as the previous solution).

Full wavelength conversion offers the greatest flexibility followed by limited conversion, because both involve the ability to reconfigure conversions. The last two in fact offer no flexibility since they are static in nature. Wavelength conversions may be done optically or electronically. The latter would require optoelectronic-optical (OEO) conversions; hence, the network would not be all-optical any more.

Figure 1.17 shows an example of a WXC node with no wavelength conversion capability [28]. The WXC has three input and three output ports, three WMUXs

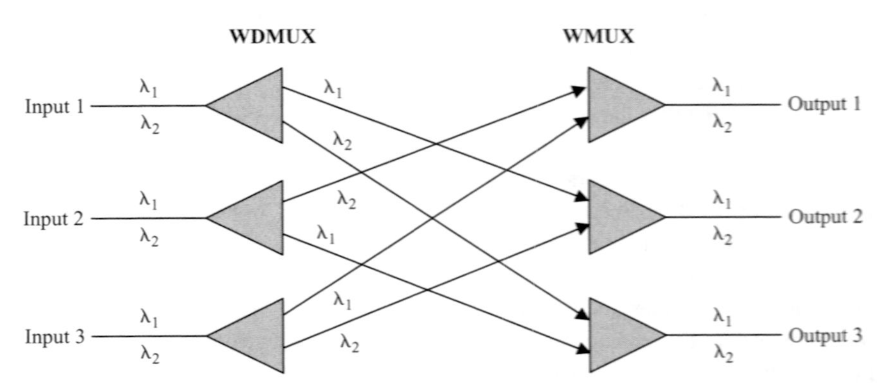

Figure 1.17 Example of a static optical WXC node with three input and three output ports. The WXC has no wavelength conversion capability [28]. (© 1993 IEEE)

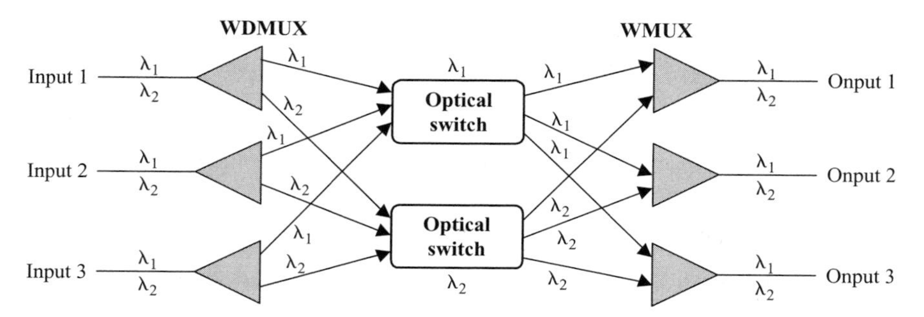

Figure 1.18 Example of a reconfigurable optical WXC node with three input and three output ports. The WXC has no wavelength conversion capability [28]. (© 1993 IEEE)

and three WDMUXs. The wavelengths, assumed to be just two for simplicity, are switched in a fixed manner. Such a WXC node is obviously static in that every signal on some wavelength entering through a specific port will always have to leave from a predetermined port on the same wavelength. For example, λ_2 of input port 1 is always switched to output port 3 (and remains λ_2). Such static WXCs are passive devices and are also called *passive routers* or *waveguide grating routers* (*WGRs*).WGRs have been commercially available for quite some time now and are proposed for architectures regarding another class of WDM optical networks[12] that is examined in a following subsection.

Apart from wavelength converters and the necessary WMUXs/DMUXs, a WXC node may also contain *optical* or *electronic switches*.[13] These would enable it to dynamically route lightpaths under the necessary control which is typically done electronically. Such a *reconfigurable* WXC node is shown in Figure 1.18 [28]. Its only difference from the static node of Figure 1.17 is that it contains a 3×3 optical switch for each wavelength. In the general case of a WXC node with N pairs of input/output ports and W wavelengths, $W N \times N$ optical switches would be required. Note that the whole set of switches can be viewed as a single $NW \times NW$ optical switch (in our example a 6×6 optical switch). A signal on a certain wavelength entering the node from some input port can leave on any output port depending on the decision of the node's electronic control (not shown in the figure).

The same WXC with full wavelength conversion capability is depicted in Figure 1.19. This offers the highest possible flexibility, since it is capable of taking an incoming wavelength from any input port and switching it to any output port on any wavelength. As an example, the WXC may take wavelength λ_2 from input port 1, convert it to λ_1 and switch it to output port 3. However, the implementation cost of WXCs offering this great flexibility is analogously much higher.

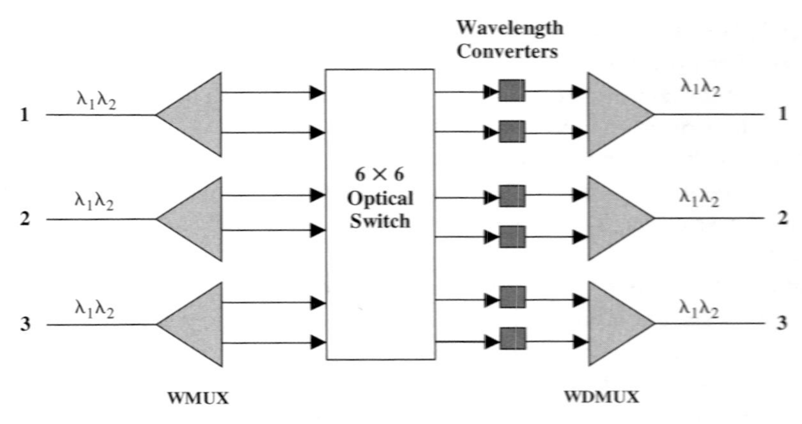

Figure 1.19 Example of a reconfigurable optical WXC node with three input and three output ports. The WXC has full wavelength conversion capability.

All the three previous examples are optical WXCs. Note that in the case of electronic WXC nodes, i.e. nodes comprising electronic switches, the necessary transceivers for optoelectronic-optical (OEO) conversions must be included as well.

When a WXC has no optical switches and involves at most fixed wavelength conversion, it is said to be *static*. As we saw, Figure 1.17 illustrates such a static WXC node incapable of converting wavelengths. If we suppose a fixed wavelength conversion was involved, the WXC would remain static. A wavelength routing network comprising *only* static WXCs is called static as well. On the other hand, when a WXC at least either includes switches or has limited or full wavelength conversion capability, it is said to be *reconfigurable* or *dynamic*. The last two examples concern reconfigurable WXC nodes, as was explained. Accordingly, a network with *at least* one reconfigurable WXC is called reconfigurable or dynamic too, since it is capable of changing its state of operation dynamically. In reconfigurable networks the set of lightpaths that can be established between end-nodes may change by reconfiguring the switches and/or wavelength converters inside some of the WXCs. On the contrary, static networks offer a fixed set of lightpaths to their users. Typically, it is much more common for wavelength routing networks to be reconfigurable, since it is very likely for optical switching to exist within some WXCs; besides this, limited wavelength conversion capability is also likely to be found in a few WXC nodes. Note that considerable amount of work has been done in how and where to optimally place wavelength conversion-capable nodes in a wavelength routing network.

1.5.3.4 Waveband Routing Networks

In WDM broadcast-and-select networks with the common star topology, the optical part of the network, i.e. excluding the end-nodes which have some electronic circuits to perform optoelectronic-optical (OEO) conversions, actually comprises only one node, the optical star coupler. This is a passive device that has *no independent control of each wavelength*; it just sends to each of its outputs a fraction of the power of the signal it gets by combining the signals received through all inputs.

In wavelength routing networks the nodes of the optical part of the network are the WXCs, since OEO conversions take place at the end-nodes. Notice that WXCs have *independent control of each wavelength*, e.g. in the example of Figure 1.18 the WXC has two optical switches, one for each wavelength, so that it can independently control and switch each wavelength. Generally the number of optical switches in a WXC must be equal to the number of available wavelengths, as we saw when we described dynamic WXCs. As technology increases the number of wavelengths that a fibre can carry, the cost of having independent control of each wavelength in WXCs and WADMs becomes considerably high. For instance, assume a fibre carrying 150 wavelengths. In the simpler case of a WADM where two pairs of unidirectional (i.e. a total of four) fibres are attached to it (excluding any local ports), two 1×150 WMUXs, two 150×1 WDMUXs and 150 2×2 optical switches would be required; in dynamic WXCs with more input/output ports the implementation would in effect be more costly.

Waveband routing networks lie somewhere in between, as far as independent control on wavelengths is concerned. The key idea is to group wavelengths in to wavebands and to have nodes do a coarser handling of wavebands instead of controlling wavelengths. Each switching component (which is different from the optical switches used in the case of WXC nodes) operates at a different waveband now, thus the number of switching components must be equal to the number of wavebands (much smaller). The cross-connect nodes of such a network can do a (coarse) demultiplexing, switching and multiplexing of wavebands, but not wavelengths within a waveband. Thereby the optical part of the network cannot distinguish between different wavelengths in a waveband; this is called *inseparability of channels* (*wavelengths*) belonging to the same waveband and is a routing constraint unique to this type of optical networks. The two constraints that were mentioned for wavelength assignment (WA) in wavelength routing networks, i.e. the wavelength continuity (assuming no wavelength conversion) and the distinct wavelength assignment constraint, also apply here. Individual wavelengths within a waveband can be separated from each other only at the end-nodes [28]. Wavelength and wavebands can be reused in different parts of the network.

In waveband routing networks, routing (cross-connect) nodes are implemented in a different way from the one used for WXCs. In fact, the basic optical elements used for the implementation are *couplers* and not optical switches. Couplers are devices that are used to split and combine optical signals and will be studied in Chapter 2. Actually, *variable* couplers that allow adjustment of the power splitting ratio in the range from 0 to 1 [28] are used. These devices are linear and the power at an output port is the linear combination of the powers at the input ports. That is why the networks incorporating such routing nodes have been called *Linear Lightwave Networks* (*LLNs*). Waveband routing networks are considered an extension of LLNs, for which routing nodes are assumed to be waveband-selective. Even though the idea of reducing the switching components required in a routing node (as compared to WXCs and WADMs) is quite attractive, it has been realized that the concept of waveband routing networks is not very practical because of the more complex routing required and the considerable splitting and combining losses introduced by using couplers. However, the couplers used inside routing nodes of LLNs provide the network with the ability to offer multicasting. This advantage could be transferred to wavelength routing networks too, if routing nodes of LLNs containing couplers were used instead of WXCs in a few selected positions of the network.

1.5.4 WDM PASSIVE OPTICAL NETWORKS (PONs)

This class of optical WDM networks has been proposed as a high speed alternative to the (*local*) *access part* of the network, i.e. the part that reaches individual homes or businesses, which is also known as the 'last mile' or 'last leg' of the network. The transmission medium that dominates the access part is definitely twisted-pair copper wire, which implements the 'last leg' of telephone networks. In countries where cable television (CATV) networks have been developed as well, one certain part of the access infrastructure (i.e. besides this belonging to the telephone network) typically comprises fibre up to a certain point (without reaching individual homes) and coaxial cable links going all the way to homes. Of course many business customers are already making use of high speed connections provided to them e.g. by first generation optical networks like SDH/SONET, but the access part for residential customers is still either the telephone or the cable network.

Figure 1.20 shows the standard architecture for an access network. The *hub* contains terminal equipment which sources the signals (but not information) sent downstream to subscribers [10]. Information is fed through a number of links to several *remote nodes* (*RNs*) placed near subscribers. Each RN in turn distributes information to several *network information units* (*NIUs*) each serving one or more

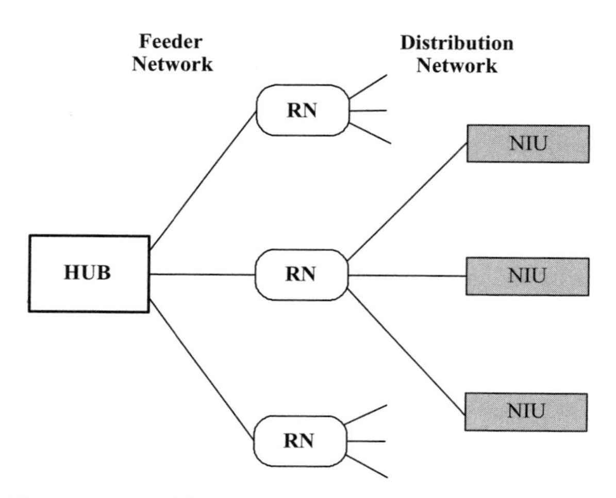

Figure 1.20 The standard architecture for an access network. The hub station connects to several remote nodes (RNs) placed near subscribers. Each RN, in turn, is connected to several network interface units (NIUs) each of them serving one or more subscribers. The part from the hub to the RNs is the feeder network, while from RNs to NIUs we have the distribution network.

subscribers. The part of the network from the hub down to the RNs is the *feeder* network. From the RNs to subscribers we have the *distribution* network, as shown in the figure. In the case of a telephone network the hub is a central office, while for cable networks it is the head end.

Considering that even among individual residential users, the demand for new types of applications and services requiring large amounts of bandwidth is constantly increasing and pushing for progress, fibre-based (optical) access networks have quite reasonably been proposed as a solution. It is expected that, on account of the great potentialities of the medium, service providers will eventually be able to offer such advanced services to individual customers. Optical access networks generally follow the same architecture depicted in Figure 1.20, where NIUs are usually called optical network units (ONUs).

Since we are dealing with the access part of the network, it would be wise not to use any complicated devices requiring special network control, such as the switches of Figure 1.18 and Figure 1.19. The installation and usage costs of such devices would surely be unjustifiable considering that here we are not dealing with a backbone network as in wavelength routing networks. Thereby, the optical networks proposed for the access part must preferably comprise passive

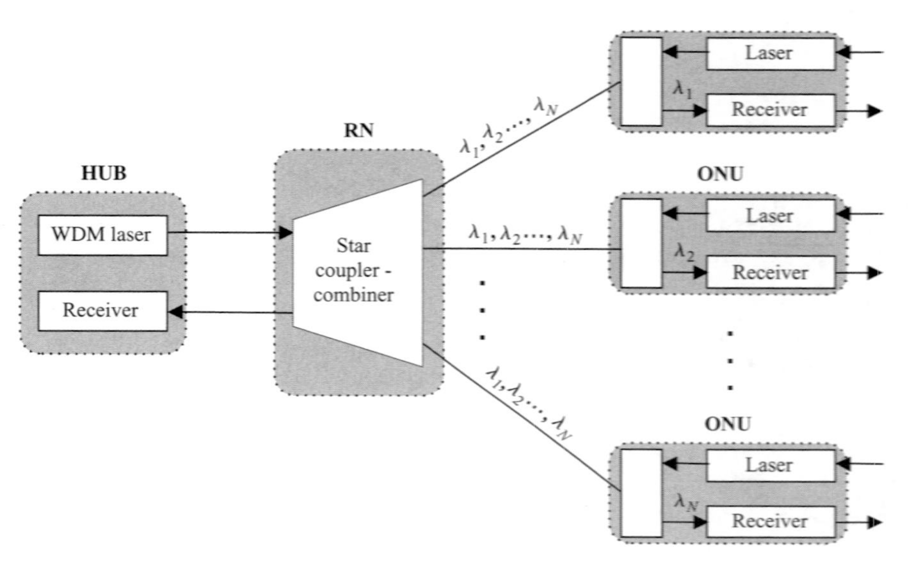

Figure 1.21 The broadcast-and-select PON architecture. (Reprinted from [29], copyright 1998, with permission from Elsevier)

devices like couplers (e.g. passive star couplers) and waveguide grating routers (WGRs).[14] Such networks are called *Passive Optical Networks* (*PONs*), while the more advanced of them incorporating WDM are called *WDM PONs*. In order for a PON (WDM or not) to be practicable and cost-effective, we must ensure that the ONUs are as simple and inexpensive as possible (since they involve a large number of customers) while the RNs, and even more the hub, can be somewhat more complicated.

Next we briefly describe the two basic architectures for WDM PONs [29]. Figure 1.21 illustrates the *broadcast-and-select WDM PON architecture*. A tunable transmitter (over the entire range of available wavelengths) is located inside the hub along with a receiver for data coming from ONUs (upstream). Instead of the tunable transmitter, an array of fixed-tuned transmitters could be used. The multichannel traffic directed from the hub to the subscribers, i.e. the downstream traffic, is broadcast through a passive star coupler located inside the RN to all ONUs. The links between RN and ONUs carry the same traffic and thus we have a broadcast distribution network. ONUs can therefore be identical, which is an advantage of this broadcast feature. Since by the use of WDM, each ONU has one channel for exclusive service, we say that we have *dedicated bandwidth* assigned

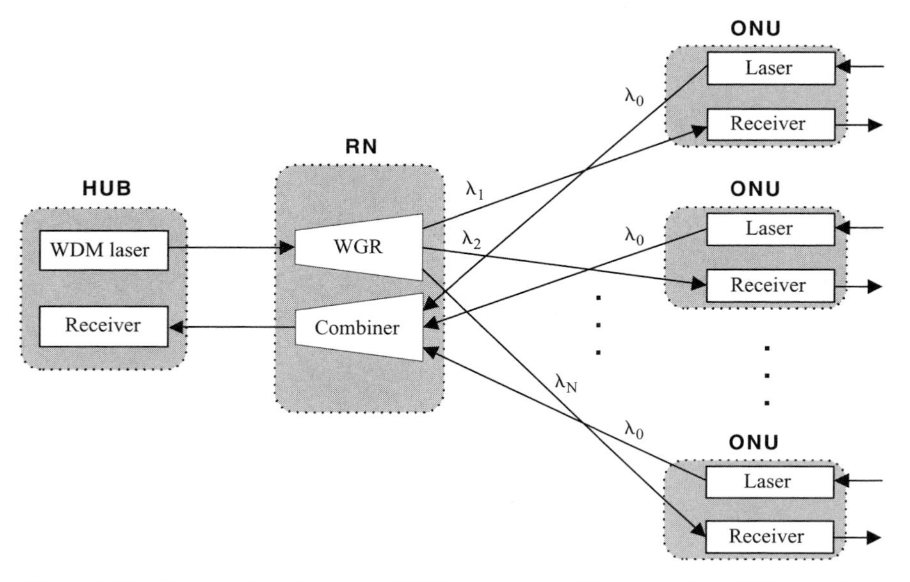

Figure 1.22 The wavelength routing PON architecture. (Reprinted from [29], copyright 1998, with permission from Elsevier)

by the feeder network to each ONU.[15] As in a WDM broadcast-and-select LAN, all ONUs receive exactly the same WDM signal and have to utilize some form of filtering so as to select their exclusive wavelength. The upstream traffic of all ONUs is carried over one channel, shared by ONUs via an appropriate multi-access method like e.g. synchronous TDM. The reasoning behind this is that traffic is largely asymmetric in PONs and one channel would be adequate for upstream transmission. Inside the RN there is a coupler functioning as a combiner of the upstream traffic which subsequently heads towards the receiver of the hub. This broadcast-and-select WDM PON is a nice solution that can be implemented with relatively low-cost optical components. However, the use of a passive star coupler in the RN introduces splitting losses ($1/N$ of the signal power is delivered to each ONU) and therefore hinders the support of many subscribers.

The *wavelength routing PON architecture* overcomes the splitting loss trouble by replacing the star coupler by an $N \times N$ WGR (passive router) inside the RN, as shown in Figure 1.22 [29]. The distribution links carry now only the appropriate channel for each ONU on account of using the WGR. This results in the distribution network becoming switched in this case and inherently more appropriate for switched services like e.g. interactive video and Internet browsing. Several different implementations of this general architecture have been demonstrated; see [15] and [16] for example.

The above two WDM PONs are subsumed in the *fibre-to-the-curb* (*FTTC*) and *fibre-to-the-home* (*FTTH*) architectures (differing in how close to homes ONUs are considered, where in FTTC copper wires connect one ONU to several homes), which were conceived and proposed before quite some time now. Thus, the idea of using fibre up to the home (or really close to it in the FTTC case) and preferably exploiting it well by means of WDM is not that new. In practice, bringing WDM technology to the access part can be considered the logical next step after introduction of WDM in the long-haul inter-exchange core and then into the metropolitan local exchange [12]. But this will happen only when optical component technology becomes more mature and less expensive.

1.5.5 PHOTONIC PACKET SWITCHING NETWORKS

OTDM networks appear fairly futuristic for today's technology in that there are still some serious problems that have to be overcome before their commercial deployment. For example, one major problem is that some part of the end user's network interface must execute certain control and processing operations at much higher speed than the potentiality of electronic circuits. These operations, though, cannot be performed in the optical domain either, at least with current optical technology. Consequently, it seems that OTDM has to wait in laboratories for several technological advancements before breaking through to commercial reality. However it is worth mentioning the fundamentals of OTDM networks as they seem capable of offering advanced services not provided by WDM networks alone. Furthermore, taking into account that this book focuses on the WDM broadcast-and-select architecture, we will notice that the latter is crucial in implementing several switches used inside OTDM networks.

As we have already seen, using pure WDM only provides granularity at the level of one wavelength. If data at a capacity of a fraction of a wavelength's granularity is to be carried, capacity will be wasted. For example, if we assume that one wavelength can carry traffic up to 40 Gb/s and we had four users, asking for a 10 Gb/s connection each, with pure WDM we would have to use four distinct wavelengths, since the lowest level of granularity is the wavelength. Obviously, it would be more cost-effective to deliver service to all of them using only one wavelength, offering in essence virtual circuit services as in the case of ATM for example. With *photonic* (*or optical*) *packet switching*, packet streams can be multiplexed together statistically by means of the OTDM technique (described in Section 1.2), making more efficient use of capacity and providing increased flexibility over pure WDM. Note, however, that photonic packet switching can be combined with WDM either within the switch itself or as a means of transferring information between switches [14].

Packet switches analyze the information contained in the packet headers and thus determine where to forward the packets. Photonic packet switching technologies enable the fast allocation of WDM channels in an on-demand fashion with fine granularities (microsecond time scales). An optical packet switch can cheaply support incremental increases of the transmission bit rate, so that frequent upgrades of the transmission layer capacity can be envisaged to match increasing bandwidth demand with a minor impact on switching nodes [9]. In addition, photonic packet switching offers high-speed, data rate/format transparency, and configurability, all of which are some of the important characteristics needed in future networks supporting different forms of data [36].

1.5.5.1 Issues Concerning Photonic Packet Switching

Photonic packet switched networks can be divided into two categories: *slotted* (*synchronous*) and *unslotted* (*asynchronous*). In a slotted network all packets have the same size. They are placed together with the header inside a fixed time slot, which has a longer duration than the packet and header to provide guard time. In a synchronous network, packets arriving at the input ports must be aligned in phase with a local clock reference [36]. Maintaining synchronization is not a simple task in the optical domain. Assuming an Internet environment, fixed-length packets imply the need to segment IP datagrams at one edge of the network and reassemble them at the other edge. This can be a problem at very high speeds. For this reason, it is worth considering asynchronous operation with variable-length packets [14].

Packets in an asynchronous network do not necessarily have the same size. In this case also, packets arrive and enter the switch without being aligned. Therefore, the packet switching function could take place at any point in time. The behaviour of packets in an unslotted network is more unpredictable than in a slotted one. This leads to an increased possibility of *packet contention*, that is to say packets e.g. intending to use the same switch ports at the same time, and therefore impacts negatively on the network throughput. Asynchronous operation also leads to an increased packet loss ratio. Then again, unslotted networks feature a number of advantages over slotted ones, such as increased robustness and flexibility, as well as ease of set-up and lower cost. The use of some methods for *contention resolution* can lead to a fairly good traffic performance, while the use of complicated packet alignment units of synchronous networks is avoided [7].

Packets travelling in a packet switched network experience variant delays. Packets travelling on a fibre can experience different delays depending on factors such as fibre length, temperature variation and chromatic dispersion. The packet propagation speed is also affected by temperature variations. The sources of delay variations described so far can be compensated for statically and not dynamically (on a packet-by-packet basis) [36].

The delay variations mentioned above are delays that packets experience while they are transmitted between network nodes. The delays that packets experience in switching nodes are also not fixed. The contention resolution scheme and the switch fabric greatly affect the packet delay. In a slotted network that uses *fibre delay lines* as optical buffers, a packet can take different paths with unequal lengths within the switch fabric [36].

Packets that arrive in a packet switching node are directed to the switch's input interface. The input interface aligns the packets so that they will be switched correctly (assuming the network operates in a synchronous manner) and extracts the routing information from the headers. This information is used to control the switching matrix. The switching matrix performs the switching and buffering functions. The control is electronic, since optical logic is in too primitive a state to permit optical control currently. After the switching, packets are directed to the output interface, where their headers are rewritten.

The header and payload of a packet can be transmitted serially on the same wavelength. Guard times must account for payload position jitter and are necessary before and after the payload to prevent damages during header erasure or insertion. Although there are various techniques to detect and recognize packet headers at Gb/s speed either electronically or optically, it is still difficult to implement electronic header processors operating at such high speed to switch packets on the fly at every node [36]. Several solutions have been proposed for this problem. One of these suggests employing *subcarrier multiplexing*. In this approach, the header and payload are multiplexed on the same wavelength, but the payload data is encoded at the baseband, while header bits are encoded on a properly chosen *subcarrier frequency* at a lower bit rate. This enables header retrieval without the use of an optical filter. The header can be retrieved using a conventional photodetector. This approach features several advantages, such as the fact that the header interpretation process can take up the whole payload transmission time, but also puts a possible limit on the payload data rate. If the payload data rate is increased, the baseband will expand and might eventually overlap with the subcarrier frequency, which is limited by the microwave electronics.

According to another approach, the header and the payload are transmitted on separate wavelengths. When the header needs to be updated, it is demultiplexed from the payload and processed electronically. This approach suffers from *fibre dispersion*, which separates the header and payload as the packet propagates through the network. Subcarrier multiplexed headers have far less dispersion problems since they are very close to the baseband frequency.

Two major difficulties prevail in optical packet switching, namely there is currently no capability of bit level processing in the optical domain, and there is no efficient way to store information in the optical domain indefinitely. The former issue concerns the process of reading and interpreting packet headers, while the latter concerns the way packet contentions are resolved in an optical network. *Contentions* occur in the network switches when two or more packets have to exploit the same resource, for example when two packets must be forwarded to the same output channel at the same time. The adopted solutions to solve these contentions are a key aspect in packet-switched networks, and they can seriously affect the overall network performance. The optical domain offers new ways of resolving contentions, but does not allow the implementation of methods that have been widely used in traditional networks. Three methods for contention resolution are described in the following: *buffering*, *deflection routing* and *wavelength conversions*.

1.5.5.2 Buffering, Deflection Routing and Wavelength Conversions for Contention Resolution

First of all, the simplest solution to overcome the contention problem is to *buffer* contending packets (hence exploit the time domain). This technique is widely used in traditional electronic packet switches, where packets are stored in the switch's random access memory (RAM) until the switch is ready to forward them. Electronic RAM is cheap and fast. On the contrary, optical RAM does not exist. *Fibre delay lines* (*FDLs*) are the only way to 'buffer' a packet in the optical domain. Contending packets are sent to travel over an additional fibre length and are thus delayed for a specific amount of time.

Optical buffers are either *single-stage* or *multi-stage*, where the term stage represents a single continuous piece of delay-line [36]. Optical buffer architectures can be further categorized into *feed-forward* architectures and *feedback* architectures [29]. In a feedback architecture, the delay lines connect the outputs of the switch to its inputs. When two packets contend for the same output, one of them can be stored in a delay line. When the stored packet emerges at the output of the fibre delay line, it has another opportunity to be routed to the appropriate output. If contention occurs again, the packet is stored again and the whole process is repeated. Although it would appear so, a packet cannot be stored indefinitely in a feedback architecture, because of unacceptable loss. Also, arriving packets can pre-empt packets that are already in the switch. This allows the implementation of multiple quality-of-service (QoS) classes. In the feed-forward architecture, a packet has a fixed number of opportunities to reach its desired output [29]. Almost all loss a signal experiences in a switching node is related to the passing through the switch. The feed-forward architecture attenuates all signals almost equally, because every packet passes through the same number of switches.

The implementation of optical buffers using FDLs entails several disadvantages. Fibre delay lines are bulky and expensive. A packet cannot be stored indefinitely on an FDL. Generally, once a packet has entered an FDL, it cannot be retrieved before it emerges on the other side, after a certain amount of time. In other words, fibre delay lines do not have random access capability. Another disadvantage of FDLs is that they introduce noise to the optical signal.

Note that optical buffering can be used in combination with wavelength conversions (see below). The use of wavelength conversion minimizes the number of FDLs. Assuming n wavelengths can be multiplexed on a single FDL, each FDL has a capacity of n packets. (The more wavelengths can be multiplexed on each FDL, the more packets the FDL can store.) Tunable optical wavelength converters[16] (TOWCs) can be used to assign packets to unused wavelengths in FDLs; this reduces the total number of FDLs [6].

Generally, it is desirable that the need for buffering is minimized. If the network operates in a synchronous manner, the need for buffering is greatly reduced, since contention occurs only when two packets need to be forwarded to the same output channel.

Second, and apart from buffering, contentions may be resolved by *deflection routing* through exploitation of the space domain. If two or more packets need to use the same output link to achieve minimum distance routing, then only one is routed along the desired link, while others are forwarded on paths possibly leading to greater than minimum distance routing [36]. The deflected packets may follow long paths to their destinations, and thus suffer high delays. In addition, the sequence of packets may be disturbed.

Deflection routing can be combined with buffering in order to keep the packet loss rate under a certain threshold. Deflection routing without the use of optical buffers is often referred to as *hot potato routing*. When no buffers are employed, the packets' queuing delay is absent, but the propagation delay is larger than in the buffering solution, because of the longer routes that packets take to reach their destination. Simple deflection methods without buffers usually introduce severe performance penalties in throughput, latency, and latency distribution [35].

The most important advantage of the deflection routing method is that it is implemented relatively easily both in terms of hardware components and control algorithms. The effectiveness of this technique critically depends on network topology; mesh topologies with high number of interconnections greatly benefit from deflection routing, whereas minor advantages are gained in simpler topologies [9]. Moreover, clever deflection rules can increase network throughput. These rules determine which packets will be deflected and where to.

Finally, besides the two previous methods, we could utilize the additional dimension that is unique in the field of optics, i.e. wavelength, in order to resolve contentions by *wavelength conversions*. If two packets having the same wavelength are addressing the same switch outlet, one of them can be converted to another wavelength using a tunable optical wavelength converter (TOWC). Only if we run out of wavelengths is it necessary to resort to optical buffering.

By splitting the traffic load on several wavelength channels and by using tunable optical wavelength converters, the need for optical buffering is minimized or even completely eliminated. In order to reduce the number of converters needed, while keeping the packet loss rate low, wavelength conversion must be optimized. Not all packets need be shifted in wavelength. Decisions must be made concerning the packets that need conversion and the wavelengths they will be converted to.

1.5.5.3 General Optical Packet Switch Architectures

Next we investigate several *optical packet switches*. Certain examples that are based on the WDM broadcast-and-select architecture and therefore are of special interest to us are included. Before proceeding to examples, however, we need to know some fundamental and more general things concerning optical packet switches.

First, we examine a general architecture. A *general optical WDM packet switch*, shown in Figure 1.23, is employed in a synchronous network and consists of three main blocks [7]:

- *Packet/Cell encoder*: packets arriving at the switch inputs are selected by a demultiplexer, which is followed by a set of tunable optical wavelength converters that address free space in the fibre delay line output buffers. Optical to electrical interfaces situated after demultiplexers extract the header of each packet (where the packet's destination is written) and thus determine the proper switch outlet. This information is used to control the switch. Additionally, optical packet synchronizers must be placed at the switch inlets to assure synchronous operation.
- *Non-blocking space switch*: this switch is used to access the desired outlet and the appropriate delay-line in the output buffer. The size of the space switch is

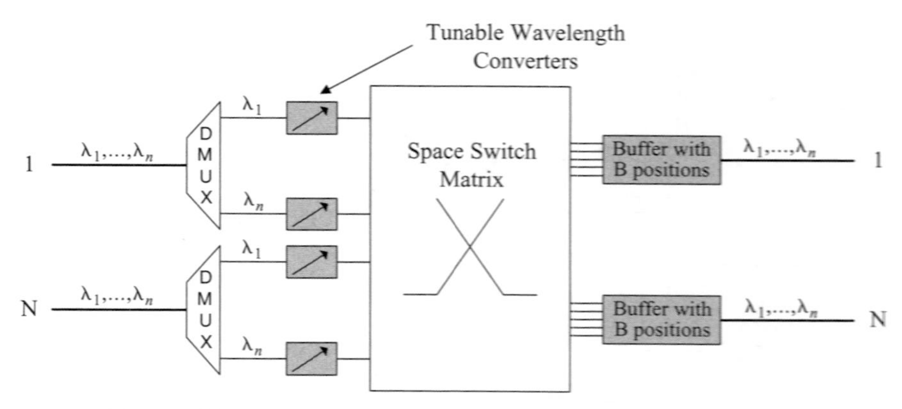

Figure 1.23 A general optical WDM packet switch [7]. (© 1998 IEEE)

$N \cdot n \times N \cdot (B/n + 1)$, where N is the number of input and output fibres and B is the number of positions in the buffer.

- *Buffers*: the switch buffers are realized using fibre delay lines.

The size of the space switch is nearly independent of the number of wavelengths. As a result, it is possible to increase the switch capacity by increasing the number of wavelength channels per inlet and outlet, while keeping the complexity in terms of number of components almost constant. The use of tunable optical wavelength converters significantly increases network throughput and higher allowed burstiness.

In the general WDM packet switch architecture of Figure 1.23 described above, tunable optical wavelength converters are used to handle packet contentions and efficiently access packet buffers. The use of TOWCs greatly improves switch performance, but implies more components and thus higher cost. In the scheme discussed above, a wavelength converter is required for each wavelength channel. That is to say for N inputs, each carrying n wavelengths, $n \cdot N$ wavelength converters will have to be employed. However, as noted in [9], only a few of the available TOWCs are simultaneously utilized; this is mainly due to the following two reasons:

- Unless a channel load of 100% is assumed, not all channels contain packets at a given instant.
- Not all packets contending for the same output line have to be shifted in wavelength because they may be already carried on different wavelengths.

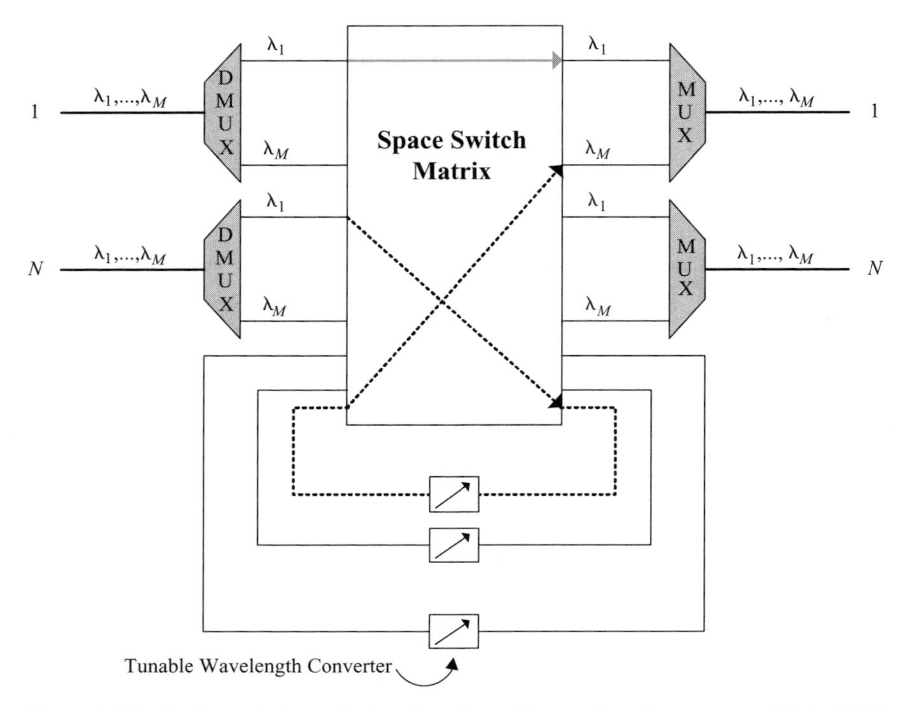

Figure 1.24 Packet switch employing shared tunable wavelength converters [9]. (© 2000 IEEE)

These observations suggest an architecture where TOWCs are shared among input channels and their number is minimized, so that only those TOWCs that are required to achieve a given performance level are employed.

A packet switch with *shared tunable wavelength converters* is shown in Figure 1.24. This switch is equipped with a number r of TOWCs, which are shared among input channels. At each input line, a small portion of the optical power is tapped to the electronic controller, which is not shown in the figure. The switch control unit detects and reads packet headers and drives the space switch matrix and the TOWCs. Incoming packets on each input are wavelength demultiplexed. Electronic control logic processes the routing information contained in each packet header, handles packet contentions and decides which packets have to be wavelength shifted. Packets not requiring wavelength conversion are directly routed towards the output lines; on the contrary, packets requiring wavelength conversions will be directed to the pool of r TOWCs and reach the output line after a proper wavelength conversion.

The switch control unit adopts a simple and fair technique in assigning the various output channels to TOWCs in each time slot. If in a time slot there are conversions to be made, the control unit randomly selects an output fibre, among those having packets to be shifted and determines the required conversions by allocating output channels to TOWCs. This operation is repeated by the control unit so long as at least one packet conflict remains unsolved and TOWCs are still available. If all of the r TOWCs are allocated but one or more packet conflicts are not solved, the packets involved in the conflict(s) are lost.

The issue of estimating the number of TOWCs needed to satisfy predefined constraints on packet loss is addressed in [9]. Packet arrivals are synchronized on a time slot basis; therefore the number of required converters in a given time slot depends only on the number of arriving packets during that slot. The performance of the switch, expressed in terms of packet loss probability, depends only on traffic intensity. Thus, both converters dimensioning procedures and switch performances hold for any type of input traffic statistic. The dimensioning of converters does not depend on the considered traffic type, but only on its intensity.

Converter sharing allows a remarkably reduced number of TOWCs as compared to other switch architectures, where it is equal to the number of input channels. The drawbacks involved in sharing TOWCs are first, the enlargement of the switching matrix by a factor equal to the number of converters used (r) and, second, an additional attenuation of the optical signal caused by crossing the switching matrix twice.

It should be noted here that the wavelength converters employed in the switch architectures discussed above were assumed to be capable of conversions over the entire range of available wavelengths. In practical systems, however, a wavelength converter normally has a limited range of wavelength conversion capability. Moreover, a wide wavelength conversion range might slow down switching speed, since it generally takes a longer time to tune to a wavelength over a wider range. Limited wavelength conversion range in optical packet switches has been studied e.g. in [29].

1.5.5.4 KEOPS and the Knockout Switch: Two Examples of Optical Packet Switches

The *KEOPS* (*KEys to Optical Packet Switching*) project in 1995 extended the study of the packet switched optical network layer [7]. The KEOPS proposal defined a multi-Gb/s interconnection platform for end-to-end packetized information transfer that supports any dedicated electronic routing protocols and native WDM optical transmission.

In KEOPS, the duration of packets is fixed; the header and its attached payload are encoded on a single wavelength carrier. The header is encoded at a low fixed bit rate, to allow the utilization of standard electronic processing. The payload duration is fixed regardless of its content; the data volume is proportional to the user-defined bit rate, which may vary from 622 Mbit/s to 10 Gbit/s, with easy upgrade capability. The fixed packet duration ensures that the same switch node can switch packets with variable bit rates. Consequently, the optical packet network layer proposed in KEOPS can be considered both bit rate and (to some degree also) transfer mode *transparent*; e.g., both ATM cells and IP packets can be switched.

The final report on the KEOPS project [5] suggests a 14-byte header. Eight bytes are dedicated to a two-level hierarchy of routing labels. Three bytes are reserved for functional such as identification of payload type, flow control information, packet numbering for sequence integrity preservation and header error checking. A 1-byte pointer field flags the position of the payload relative to the header. Finally, two bytes are dedicated to the header synchronization pattern.

Each node in the KEOPS network has the following sub-blocks:

- an input interface, defined as a 'coarse-and-fast' synchronizer which aligns the incoming packets in real time against the switch master clock;
- a switching core which routes the packets to their proper destination, solves contention, and manages the introduction of dummy packets to keep the system running in the absence of useful payload;
- an output interface which regenerates the data streams and provides the new header.

Two architectural options for the implementation of the switching fabric were evaluated exhaustively. The first one, i.e. the *Wavelength Routing Switch*, utilizes WDM to execute switching, while the second one, i.e. the *Broadcast-and-Select Switch*, achieves high internal throughput thanks to WDM. Figure 1.25 shows the broadcast-and-select switch suggested in KEOPS.

The principle of operation for the broadcast-and-select switch can be described as follows. Each incoming packet is assigned one wavelength through wavelength conversion identifying its input port and is then fed to the packet buffer. By passive splitting, all packets experience all possible delays. At the output of each delay line, multiwavelength gates select the packets belonging to the appropriate time slot. All wavelengths are gated simultaneously by these gates. Fast wavelength selectors are used to select only one of the packets, i.e. one wavelength. Multicasting can be achieved when the same wavelength is selected at more than one output. When the same wavelength is selected at all outputs, broadcasting is achieved.

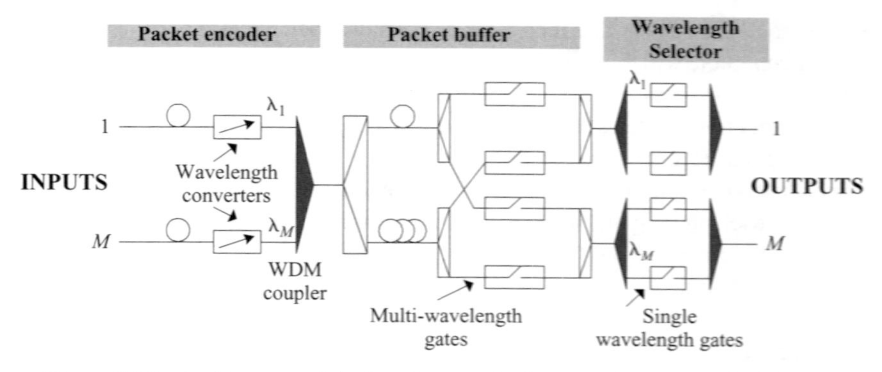

Figure 1.25 Architecture of the broadcast-and-select switch suggested in KEOPS.

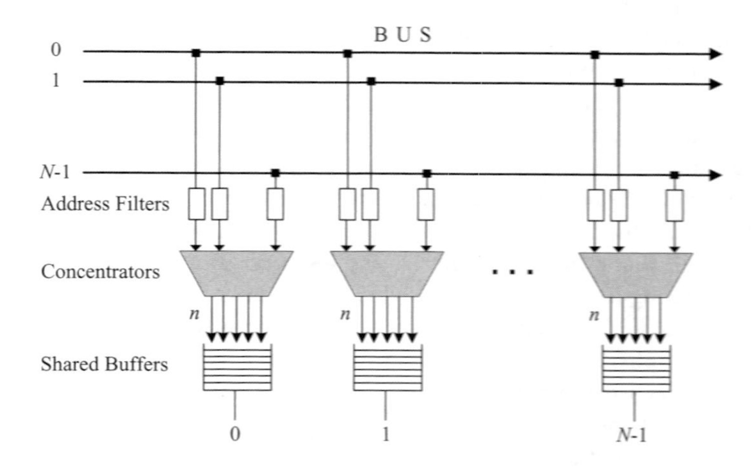

Figure 1.26 The original knockout switch.

Another example worth mentioning of a packet switch employing the WDM broadcast-and-select and wavelength routing architectures is the *Knockout Switch*. A schematic of the original knockout switch [22] is shown in Figure 1.26. This switch has N input and N output ports. Each output port can simultaneously accept n packets. The switch is based on a fully interconnected approach, i.e. every input port can directly reach every output port through a bus (thus, there are N^2 interconnections). Using this bus, all packets are distributed to the output ports of the switch. Each output port has three parts:

- a set of address filters that recognize packets destined for that port;
- a concentrator, which is an $N \times n$ switch that selects the packets that will be forwarded towards the output. If more than n packets are sent to the concentrator, the concentrator selects n packets and discards the rest. In order to select the winning packets, the concentrator performs a 'knockout tournament' among packets contending for the same output. The tournament is organized in several rounds, each picking one winner. The losers from one round continue to the next round; this enhances the fairness of the switch;
- an n-input 1-output queue that is used to smooth data output.

As a packet switch architecture, the knockout switch features high performance with simple switch fabric and packet contention control, and is suitable for high-speed packet switching. By applying optical technologies to the bus, the address filters the concentrators, the knockout switch can be made to handle bit streams in the order of Gb/s. In [22], a photonic knockout switch is proposed that uses wavelength division multiplexing. Since packets are transported using WDM, the N^2 interconnections of the bus output can be physically reduced to N interconnections.

The proposed switch is designed for fixed-length packets. Concerning the implementation of the optical bus, two types of switches are proposed: a broadcast-and-select switch and a wavelength routing switch.

The *broadcast-and-select WDM knockout switch* (Figure 1.27) [22] consists of fixed wavelength channel transmitters and a star coupler. All packets are copied to the wavelength channel selectors, which operate as concentrators. Routing and contention control are implemented by electronic circuits, which are referred to as routing controller and contention controller, respectively. Each transmitter sends its destination to the routing controller, which extracts the routing information (i.e. the wavelength channel to be selected) and feeds it to the contention controller. The contention controller selects the n winning packets (if necessary) and forwards them to the wavelength channel selectors. The speed of contention control limits switch throughput because control should be completed within one time slot. As the switch size increases, the contention control response becomes slower.

The *wavelength routing WDM knockout switch* shown in Figure 1.28 [22] consists of tunable wavelength channel transmitters and a wavelength router. The wavelength channel on which a packet is transmitted corresponds to its destination, i.e. *self-routing* is established. Apart from replacing the bus in the knockout switch, the wavelength router also acts as the address filter, since it eliminates the possibility of conflict between packets and the need for negotiations between transmitters. There is no need for contention control, since only address-matching packets are sent to the wavelength channel selectors. The wavelength channel selectors will most likely be implemented using wavelength routers and optical gates.

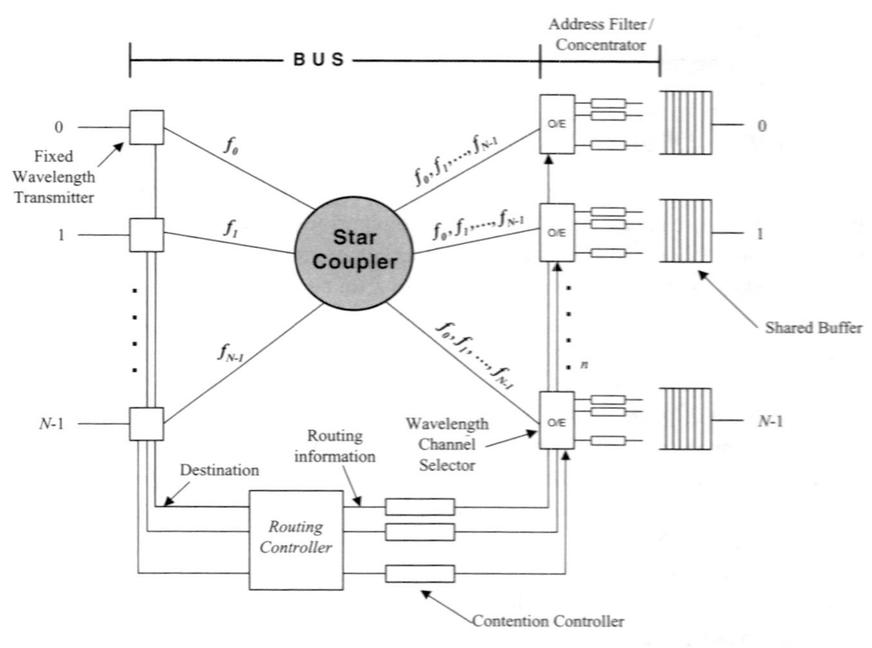

Figure 1.27 Broadcast-and-select WDM knockout switch [22]. (© 1998 IEEE)

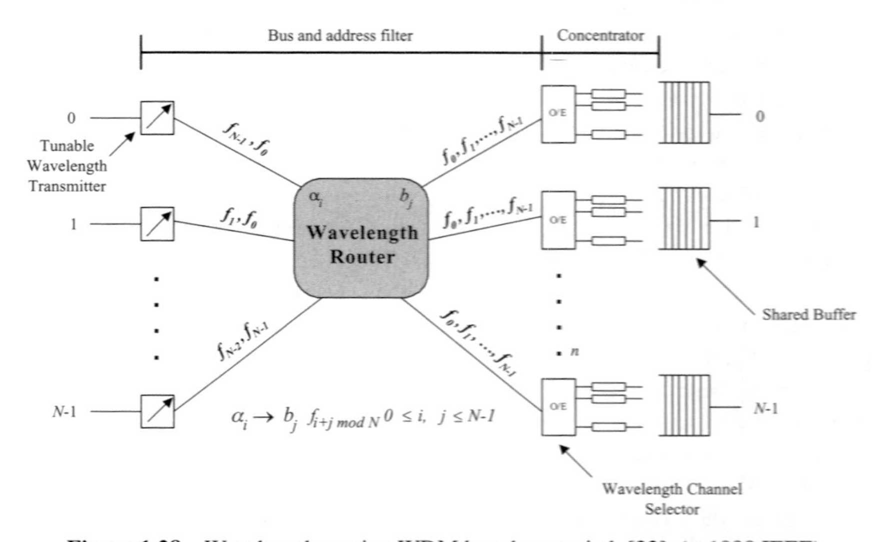

Figure 1.28 Wavelength routing WDM knockout switch [22]. (© 1998 IEEE)

This implementation satisfies the system requirements for high-speed switching and wide tuning range.

1.5.6 OPTICAL BURST SWITCHING NETWORKS

Optical burst switching (*OBS*) is an attempt at a new synthesis of optical and electronic technologies seeking to exploit the tremendous bandwidth of optical technology, while using electronics for management and control [34]. Burst switching is designed to facilitate switching of the user data channels entirely in the optical domain.

This approach divides the entire network in two regions: edge and core. At the edge the usual packets are assembled with some procedures to form *bursts*, i.e. collections of packets that have certain features in common (e.g. destination). Bursts are assigned to wavelength channels and are switched through transparently without any conversion. Bursts travel only in the core nodes, and when they arrive at the network edge they are disassembled into the original packets and delivered with the usual methods.

A burst switched network that has been properly designed can be operated at reasonably high levels of utilization, with acceptably small probability that a burst is discarded due to lack of an available channel or storage location; thus, very good statistical multiplexing performance is achieved. When the number of wavelength channels is large, reasonably good statistical multiplexing performance can be obtained with no burst storage at all [33].

The transmission links in a burst switched system carry multiple channels, any one of which can be dynamically assigned to a user data burst. The channels are wavelength division multiplexed. One channel (at least) on each link is designated as the *control channel* used to control dynamic assignment of available channels to user data bursts. It is also possible to provide a separate fibre for each channel in a multi-fibre cable.

In principle, burst transmission works as follows. Arriving packets are assembled to form bursts at the edge of the OBS network. The assembly strategy is a key design issue and is discussed later. Shortly before the transmission of a burst, a control packet is sent in order to reserve the required transmission and switching resources. Data is sent almost immediately after the reservation request without receiving an acknowledgement of successful reservation. Although there is a possibility that bursts may be discarded due to lack of resources, this approach yields extremely low latency as propagation delay usually dominates transmission time in wide area networks [8].

The reservation request (control packet) is sent on the dedicated wavelength some offset time prior to the transmission of the data burst. This basic offset has to be large enough to electronically process the control packet and set up the switching matrix for the data burst in all nodes. When a data burst arrives in a node, the switching matrix has been already set up, i.e. the burst is kept in the optical domain.

The format of the data sent on the data channels is not constrained by the burst switching system. Data bursts may be IP packets, a stream of ATM cells or frame relay packets. However, since the burst switching system must be able to interpret the information on the control channel, a standard format is required here [33]. The control packet includes a length field specifying the amount of data in the burst, as well as an offset field that specifies the time between the transmission of the first bit of the control packet and the first bit of the burst.

The way in which packets are assembled to form bursts can seriously affect network performance. The assembly method determines the network traffic characteristics. The process according to which bursts are assembled must take several parameters into account, such as destination of packets and quality of service requirements.

There must be a minimum requirement on the burst size. A burst must be sufficiently long in order to allow the node receiving the control packet to convert it into electronic form, process and update it (if necessary), and prepare the switching fabric. This requirement is also dictated by the limited capacity of the control channel. If bursts are made too small, the corresponding control packets may exceed the capacity of the control channel. Conversely, if data traffic is not intense, the burst generator must not delay bursts indefinitely until the minimum size requirement is met. After a specified time period has elapsed, existing packets must be joined to form a burst and, if necessary, the burst will be padded to achieve a minimum length.

After briefly studying optical burst switching and photonic packet switching, along with several examples of packet switches, some of which were actually based on the *WDM broadcast-and-select architecture* (thus relevant to the subject of our book), we have finished the general introduction to the various classes of (second generation) optical networks. In the next subsection, we examine in more detail WDM broadcast-and-select LANs, as stated in the Preface.

1.6 A CLOSER LOOK AT WDM BROADCAST-AND-SELECT LOCAL AREA NETWORKS

We have already discussed some of the basic features concerning WDM broadcast-and-select LANs when we examined the various classes of optical networks. In this subsection we give a closer look and attempt to describe in more detail the main issues, i.e. characteristics that differentiate one broadcast-and-select WDM network from the other.

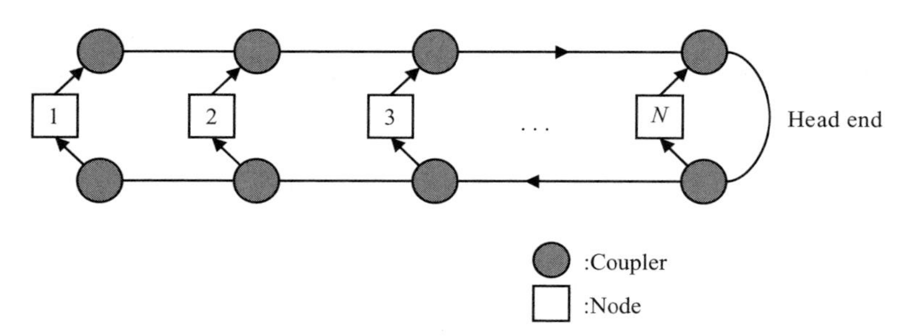

Figure 1.29 Folded bus topology.

1.6.1 PHYSICAL TOPOLOGY

The way nodes are connected with each other through (physical) fibre links is obviously a characteristic that differentiates the examined networks. Broadcast-and-select WDM networks can be implemented using various physical topologies, the most popular of which are *star*, *bus*, *tree* and *ring*.

Recall that we had discussed a star topology in Figure 1.11, while a (folded) bus topology for a WDM broadcast-and-select network is illustrated in Figure 1.29. The network has N nodes and the shared medium where broadcasting and selection take place is a folded bus. The nodes transmit through couplers (in fact functioning as 2×1 combiners) on the upper part of the topology and receive via couplers (functioning here as 1×2 splitters) on the lower part. Comparing star and bus we must notice that the number of couplers used in the case of bus changes linearly with N, while for constructing a $N \times N$ star coupler it can be shown that it only grows logarithmically with N. Since the non-ideal (of course) couplers used are accompanied by a relevant excess loss in signal power, it can be concluded that a star network can support more nodes without such significant power losses as in the case of a bus. However, since EDFAs are getting quite cheap, this drawback of bus is eliminated; anyway it is still recognized that the star topology is generally better than the bus for most cases of broadcast-and-select WDM networks.

An example of a tree physical topology is shown in Figure 1.30. The presented network consists of six nodes. Signals from nodes (1, 2) are combined and broadcast to nodes (3, 4, 5, 6) and, vice versa, signals transmitted from nodes (3, 4, 5, 6) are combined and sent to nodes (1, 2). Each node is equipped with a receiver to select the appropriate channel. Since a transmitted signal does not reach all nodes including the source itself as in a classical broadcast architecture, it would be more precise to use the term 'multicasting' instead of 'broadcasting' in this case.[17] A straightforward consequence is that the distinct channel assignment constraint is

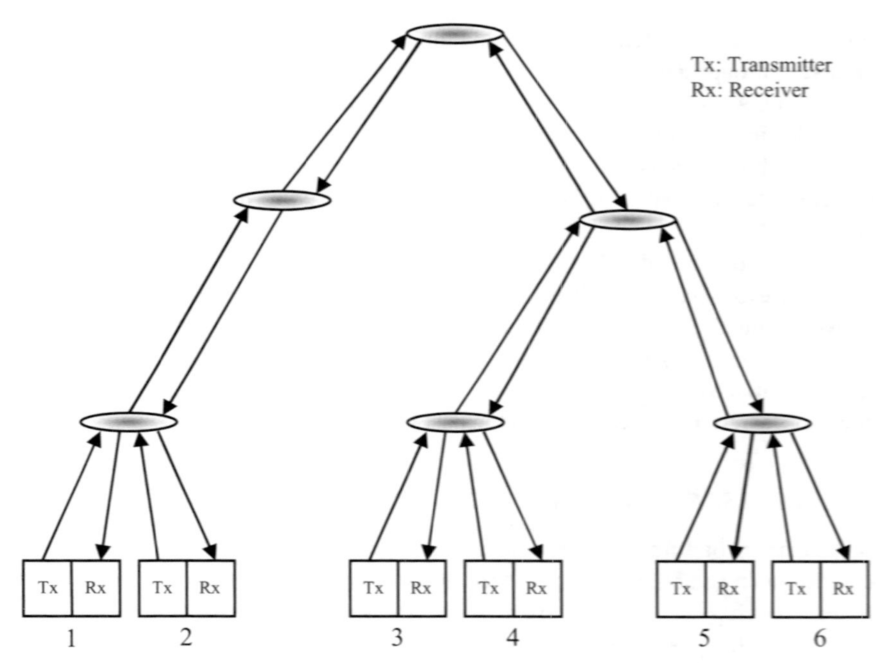

Figure 1.30 A tree physical topology.

not valid here as far as the two latter sets of nodes are concerned: for example, if node 2 transmits using wavelength λ, it is possible for node 4 (from the other set) to transmit on the same wavelength at the same time.

Another topology proposed for broadcast-and-select networks is the ring. Ring topologies have been demonstrated in a couple of experiments (far less than star of course). In comparison to star topologies, they generally suffer from greater losses in power for a given number of nodes, while, on the other hand, they offer better failure management, especially when the topology is a dual ring.

For example, a certain proposal for a broadcast-and-select ring network considers a two-fibre ring connection of nodes, with each node having two fixed-tuned transmitters to a unique pre-assigned wavelength and two arrays of receivers so that it can 'hear' from all other stations. EDFAs can greatly help the situation when an increase in the number of connected nodes is desired. More details about this network may be found in [31].

In [20] an experimental ring network is described. Specifically, a unidirectional WDM ring coming with a broadband light source and a passive star coupler is implemented. First, light from a single multi-wavelength source is distributed to all nodes through a passive star coupler and then data is transmitted between nodes on the ring. A node selects the wavelength corresponding to the desired destination (each destination is pre-assigned a unique wavelength) from the distributed light via a tunable bandpass filter and then transmits the data on the ring. The nodes receive data using add-drop filters that are fixed-tuned to their unique receiving wavelengths.

The development of EDFAs has recently revived interest in bus and ring topologies, however, here we will have to deal with star WDM broadcast-and-select networks mostly, in full accordance with the relevant literature which mainly focuses on star too. At any rate, the problem of designing the optimal physical topology, in terms of low (or at least working) losses in optical signal power, has been addressed quite frequently in the past [26].

1.6.2 TRANSCEIVERS PER NODE

Another point of differentiation among broadcast-and-select WDM networks is the number and type of transmitters and receivers assumed for each node in the system. More explicitly we could distinguish the following four possible combinations concerning the type of transceivers used by user nodes:

- Fixed-tuned Transmitter(s)—Fixed-tuned Receiver(s) (FT-FR);
- Fixed-tuned Transmitter(s)—Tunable Receiver(s) (FT-TR);
- Tunable Transmitter(s)—Fixed-tuned Receiver(s) (TT-FR);
- Tunable Transmitter(s)—Tunable Receiver(s) (TT-TR).

On the right of each combination the symbols for the corresponding WDM systems are quoted, where for instance TT-FR indicates a system with one tunable transmitter and one fixed-tuned receiver per node. As an example, the broadcast star network of Figure 1.11 is clearly a FT-TR system.

In general, each node could be equipped with one or more transmitters and one or more receivers, each one of either the fixed-tuned or tunable type. In these symbolizations the exact number of transceivers per node can easily be introduced as a superscript in the proper position resulting in the general representation of a WDM system as: $FT^iTT^j—FR^kTR^l$. Consequently, a sample system with three tunable transmitters and N fixed-tuned receivers per node would be symbolized as a TT^3-FR^Nsystem. Let us note for completeness that, if M separate channels are used by nodes for control purposes, 'CC^M' is usually added at the start of the symbol (to denote the 'M Control Channels'). As we will see, a protocols category of great size and importance has the key feature of using one or more separate (or just time-multiplexed with data) control channels for coordination among network nodes. If one separate control channel was employed in the previous example, the system would be said to be CC-TT^3-FR^N.

1.6.3 SINGLE-HOP AND MULTI-HOP NETWORKS

Yet another fundamental distinction of broadcast-and-select WDM networks is between the so-called *single-hop* and *multi-hop* systems. What is considered here is the actual number of intermediate nodes that a signal needs to pass through in order to reach its final destination, for all sources and destinations. In other words it is examined whether for every possible source-destination pair in the network the number of hops made by a prospective transmitted signal is exactly one (i.e. there are no intermediate nodes) or more (there are intermediate nodes along the way). In the former instance of zero intermediate nodes the broadcast-and-select network is said to be single-hop and in the latter case multi-hop. Apparently, when a signal has to pass through intermediate nodes, it has to be converted into electronic form at each intermediate station, and then back to optical form, before it is routed through the network towards the intended final recipient. On the other hand, in single-hop networks the light signal remains in optical form all the way from one end to the other; this implies high speeds and data transparency.

An example of a single-hop WDM broadcast-and-select star network was presented in Figure 1.11. In this FT-TR network, tunability on the receivers' side over the entire range of system's channels ensures the capability of communication between any source-destination pair in exactly one hop: so long as a receiver tunes to the appropriate wavelength λ_i it can achieve single-hop communication with any source i, for $i = 1, \ldots, N$. Indeed, a broadcast-and-select network can generally be single-hop, either if in each node there is tunability at least on one side (transmitters or receivers) or if at least one side is equipped with an array of exactly W fixed-tuned devices, where W denotes the number of wavelengths in the system. Obviously, as W grows a bit, the latter solution turns out to be rather too expensive, if we keep in mind that transceivers are generally costly. On the other hand, even though there are devices with a wide tuning range, these usually have other important deficiencies, i.e. they have slow (millisecond) tuning times or they are not acceptably narrowband[18] [28]. In this context and until rapidly tunable narrowband transceivers with wide tuning ranges are available at a reasonable cost, single-hop networks seem more appropriate for circuit-switching services where slower tuning times could be accepted, contrary to packet-switching. They are also good candidate solutions when the number of nodes and wavelengths in the system is relatively small.

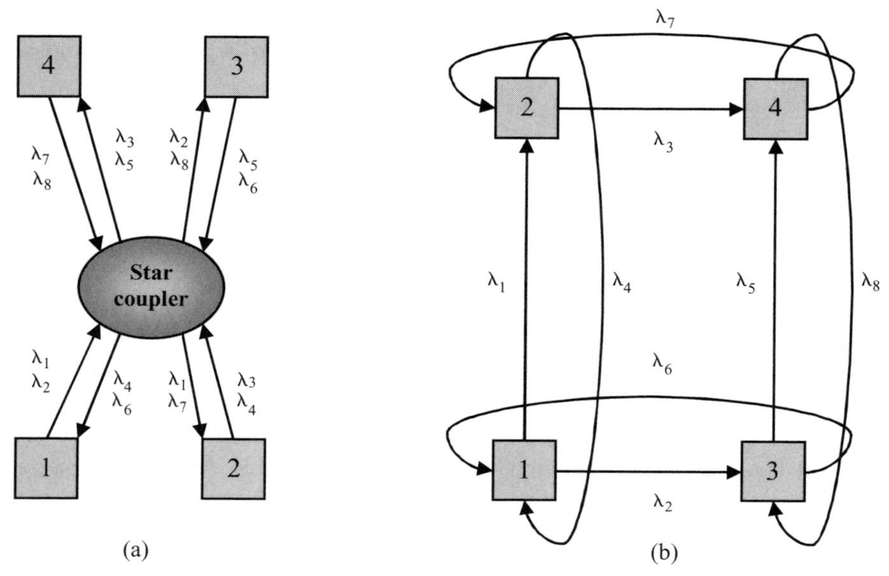

Figure 1.31 (a) A physical star multi-hop topology and (b) the corresponding virtual torus topology.

A way of coping with the aforementioned problems is to use a multi-hop network, like the one shown in Figure 1.31. Figure 1.31(a) illustrates the physical topology of a multi-hop WDM broadcast-and-select network including four nodes interconnected via an optical passive star coupler. Each node can transmit and receive in exactly two fixed wavelengths, different for transmission and reception, as depicted in the figure. Using the terminology of the previous paragraph, this is obviously a FT^2-FR^2 network. Multi-hopping is the case here: it is easy to notice for example that node 2 can transmit to node 3 either via node 4 through λ_3 and λ_8, or via node 1 through wavelengths λ_4 and λ_2. In general, multi-hop systems have a small number of fixed-tuned transmitters and fixed-tuned receivers per node. However, multi-hopping may also be the case when hardware equipment of nodes includes tunable devices with limited tuning capability. We will refer to such a case later.

The physical topology is in practice transparent to end users, who only 'see' a set of logical connections provided to them by the optical infrastructure. The set of logical connections that are realized over a given transparent optical physical topology constitute the so-called *logical* (*or virtual*) *topology* which is said to be embedded over the underlying physical topology. In the case considered, a virtual *torus* topology (Figure 1.31(b)) is embedded over the supporting physical star topology of Figure 1.31(a).

In multi-hop networks the assignment of wavelengths to transmitters and receivers of the various nodes is generally static in nature. Even when tunable devices are employed, rapid tuning times are not that critical, as in single-hop systems. That is because in multi-hop networks the assignment of channels need not change that frequently; global changes may be decided only sporadically when some new set of assignments seems more advantageous for the overall network performance. What *is* really important in multi-hop systems is the choice of a virtual topology (to be embedded over the physical infrastructure) ensuring the best possible network performance, for example, in terms of minimal average number of hops that a packet needs to traverse in its way from source to destination. In addition to this, it is also crucial that by adopting this topology the routing mechanism needed does not become too complex and thus wasteful of time, which is generally unacceptable in high-speed communication systems like optical networks. As expected, the problem of designing good virtual topologies has been extensively addressed in the relevant literature, since the time when multi-hopping in broadcast-and-select networks was proposed as a solution to the already discussed problems of single-hop systems in employing packet-switched services, around 1987.

The virtual topology design efforts were basically focused on two different directions: *irregular* and *regular* topologies. Topologies with several well-known properties and the special feature that all nodes have the same degree are termed ‘regular’, while the remaining ones are called ‘irregular’.

The design of irregular topologies requires development of special heuristic methods that usually are not that simple. The proposed heuristics typically concentrate on achieving the best possible performance, i.e. straightforwardly seek optimality in terms of (for example) minimum average number of hops. They do not presuppose any type of constraint in the supported number of nodes and furthermore their effectiveness is not affected by the various traffic characteristics, even though they take traffic demands into account in their effort to maximize performance. On the other hand, their main deficiency regards the necessary accompanying routing mechanism, which usually tends to be of high complexity and therefore imposes analogously high processing delays at nodes. Many studies have dealt with the logical topology design problem following this approach, e.g. [1] and [25].

In the case of regular topologies, such as torus with degree equal to two in Figure 1.31(b), the traffic demands are not taken into account and optimal performance is not the highest priority. Of course a high performance is always desired but in combination with simple routing strategies and, therefore, acceptable processing delays. All these are achieved as a consequence of their structure. Their main drawback, however, is that performance is not satisfactory under non-uniform traffic (commonly encountered in practice). Moreover, these topologies typically impose some constraints on the number of nodes, i.e. they may not support arbitrary number of nodes; note that one remarkable exception to this rule is the *GEMNET* topology [17]. Other important and well-studied regular topologies proposed as embedding options over broadcast-and-select WDM networks include the popular *Shufflenet, de Bruijn graph, Manhattan Street Network* or *torus* (used in our example), *hypercube* and *linear dual-bus*. A good starting point with references for all of these is [23].

1.6.3.1 Regular Topology Embedding Issues over Broadcast-and-Select Networks

In this section we underline the main aspects of how a regular topology can be embedded over a WDM broadcast-and-select network with N nodes. We consider that the underlying physical topology is a star.

Let d denote the degree of the virtual topology. This implies that each node would generally need d transmitters and d receivers. According to the digital channel assignment constraint for WDM broadcast-and-select networks, no two transmitters should transmit signals on the same channel simultaneously. Analogous constraint of course stands for receivers too. This dictates tuning each transmitter in the network to a different wavelength, i.e. a total of $\mathrm{M} = N \cdot d$ wavelengths would be required. The receivers would be tuned similarly on those $N \cdot d$ wavelengths. The critical point here is the particular way these wavelengths are assigned to transceivers in every instance. Exactly this will determine the virtual topology embedded over the physical star infrastructure each time.

As an example let us consider again the network of Figure 1.31. Since the virtual topology desired was torus of degree two, each node was equipped with two transmitters and two receivers, and the assignment of wavelengths shown in Figure 1.31(a) was such that in fact the torus virtual topology of Figure 1.31(b) was implemented.

Note that it would be really advantageous for a multi-hop system if it were capable of *changing its global channel assignments* every now and then (not too frequently of course) according to the specific traffic characteristics, so that a more appropriate virtual topology is embedded for each case and performance is improved. When this happens, we say that the multi-hop system has the capability of *(virtual) topology reconfiguration*. Apart from conforming to varying traffic

Table 1.2 Wavelengths assigned to transmitters and receivers for each node of the network in Figure 1.31(a) so that it embeds the virtual topology of Figure 1.32.

Node #	Transmitters	Receivers
1	λ_1,λ_8	λ_4, λ_7
2	λ_2,λ_7	λ_1, λ_6
3	λ_3,λ_6	λ_2, λ_5
4	λ_4,λ_5	λ_3, λ_8

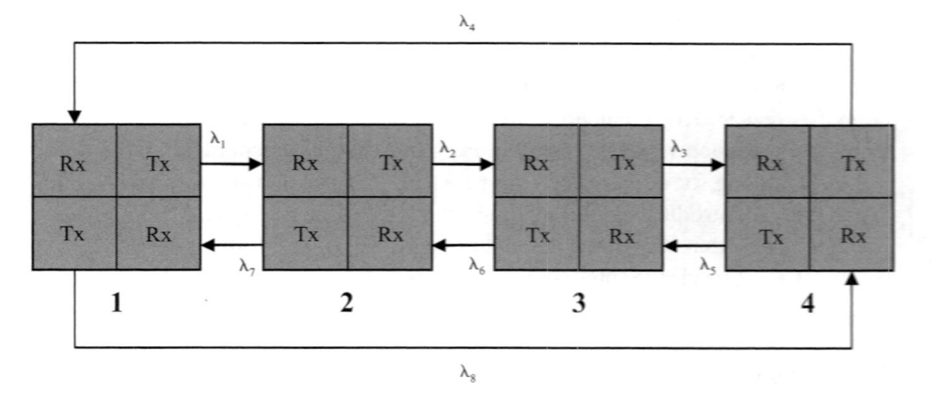

Figure 1.32 A simple virtual topology with 4 nodes and degree 2 (dual-ring).

patterns, topology reconfiguration is really helpful as a response to potential failures in the network.

Let us assume for instance that at some point of network operation, it is decided that (for some reason) the network of Figure 1.31(a) should embed another regular topology, say, for example the one demonstrated in Figure 1.32. Four nodes are shown in full accordance to the specific example. This topology resembles a linear dual-bus topology with additional links between the first and last node. Thus, a dual-ring is essentially formed. It is quite interesting to see that in order to embed this topology, the nodes of the network in Figure 1.31(a) should tune their transmitters and receivers exactly as Figure 1.32 dictates, i.e. as shown in Table 1.2.

It is worth noting that several studies have been reported on the subject of constructing regular virtual topologies (having, of course, simple routing mechanisms as discussed) that additionally pursue some kind of optimality in performance (or at least a near-optimal operation in some sense), which is one point where regular topologies are generally deficient in. In the previously referred study [23] one could find information on studies dealing with this problem for various structures including linear dual-bus. For this particular case, heuristic algorithms that determine how the nodes of a linear dual-bus network should near-optimally be placed are proposed.

Another important observation that has to be made finally concerns the number of wavelengths required in multi-hop systems with broadcast star physical topologies. The $W = N \cdot d$ wavelengths needed (in our example $N = 4, d = 2$ and thus $W = 8$) may be quite a serious concern as compared to the single-hop case where only N wavelengths suffice (Figure 1.11, $W = N$). The solution in this case is to allow more than one link of the virtual topology to *share* the same channel. That is to say, within each channel we could have another level of electronic frequency-division multiplexing where the channel bandwidth can be subdivided into many radio frequency (RF) channels, each at a different microwave frequency. This technique is termed *subcarrier multiplexing* (*SCM*) (also encountered in photonic packet switching for header processing). An alternative (and at least this familiar) approach is to introduce electronic time-division multiplexing (TDM) within each wavelength, so that it is time-shared among virtual links. Both solutions can decrease (if so desired) the total number of required channels in a WDM network.

1.6.3.2 Notes on Related Work

The authors in [4] examine how three graph-based topologies can be embedded over a WDM optical passive star network. The topologies considered are the *complete-graph, mesh* and *hypercube*, which happen to be among the most widely used ones in networks and parallel computing. The key transceiver feature taken into account on this work is the *limited tuning range* of transmitters which happens to be quite realistic i.e. in accordance to the devices encountered in practice. The system model is a WDM passive star network such as the ones described here with one tunable transmitter and one fixed-tuned receiver per node, i.e. it is a TT-FR WDM network. The limited tunability of transmitters results in the need for multi-hopping within the network. It also has several effects on the maximum number of wavelengths that can be used in different topological embeddings and on the maximum delay, which is an important performance metric in computer networks.

For the complete-graph topology, first, a lower bound on the maximum delay (measured in number of hops) depending on the tuning range of transmitters and the number of channels is determined. Then, an algorithm that embeds a complete-graph topology over a WDM passive star network is described. This algorithm assigns channels to transmitters and receivers in a way guaranteeing both that topology is embedded and an optimal performance is achieved (by ensuring the maximum delay meets the previously derived lower bound). That is to say, the algorithm minimizes the maximum number of hops needed in a complete-graph topology.

In addition, for both the mesh and hypercube topologies the authors in [4] derive a relationship between the total number of wavelengths that can be used and parameters of each specific structure. In each case, efficient embedding algorithms capable of hosting the maximum possible number of wavelengths are provided.

Another noteworthy study [18] considers a FT-TR architecture, in which, however, the system retains its multi-hop character (the tunability range may be limited, for example). Here the main goal of using tunable receivers is to provide the network with the capability of virtual topology reconfiguration. A new class of protocols is proposed for this multi-hop system, namely *Multiconfiguration Multihop Protocols* (*MMPs*). In the *static version* of these protocols the network cycles through certain configurations (i.e. virtual topologies determined by the state of tunable receivers) in a TDM way. The selection of the specific topologies is based on known *a priori* (or expected) traffic patterns. On the other hand, the *dynamic version* of MMPs attempts to be more adaptive in terms of employing a specific configuration on the basis of current traffic conditions. So, in this case, the virtual topology selection is in fact a reply by the system to the prevailing traffic conditions, which aims to be profitable for the overall performance.

1.6.3.3 Hypergraph Topologies

So far we have considered only the case where the virtual topology embedded over a broadcast-and-select WDM network is graph-based. In graph-based topologies there are only point-to-point links between nodes, i.e. each virtual link connects exactly two nodes. However, it has been noticed that these topologies usually suffer from one of the following problems [2]:

- need for quite a lot of transceivers;
- need for supporting many channels;
- large diameter.[19]

An interesting concept that has emerged over the last years is to use an alternative type of topologies, namely *hypergraph topologies*, which are a generalization of simple graph-based structures. In such topologies a virtual 'link' connects

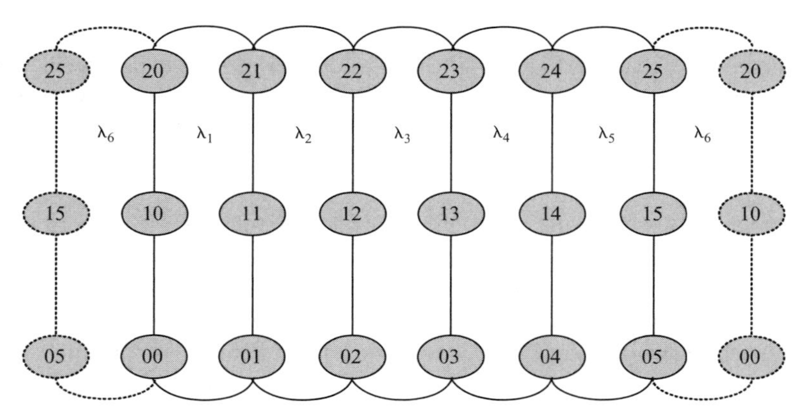

Figure 1.33 A stack-ring hypergraph topology of size 6 and order 3 [30]. (© 1998 IEEE)

more than just two nodes and therefore hypergraphs directly support one-to-many (multicast) transmissions. As far as WDM broadcast-and-select star networks are concerned, this keeps up well with the internal operation. Such a virtual 'link' is termed *hyperedge* and is implemented as a specific channel shared by all nodes within the hyperedge. Thus, each hyperedge can be seen as a simple shared medium like a bus, i.e. it essentially corresponds to a logical bus that can be accessed by all nodes included in the hyperedge.

A relatively simple example of a hypergraph topology is shown in Figure 1.33 [30]. This hypergraph topology is called a *stack-ring*, because it is formed by piling up several simple ring graphs. In Figure 1.33, as an example, nodes 20, 21, 11, 01, 00 and 10 are within the same hyperedge. A hyperedge models a logical bus in the sense that if, for example, node 11 transmits a message on wavelength λ_1 nodes 20, 21, 01, 00 and 10 will also receive it. If a source node wants to send to a destination node of another hyperedge, the message will have to pass through several virtual links (i.e. other hyperedges) before reaching the final destination, as in every multi-hop network. In accordance with the above description, a single wavelength is assigned to each hyperedge in the figure. The dashed lines suggest that the first triple of nodes (20, 10, and 00) connects to the last triple (25, 15, and 05) by means of a hyperedge that is assigned wavelength λ_6.

In the same way as a stack-ring can be constructed by piling up multiple rings, a *stack-graph* hypertopology can be constructed as a generalization of a graph topology. More complex hypergraphs can also be constructed by the *Cartesian product of stack-graphs*. In [3] a stack-ring is defined in more detail along with *stack-torus* and *hypertorus*. The stack-graphs are regular virtual (hyper) topologies, while this is not the case (at least globally) for hypertorus. These three hypergraph topologies are compared in [3] with two very popular graph-based ones: hypercube and torus. To be exact, the comparison regards performance achieved when the same star WDM broadcast-and-select network embeds each virtual topology. Five performance metrics are considered, namely *packet delivery time*, *waiting time*, *link utilization*, *throughput* and *number of hops*. The results show that hypergraph topologies outperform graph-based ones in almost all aspects. They are especially suited for multicast communications, but generally do better than graph-based topologies even in the case of one-to-one communications. Work on this relatively recent concept of hypergraph topologies embedding over WDM broadcast-and-select networks, is of course still in progress.

1.6.4 MEDIA-ACCESS CONTROL (MAC) PROTOCOLS

So far we have investigated several features that distinguish broadcast-and-select WDM local area networks from one another. Nevertheless, we have not yet examined a characteristic with major significance for multiwavelength optical LANs: the common strategy (rules) implemented in the network and adopted by all nodes in order to communicate with each other, that is to say the *media-access control (MAC) protocol* used.

As noted in the overview of the book presented in the Preface, a thorough investigation of MAC protocols is carried out in Chapter 3. Thus, we defer the discussion on this crucial point of optical LANs until then. Before this, however, the main features of optical devices used as building blocks for multiwavelength optical networks are presented and described in detail next in Chapter 2 which, therefore, completes the necessary background knowledge for the rest of the book.

NOTES

1 Time-division multiplexing (TDM) will be explained later on a separate section about multiplexing techniques.
2 Of course such problems do not exist when communication takes place outside the Earth's atmosphere.
3 Wavelength-division multiplexing (WDM) will be explained later in a separate section about multiplexing techniques.
4 An end-user can be an individual workstation or a gateway that interfaces with lower-speed subnetworks.

5 Small channel spacing accounts for the term 'dense', yet in this book by WDM we will often include (or even refer only to) DWDM systems. The details on when systems should be called DWDM or not, are beyond the scope of this chapter; WDM will primarily be used from now on.
6 Of course, the actual distances are shorter, if we think that optical fibres were installed in groups of fibre cables containing e.g. 20 fibres each, while many cables were laid all along a certain route.
7 ITU was called CCITT back in 1986 when the attempts at developing SDH began.
8 Examples of such high demand services were mentioned in the beginning of this chapter.
9 From now on when we simply mention 'optical networks' we will refer to second generation optical networks.
10 There are certain other types of optical networks such as Optical Code-Division Multiplexing (OCDM); these are considered beyond the scope of this introduction as rather futuristic. For OCDM the interested reader could refer to [19] for example.
11 The optical network component devices performing wavelength conversion (i.e. wavelength converters) will be explored in Chapter 2 on enabling technology. Wavelength conversion will also be discussed briefly in a following subsection.
12 WGRs are used in a type of WDM Passive Optical Networks (PONs).
13 Optical switches are typically included in the implementation of full or limited wavelength conversion. Here we refer to optical switches that are not part of wavelength converters and are used inside the WXC node to offer dynamic wavelength routing capability.
14 Recall that Figure 1.17 shows a static WXC which in fact is a passive router also called a waveguide grating router (WGR). Passive star couplers are used in broadcast-and-select networks.
15 Contrary to this, without WDM the total bandwidth of the feeder network would have to be *shared* by means of a proper media-access control protocol among the ONUs.
16 See Chapter 2 on enabling technology.
17 Maybe the unproved term 'multicast and select' better describes the architecture of this WDM tree network.
18 Of course significant progress has been made since [28] and is constantly being made in laboratories. At any rate the cost usually remains high.
19 The *diameter* of a virtual topology is defined as follows: consider a set of values, with each value being equal to the minimum possible path length

(in hops) between two nodes, for all node pairs. The maximum value in this set is the diameter of the topology.

REFERENCES

[1] Bannister A, Fratta L and Gerla M, "Topological design of the wavelength-division optical network", in *Proceedings IEEE INFOCOM'90*, San Francisco California vol. 3, pp. 1005–1013, June 1990.

[2] Bogineni K and Dowd PW, "Collisionless media access protocols for high speed communication in optically interconnected parallel computers", in *Proceedings SPIE* vol. 1577, pp. 276–287, September 1991.

[3] Bourdin H, Ferreira A and Marcus K, "A performance comparison between graph and hypergraph topologies for passive star WDM lightwave networks", *Elsevier Computer Networks and ISDN Systems*, vol. 8, no. 30, pp. 805–819, 1998.

[4] Cao F, Du DHC and Pavan A, "Topological embedding into WDM optical passive star networks with tunable transmitters of limited tuning range", *IEEE Transactions on Computers*, vol. 47, no. 12, pp. 1404–1413, December 1998.

[5] Chiaroni D, Janz C, Renaud M, Gravey P, Guillemot C, Gambini P, Hansen PB, Schilling M, Corazza G, Melchior H, Andonovic I and Renaud M, "Keys to optical packet switching", *KEOPS Project (European Union Programme ACTS) Final Report*, December 1998.

[6] Danielsen SL, Mikkelsen B, Joergensen C, Durhuus T and Stubkjaer KE, "WDM packet switch architectures and analysis of the influence of tunable wavelength converters on the performance", *IEEE Journal of Lightwave Technology*, vol. 15, no. 2, pp. 219–227, February 1997.

[7] Danielsen SL, Hansen PB and Stubkjaer KE, "Wavelength conversion in optical packet switching", *IEEE Journal of Lightwave Technology*, vol. 16, no. 12, pp. 2095–2108, December 1998.

[8] Dolzer K and Gauger C, "On burst assembly in optical burst switching networks—A performance evaluation of just-enough-time", in *Proceedings of ITC-17*, Salvador da Bahia, Brazil pp. 149–160, September 2001.

[9] Eramo V and Listanti M, "Packet loss in a bufferless optical WDM switch employing shared tunable wavelength converters", *IEEE Journal of Lightwave Technology*, vol. 18, no. 12, pp. 1818–1833, December 2000.

[10] Frigo NJ, "Local access optical networks", *IEEE Network Magazine*, pp. 32–36, November/December 1996.

[11] Green Jr PE, "Optical networking update", *IEEE Journal on Selected Areas in Communications*, vol. 14, pp. 764–779, June 1996.

[12] Green Jr PE, "Optical networking has arrived", *IEEE Communications Magazine Guest Editorial*, vol. 36, p. 38, February 1998.

[13] Henderson PM, "Fundamentals of SONET/SDH", *Mindspeed Technologies, Inc.* 2001.

[14] Hunter DK, "Issues in optical packet switching", in *Business Briefing: Global Optical Communications*, World Markets Research Centre (WMRC), pp. 58–62, July 2001.

[15] Iannone PP, Reichmann KC and Frigo NJ, "Broadcast digital video delivered over WDM passive optical networks", *IEEE Photonics Technology Letters*, vol. 8, no. 7 pp. 930, July 1996.

[16] Iannone PP, Reichmann KC and Frigo NJ, "High-speed point-to-point and multiple broadcast services delivered over a WDM passive optical network", *IEEE Photonics Technology Letters*, vol. 10, no. 9, pp. 1328–1330, September 1998.

[17] Iness J, Banerjee S and Mukherjee B, "GEMNET: A generalized, shuffle-exchange-based, regular, scalable, modular, multihop, WDM lightwave network", *IEEE/ACM Transaction on Networking*, vol. 3 no. 4, pp. 470–476, August 1995.

[18] Jue JP and Mukherjee B, "Multiconfiguration multihop protocols: A new class of protocols for packet-switched WDM optical networks", *IEEE/ACM Transactions on Networking*, vol. 8, no. 5, pp. 631–642, March 1998.

[19] Kitayama K, Sotobayashi H and Wada N, "Optical code division multiplexing (OCDM) and its applications to photonic networks", *IEICE Transactions on Fundamentals of Electronics, Communications and Computer Sciences*, vol. E82-A, pp. 2616–2626, December 1999.

[20] Lauder RDT, Badcock JM, Holloway WT and Sampson DD, "WDM ring network employing a shared multiwavelength incoherent source", *IEEE Photonics Technology Letters*, vol. 10, no. 2, pp. 294–296, February 1998.

[21] Melle S, Pfistner CP and Diner F, "Amplifier and multiplexing technologies expand network capacity", *Lightwave Magazine*, pp. 4246, December 1995.

[22] Misawa A, Sasayama K and Yamada Y, "WDM knockout switch with multi-output-port wavelength-channel selectors", *IEEE Journal of Lightwave Technology*, vol. 16, no. 20, pp. 2212–2219, December 1998.

[23] Mukherjee B, "WDM-based local lightwave networks—part II: Multihop systems", *IEEE Network Magazine*, vol. 6, no. 4, pp. 20–32, July 1992.

[24] Mukherjee B, "WDM optical communication networks: progress and challenges", *IEEE Journal on Selected Areas in Communication*, vol. 18, no. 10, pp. 1810–1824, October 2000.

[25] Mukherjee B, Banerjee D, Ramamurthy S and Mukherjee A, "Some principles for designing a wide-area optical network", *IEEE/ACM Transactions on Networking*, vol. 4, no. 5, pp. 684–696, October 1996.

[26] Nassehi MM, Tobagi FA and Marhic ME, "Fibre optic configurations for local area networks", *IEEE Journal on Selected Areas in Communication*, vol. SAC-3, pp. 941–949, November 1985.

[27] Papadimitriou GI, Obaidat MS and Pomportsis AS, "Advances in optical networking", *Wiley International Journal of Communication Systems*, vol. 15, nos 2–3, pp. 101–114, March/April 2002.

[28] Ramaswami R, "Multiwavelength lightwave networks for computer communication", *IEEE Communications Magazine*, vol. 31, pp. 78–88, February 1993.

[29] Ramaswami R and Sivarajan KN, '*Optical Networks: A Practical Perspective*', Morgan Kaufmann Publishers, San Francisco, CA, 1998 and 2001 editions.

[30] Senior JM, Handley MR and Leeson MS, "Developments in wavelength division multiple access networking", *IEEE Communications Magazine*, vol. 36, no. 12, pp. 28–36, December 1998.

[31] Sharma M, Ibe H and Ozeki T, "WDM ring network using a centralized multiwavelength light source and add-drop multiplexing filters", *IEEE Journal of Lightwave Technology*, vol. 15, no. 6, pp. 917–929, June 1997.

[32] Stern TE and Bala K, *Multiwavelength Optical Networks: A Layered Approach*, Addison Wesley Longman Editions, Reading, MA, 1999.

[33] Turner JS, "Terabit burst switching", *IOS Press Journal of High Speed Networks*, vol. 8, no. 1, pp. 3–16, 1999.

[34] Turner JS, "WDM burst switching for petabit data networks", in *Proceedings of Optical Fibre Communication Conference*, vol. 2, pp. 47–49, 2000.

[35] Yang O, Bergman K, Hughes GD and Johnson FG, "WDM packet routing for high-capacity data networks", *IEEE Journal of Lightwave Technology*, vol. 19, no. 10, pp. 1420–1426, October 2001.

[36] Yao S, Mukherjee B and Dixit S, "Advances in photonic packet switching: an overview", *IEEE Communications Magazine*, vol. 38, no. 2, pp. 84–94, February 2000.

[37] Zang H, Jue JP and Mukherjee B, "A review of routing and wavelength assignment approaches for wavelength-routed optical WDM networks", *SPIE Optical Networks Magazine*, vol. 1, no. 1, January 2000.

[38] Zang H, Jue JP, Sahasrabuddhe L, Ramamurthy R and Mukherjee B, "Dynamic lightpath establishment in wavelength-routed WDM networks", *IEEE Communication Magazine*, vol. 39, pp. 100–108, September 2001.

2

Enabling Technologies

Telecommunications technology is undergoing an extraordinary transformation, brought on mainly by the shift from a voice-based network to a data-based network. As discussed in the previous chapter, the network is no longer carrying only voice signal; it carries now additional traffic types such as video, data, and multimedia traffic types. In the past the design of the network was dictated by voice traffic consideration, however, this paradigm has now changed as traffic patterns have changed. This change in network traffic requires us to change the design, control and management of the network in order to meet the characteristics and requirements of the new traffic types. At the core of this next generation networks is the multiwavelength optical local area networks, among others. Optical networks are becoming very popular recently due to their great potential in terms of capacity, data rate, security, signal quality and quality of service (QoS). Future computer communication networks are expected to have dramatic increases in bandwidth demands, which can only be met by optical technology. Optical networks have moved from being a research curiosity to becoming a well-established technology with a multi-billion dollar industry, as can be seen by the many start-up companies all over the world. Over the past ten years, many telecommunications companies have installed optical cables as the backbone of their networks. Fibre optic cables are becoming the foundation of telecommunications infrastructures as they can

Multiwavelength Optical LANs, G.I. Papadimitriou, P.A. Tsimoulas, M.S. Obaidat and A.S. Pomportsis.
 ISBN 0-470-85108-2.

carry data at very high line speeds. The use of wave division multiplexing and other enabling technologies has moved optical networking from the laboratory to practice. Among the important optical networks technologies is the multiwavelength optical local area networks (LANs) [1–10].

In this chapter, we review the main aspects, characteristics, and structures of the enabling technologies of multiwavelength optical LANs.

2.1 INTRODUCTION

Since the mid-1980s, most of the telecom companies have started to migrate to fibre optic cables. The widespread use of fibre has been possible due to the industry's acceptance of some enabling technologies such as SONET/SDH, optical filters, lasers, light emitting diodes, passive start couplers, wavelength routers, wave division multiplexers/demultiplexers, combiners, splitters, wavelength converters, and directional couplers, among others.

Optical networks utilize optical signal transmission for information exchange. An optical network communications system consists of a source, a destination, enabling devices, and a communication medium. As was mentioned in Chapter 1, from the viewpoint of the selection of the communication medium, optical networks can be divided into two categories: guided and unguided systems. In the unguided systems, the optical beam transmitted from the source widens as it spreads into space. This introduces interference problems. In guided systems or the fibre-optic communications systems, this problem does not exist. An optical transmission system has three chief components: light source, transmission medium, and detector. Traditionally, a pulse of light indicates a 1-bit and the absence of the light indicates a 0-bit. The transmission medium is a thin fibre of glass. The detector generates an electrical pulse when light falls on it [1–52].

In general, optical networks can be opaque or all-optical and can be of single-wavelength or based on wavelength division multiplexing (WDM). It is known that the data rates of single channel optical networks are restricted by the limited speed of the stations' electronic circuits. On the other hand, the Wavelength Division Multiplexing technique solves this problem by dividing the available optical bandwidth into multiple channels of lower bandwidth, which can easily be supported by the stations' electronic circuits. Operations such as multiplexing and demultiplexing of the multiple channels are performed in the optical domain without the need for optical to electronic conversion and vice versa. Clearly, WDM technique allows the implementation of all-optical networks, which are capable of providing multi-gigabit per second data rates by current optical and electronic technology. Figure 2.1 shows the multi-mode and single-mode fibres [1–5].

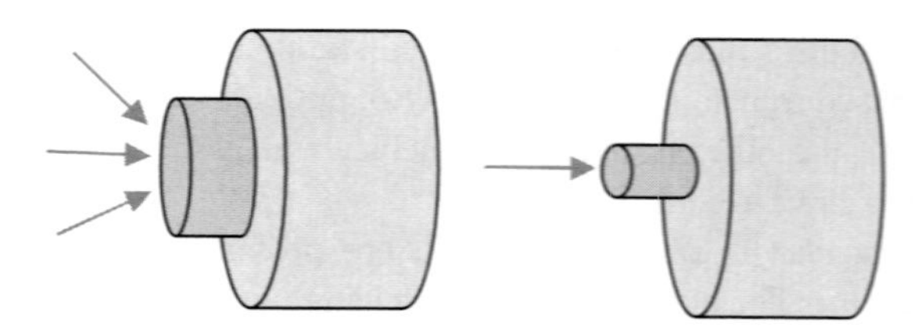

Figure 2.1 Multi-mode and single-mode fibres.

Opaque networks use optical fibre as a transmission medium; however, optical-electronic-optical (OEO) conversion occurs at the intermediate nodes of the communicating stations. Examples of such systems include the synchronous optical network/synchronous digital hierarchy (SONET/SDH), which is an opaque single-wavelength system. Many other opaque systems are based on the wavelength multiplexing systems (WDM). Despite the fact that opaque systems are popular, they do not have the power and potential of having all-optical networks. All-optical network systems have all connections at the intermediate nodes totally optical. Recent technological advances allow us to have practical all-optical cross-connects (OXCs) and all-optical add-drop multiplexers (OADMs). This has led to the emergence of a multibillion dollar industry, which is expected to dominate the marketplace of the future. The present situation in optical networking seems to be a combination of the all-optical and opaque paradigms, with all-optical islands being interconnected with OEO connections [1–5].

Optical networks have many advantages that can be summarized below [1, 3, 6, 8, 15–20]:

- *Higher bandwidths.* Optical fibres provide much higher bandwidth than other alternative communication media such as the copper twisted pairs or coaxial cables. For example, in just the 1.5 micron (1 μm is equal to 10^{-6} metre) band of each single-mode fibre, the available bandwidth is about three orders of magnitude more than the entire usable radio-frequency bandwidth on Earth.
- *Ease of deployment and maintenance.* In general, a good quality optical fibre is lighter and sometimes less fragile than a copper-based cable. Moreover, optical fibres are not subject to corrosion, making them more tolerant to environmental hazards. Furthermore, since their weight is less than that of copper wires, optical fibres become very attractive for deployment over long distances.
- *Better signal qualities.* Since optical transmission is not affected by electromagnetic fields, optical fibre exhibits superior performance when compared to other alternatives such as copper wires. At a given distance the Bit Error Rate (BER)

of a fibre-based transmission system is significantly better than that for a copper or wireless-based transmission system. Also, optical fibre provides better noise immunity than the other transmission media, which suffer from considerable electromagnetic interference.

- *Better security characteristics*. Optical fibre provides a secure transmission medium since it is not possible to read or change optical signals without physical disruption. It is possible, though not easy, to break a fibre cable and insert a tap, but that involves a temporary disruption. For many critical applications including military and e-commerce applications, where security is of the utmost importance, optical fibre is preferred over copper transmission media, which can be tapped from their electromagnetic radiations.

There are many applications for optical networks including local and metropolitan area networks and high speed networks. Among the chief applications that can be enabled by high-speed optical networks are [1–42]:

- *Broadband services to the home*. Both the telephone companies and cable TV have been trying to offer new high speed services to their residential and small business customers. The future promises to deliver applications such as interactive television, video-on-demand (VoD), e-commerce, e-government and distance learning. In order to realize this, fibre-to-the-curb (FTTC) and fibre-to-the-home (FTTH) should be installed.
- *Medical image access and distribution*. Telemedicine requires that a surgeon consults with a radiologist, especially for remote rural sites where specialists may not be available. In order to realize this, a fast link is needed between the two locations that are distant from each other. The distribution of colour images of size $2,000 \times 2,000$ for medical consultation to multiple locations will require high speed networks and optical technology is the best candidate for such links.
- *Internet and Web browsing*. Due to the fact that new Web and Internet applications are increasing in an exponential manner, annual bandwidth requirements for each user have grown by a factor of eight. Not just this, the number of new users is increasing in a super-exponential manner. Users are experiencing a response time measured in seconds instead of the few milliseconds that they like to have. Moreover, new Java-enabled active home pages have introduced the prospect of a huge increase in file sizes downloaded from any server and some researchers and developers have extended these trends to a new model of distributed and networked computing.

- *Multimedia conferencing and distance learning*. Business costs and travel time have been reduced due to the rapid introduction of fast links used for video conferencing. In order to have full utilization of the multimedia conferencing to support distance collaborative meetings and distance interactive learning sessions, the networks should be able to provide required quality of service (QoS) for each traffic type. For example, real time voice requires low delay and jitter (delay variation) while video requires low delay and high bandwidth.
- *Graphics and visualization*. Graphics and visualization applications require both data-and computer-intensive programmes. In order to extend the capabilities of these applications to a distributed mode, both low latency and high data rates will be required. Examples of such application include 3D visualization and animation in simulation.

2.2 CLASSES OF OPTICAL NETWORKS

In this section the main classes of second generation optical networks are summarized. The main classes are the following [1]: optical link networks, broadcast-and-select networks, wavelength-routed networks and photonic packet-switching networks. All of these use optical links and employ non-switching optical components. These classes can be connection-oriented or connectionless and their switching can be circuit switching or packet switching.

2.2.1 OPTICAL LINK NETWORKS

This class characterizes optical networks that consist of point-to-point or shared medium optical links. In case of point-to-point links, multiplexers/demultiplexers are used at both ends of the optical link, whereas WDM passive star couplers are used in shared-medium optical link networks. These devices are not reprogrammable, therefore prohibiting reconfiguration. The optical links interconnect the network's switches, which are of an all-electronic nature, thus optical add-drop multiplexers (OADMs) and optical cross-connects (OXCs) do not exist in such networks. Switches in optical link networks can be either based on connection-oriented or connectionless packet switching. They can also be based on circuit switching or a combination of packet and circuit switching [1].

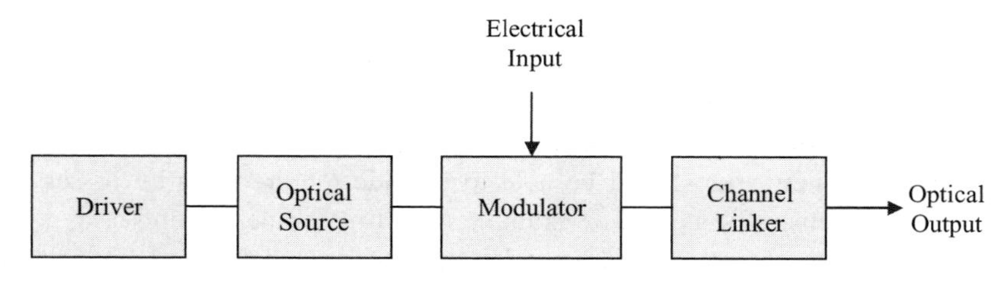

Figure 2.2 A block diagram of an optical transmitter.

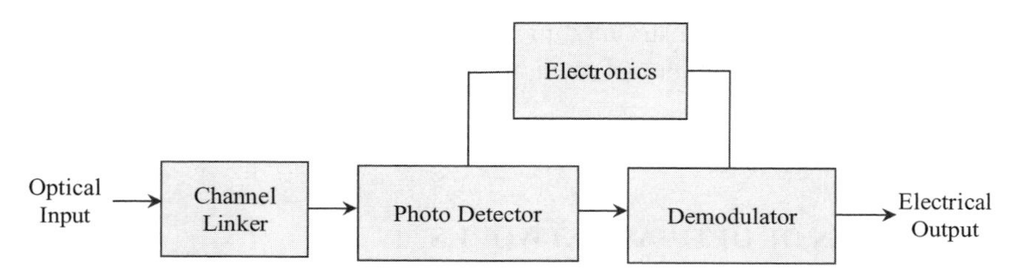

Figure 2.3 A block diagram of an optical receiver.

2.2.2 BROADCAST-AND-SELECT NETWORKS

This category of optical networks characterizes optical networks that only employ optical tunable transmitters and receivers. Figures 2.2 and 2.3 show block diagrams of an optical transmitter and optical receiver, respectively. Only optical switching components are used in such networks. Data transmissions are broadcast on all outgoing links of a node and receivers tune to the appropriate wavelength in order to receive the desired transmission. Obviously, more than one network node can tune to a specific wavelength, a fact that provides support for multicast services. Figure 2.4 shows a passive star coupler-based optical broadcast-and-select optical network [1–10].

Broadcast-and-select networks can be single-hop and multi-hop networks:

- In single-hop broadcast-and-select networks, the end-to-end data transmission only passes through optical-switching components. In single-hop networks, either packet-switches (connection-oriented or connectionless) or circuit-switches are used. A combination of different switches is not possible.
- In multi-hop broadcast-and-select networks, however, the end-to-end data transmission passes through a combination of optical and electronic switches. Since

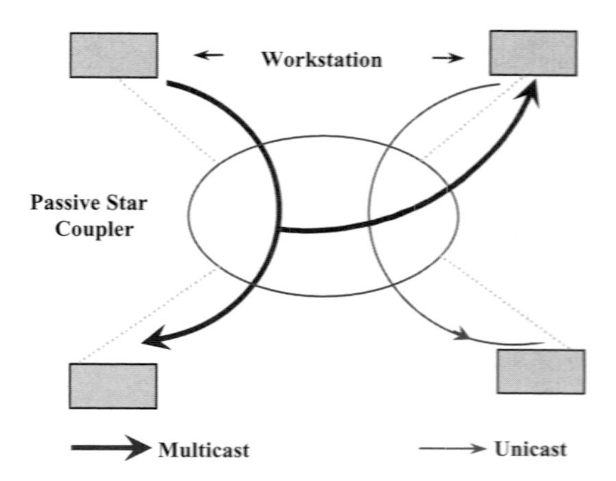

Figure 2.4 A passive star coupler-based optical broadcast-and-select network.

optical switches operate faster than electronic switches, one can see the obvious advantage of single-hop networks from the point of view of bit-rate transparency.

2.2.3 WAVELENGTH-ROUTED NETWORKS

This class characterizes optical networks that have to use OADMs/OXCs and optionally optical tunable transmitters/receivers. Wavelength-routed networks can also be single-hop or multi-hop like the broadcast-and-select networks:

- Single-hop networks that employ fixed transmitters/receivers are entirely based on circuit switching. However, if these devices are tunable, then single-hop networks can be either based entirely on circuit switching or employ a combination of circuit and packet switching.
- Multi-hop wavelength-routed optical networks use both optical and electronic switches. They can be either based on circuit switching or a combination of packet and circuit switching.

A wavelength-routed network consists of a collection of optical switches interconnected by optical links. Each network node is attached to an optical switch and consists of a set of, possibly tunable, transmitters and receivers that send and receive data to/from the network. Obviously, data transmissions can occur in parallel over the same link; however, they need to be on different wavelengths to prevent interference. In order for a data transmission to occur between two nodes, the optical signal has to be routed through intermediate switches forming the

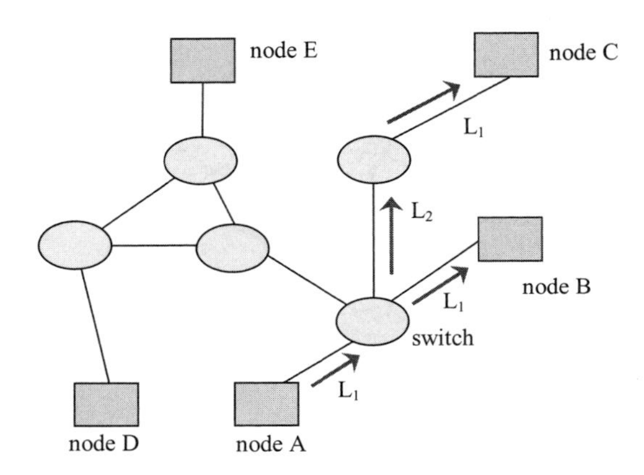

Figure 2.5 A wavelength-routed optical network.

so-called 'lightpath'. The lightpath may contain several wavelengths if wavelength converters are used in the network. In the opposite case, which is known as the wavelength continuity property of a lightpath, the latter uses the same wavelength throughout the network. Figure 2.5 shows a wavelength-routed optical network. The link from A to B is a wavelength-continuous lightpath, since it uses the same wavelength L1 throughout the network. The opposite stands for the link from A to C [1–20].

2.2.4 PHOTONIC PACKET SWITCHING NETWORKS—OPTICAL BURST SWITCHING NETWORKS

Photonic packet-switching networks must contain optical packet switches. Optical circuit switches and tunable transmitters/receivers are also an option. Among all possible switching schemes, photonic packet switching appears to be the most promising as it offers high speeds, compatibility and data rate transparency.

Optical burst switching networks try to combine the concepts of circuit and packet switching. Burst switching avoids the associated cost in time for the establishment of a circuit by sending bursts of data following a control packet in order to find allocated resources prior to the arrival of the burst to the switch. Upon success, the round trip set-up delay that exists in circuit-switching nodes is obviously reduced; however, buffers are needed in burst switching in order to handle the cases when the bursts arrive at the switch with no available resources at the switch.

From the classes considered in this and the previous three subsections, optical link networks comprise a feasible commercial alternative today. Among the remaining classes, market attention seems to focus on wavelength-routed networks, since the hardware needed for broadcast-and-select networks (tunable transmitters/receivers) and photonic packet-switched networks (optical memory) comes at a relatively high cost in today's terms [1–5]. Of course, things might very well change in the near future.

2.3 OPTICAL NETWORK COMPONENTS

Second generation optical networking, unlike SONET/SDH, does not rely on electrical data processing. As such, its development is more closely tied to optics than to electronics. In its early form, as described previously, Wave Division Multiplexing (WDM) was capable of carrying signals over two widely spaced wavelengths, and for a relatively short distance. To move beyond this initial state, WDM needed both improvements in existing technologies and the invention of new technologies. Improvements in optical filters and narrowband lasers enabled DWDM to combine more than two signal wavelengths on a fibre. The invention of the flat-gain optical amplifier, coupled in line with the transmitting fibre to boost the optical signal, dramatically increased the viability of DWDM systems by greatly extending the transmission distance. Other technologies that have been important in the development of DWDM include improved optical fibre with lower loss and better optical transmission characteristics and devices such as fibre Bragg gratings used in optical add/drop multiplexers. Figure 2.6 shows the spectrum of light bands [24–47].

Digital networks have evolved in three fundamental stages: asynchronous, synchronous, and optical [22–30].

- *Asynchronous.* The first digital networks were asynchronous networks. In asynchronous networks, each network element's internal clock source timed its transmitted signal. Due to the fact that each clock had a certain amount of variation, signals arriving and transmitting could have a large variation in timing, which

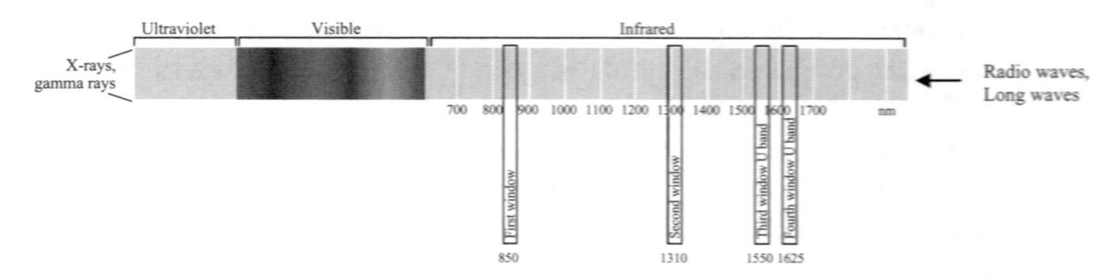

Figure 2.6 Spectrum of light.

often resulted in bit errors. More importantly, as optical-fibre deployment increased, no standards existed to control how the network elements should format the optical signal. A myriad of proprietary methods appeared, making it difficult for network providers to interconnect equipment from different vendors.

- *Synchronous.* The need for optical standards led Belcore (now Telcordia) to introduce the synchronous optical network (SONET). SONET standardized line rates, coding schemes, bit-rate hierarchies, and operations and maintenance functionality. SONET also defined the types of network elements required, the network architectures that vendors could implement, and the function that each node must perform. Network providers could now use different vendor's optical equipment with the confidence of at least basic interoperability.
- *Optical.* The one aspect of SONET that has allowed it to survive during a time of tremendous changes in network capacity needs is its scalability. Based on its open-ended growth plan for higher bit rates, theoretically no upper limit exists for SONET bit rates. However, as higher bit rates are used, physical limitations in the laser sources and optical fibre begin to make the practice of endlessly increasing bit rate on each signal an impractical solution. Moreover, connection to the networks through access rings has also increased requirements. Customers are demanding more services and options and are carrying more and different types of data traffic with different quality of service (QoS) requirements. To provide full end-to-end connectivity, a new paradigm was needed to meet all the high-capacity and different applications. Optical networks provide the required bandwidth and flexibility to enable end-to-end wavelength services.

With the development of the Internet and World Wide Web technologies and applications, the network bandwidth requirements have increased dramatically in recent years. This needs boosted research and development in WDM and progress in this field has been moving at a rapid rate. All-optical network systems that employ WDM and wavelength routing are a viable solution for the future wide area networks (WANs) and metropolitan area networks (MANs) as well as the next generation Internet(s). Such wavelength-routed WDM networks offer several advantages including protocol transparency, simplified management and processing when compared to routing in systems that use digital cross-connects [11–17].

The early 1990s saw a second generation of WDM, sometimes called narrowband WDM, in which two to eight channels were used. These channels were now spaced at an interval of about 400 GHz in the 1550-nm window. By the mid-1990s, dense WDM (DWDM) systems were emerging with 16 to 40 channels and spacing from 100 to 200 GHz. By the late 1990s DWDM systems had evolved to the

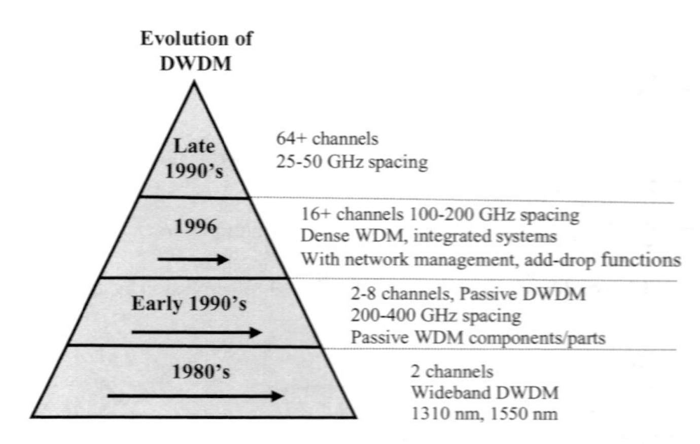

Figure 2.7 Evolution of the Dense Wavelength Division Multiplexing (DWDM).

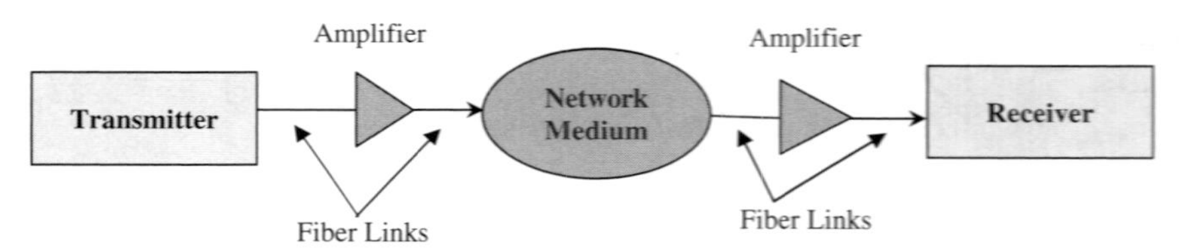

Figure 2.8 Basic structure of a transmitter and a receiver.

point where they were capable of 64 to 160 parallel channels, densely packed at 50 or even 25 GHz intervals.

Figure 2.7 summarizes the evolution of DWDM technology. As can be seen in Figure 2.7, there is an increase in the number of wavelengths accompanied by a decrease in the spacing of the wavelengths, along with increased density of wavelengths. Furthermore, there is an improvement in their flexibility of configuration, through add-drop functions, and management capabilities [5, 12, 15, 27].

The components of any optical network can be divided into switching and non-switching components. Switching components are programmable and enable networking while non-switching components are used on optical links. These components are: lasers, light-emitting diodes (LEDs), photodiodes, optical filters, passive star couplers, WDM Multiplexers/Demultiplexers, wavelength routers, combiners/splitters, wavelength converters, and directional couplers. Figure 2.8 depicts the structure of a transmitter and a receiver, while figure 2.9 depicts a block diagram of a WDM transmission system [1, 4–8].

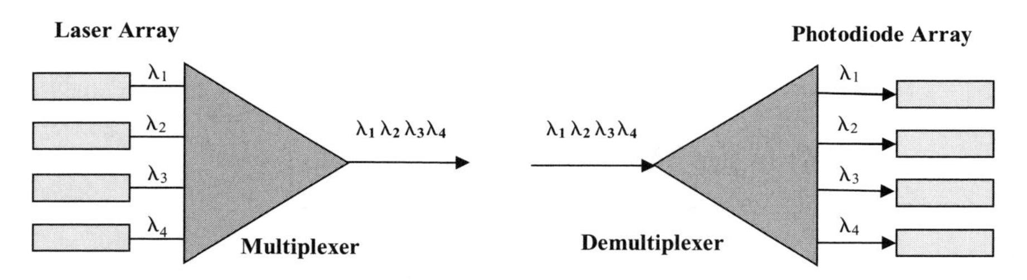

Figure 2.9 Block diagram of a WDM transmission system.

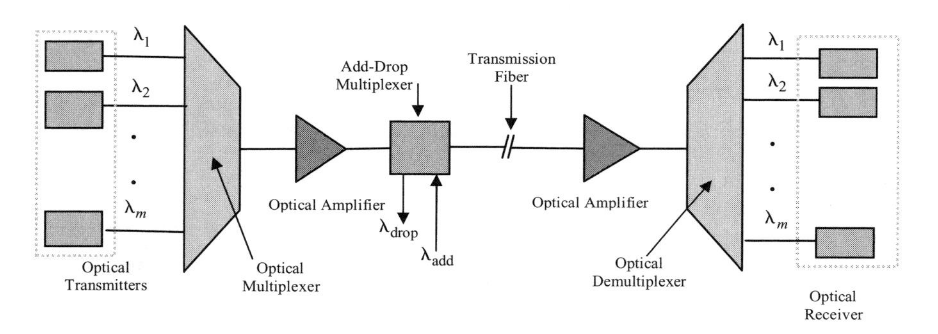

Figure 2.10 A detailed WDM system.

A typical detailed WDM system is shown in Figure 2.10.

2.3.1 OPTICAL FIBRE

Optical fibres are used to guide lightwaves with minimum attenuation. There are two general categories of optical fibre in use today, multimode fibre and single-mode fibre. Multimode, the first type of fibre to be commercialized, has a larger core than single-mode fibre. It gets its name from the fact that numerous modes, or light rays, can be carried simultaneously through the waveguide. Note that the two modes must travel different distances to arrive at their destinations. This disparity between the times that the light rays arrive is called *modal dispersion*. This phenomenon results in poor signal quality at the receiving end and ultimately limits the transmission distance. This is why multimode fibre is not used in wide-area applications.

The second general type of fibre, single-mode, has a much smaller core that allows only one mode of light at a time through the core. As a result, the fidelity of

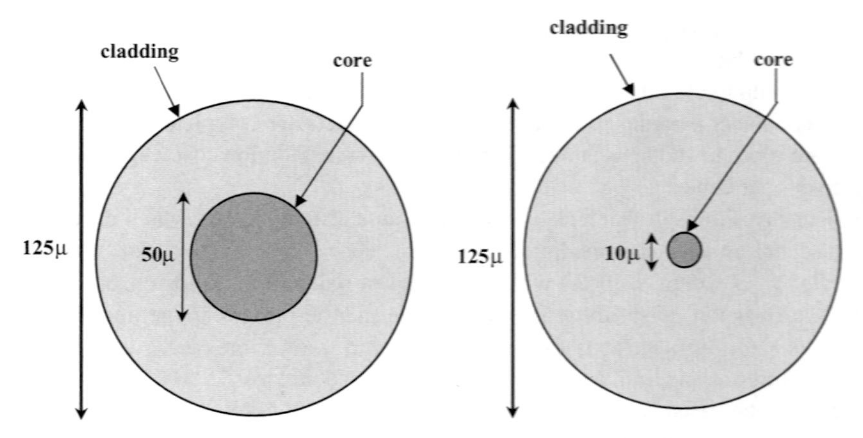

Figure 2.11 Multimode and single mode optical fibres.

the signal is better retained over longer distances, and modal dispersion is greatly reduced. These factors contribute to a higher bandwidth capacity than multimode fibres. Due to its large information-carrying capacity and low intrinsic loss, single-mode fibres are preferred for longer distance and higher bandwidth applications such as DWDM. Typical single mode fibre is 9–10 micron and multimode fibre is 50 micron. But Bells Labs announced high-power single-mode fibre lasers from 1.065 to 1.472 micrometers, using ytterbium-doped cladding-pumped and cascaded Raman lasers. Figure 2.11 shows these two types of optical fibres.

The main advantages of optical fibre are [3, 18, 29, 39]:

- *Higher bandwidth and greater capacity*. Fibre can handle much higher bandwidths than copper. The potential bandwidth and data rate of optical fibre are enormous. Data rates of a few Gbps over ten kilometres have been demonstrated. Compare these capabilities to the maximum of hundreds of Mbps over 1 km for coaxial cables.
- *Low attenuation and greater repeater spacing*. The attenuation in fibre is much less than copper, therefore, repeaters are needed only about every 50 km on long lines, versus every 5 km for copper. Fewer repeaters mean lower overall cost and fewer sources of error.
- *Electromagnetic isolation*. Optical fibre has the advantage of not being affected by power surges, electromagnetic interface, or power failures. Also, it is not affected by corrosive chemicals in the air, which makes it an ideal choice for harsh environments.

- *Small size and lighter weight.* Optical fibres are considerably thinner than coaxial cable or bundles of twisted-pair cable, at least an order of magnitude thinner for comparable capacity. These characteristics make fibres very attractive to telephone companies as many existing cable ducts are completely full, so there is no room to add new cables. Fibre has a lower installation cost when compared to copper cables.
- *Security.* Fibres do not leak light and are quite difficult to tap, which make them a secure transmission medium.
- *High work safety.* Working with fibres has low risk of fire, explosion, or ignition.
- *Relatively low cost.* Fibres are inexpensive and their prices are getting lower.
- *High reliability and ease of installation and testing.* Fibres are a very reliable transmission medium and they are easy to install and test.
- *Error-free transmission over longer distances.* Using fibres implies flexibility in network planning and possibility of taking advantage of new architectures.
- *Long-term economic benefits over copper (over the lifetime of the network).* Fibre has superior reliability which reduces operating costs by minimizing network outages. Also, higher bandwidth can produce considerable savings by eliminating the need to pull new cable when the network is upgraded to support higher bandwidth and long distance capability, which allow all hub electronics to be centrally located. Centralization reduces the cost of cabling and electronics, and reduces administration and maintenance efforts.

The challenges that face fibre transmission are shown below [3, 18, 29]:

- *Attenuation.* The decay of signal strength or loss of light power, as the signal propagates through the fibre.
- *Chromatic dispersion.* This is spreading of light pulses as they travel down the fibre. One way to keep these spread-out pulses from overlapping is to increase the distance between them, however, this can be done only by reducing the signalling rate. The good news is that it has been found that by making the pulses in a special shape related to the reciprocal of the hypercube cosine, nearly all the dispersion effects cancel out. Therefore, it is possible to send pulses for thousands of kilometres without much shape distortion. These pulses are called solitons and at time of writing these are still in the laboratory.
- *Nonlinearities.* This is the cumulative effect from the interaction of light with the material through which it travels, resulting in changes in the lightwave and interactions between lightwaves.
- *Costly fibre interface.* Fibre interfaces cost more than electrical interfaces.

The general categories of applications of optical fibre are: (a) long-haul trunks; (b) metropolitan area networks (MANs); (c) local area networks (LANs); (d) rural-exchange trunks; and (e) subscriber loops. The main scope of this book is multiwavelength local area networks which use optical fibre as the transmission medium.

As mentioned earlier, an optical transmission system consists of three chief components: the light source, the transmission medium and the photodetector. The presence of a pulse of light indicates a 1 bit and the absence of light indicates a 0 bit. The transmission medium can be made of ultra thin fibre or glass. Various glasses and plastics can be used to make optical fibres. The ones that have the lowest losses are the fibres of ultrapure fused silica. However, ultrapure fibre is difficult to manufacture. Higher loss glass and plastic fibres are more cost-effective.

An optical fibre cable has a cylindrical shape and consists of three sections: the *core*, *cladding*, and *jacket*. The core is surrounded by a glass cladding with a lower index of refraction than the core, in order to keep all the light in the core. The jacket is a thin plastic layer used to protect the cladding. Usually, fibres are grouped in bundles, protected by an outside sheath. The attenuation of light though glass depends on the physical characteristics of the glass and the wavelength of the light. The attenuation is usually measured in decibels; it is given by the formula:

$$\text{Attenuation of light through glass} = 10\log_{10}\text{(Transmitted Power)}/\text{(Received Power)}.$$

There are three wavelength bands that are used for optical communications. These are the ones centred at 0.85, 1.30 and 1.55 microns. All of these three bands are 26,000 GHz to 30,000 GHz wide. The 0.85 micron band has the highest attenuation, but at that wavelength, the lasers and electronics can be made from the same material, gallium arsenide. The 1.30 and 1.55 micron bands have very good attenuation characteristics; less than 5% loss/km [15, 18, 31]. Figure 2.12 shows the attenuation of light through fibre in the infrared region.

2.3.2 *LIGHT SOURCES*

The light source used in the design of an optical system is an important consideration since it can be one of the most costly elements. Its characteristics are often a strong factor that dictates in part the overall performance of the optical link. The

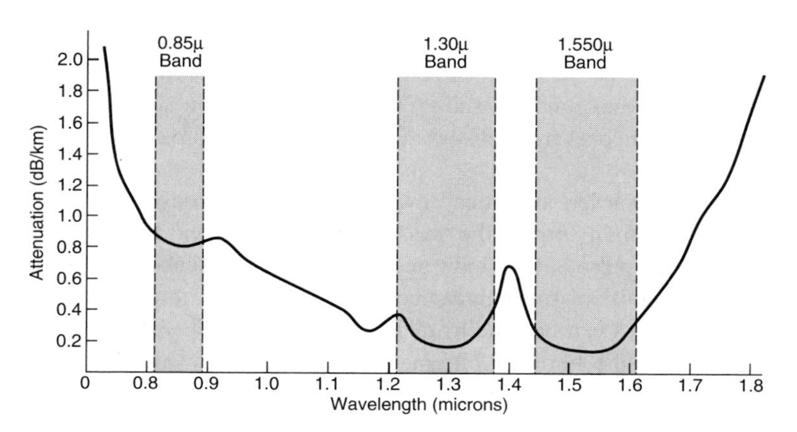

Figure 2.12 Attenuation of light in decibels versus wavelength in micron.

two possible light sources for optical networks are light emitting diodes (LEDs) and semiconductor lasers or laser diodes. They have different characteristics. Light emitting devices used in optical transmission must be compact, monochromatic, stable, and long-lasting. There are several factors that dictate the choice of light source for optical fibre systems. Among these are: (a) the wavelength of the signal source must fall within the transmission window of the optical fibre; (b) the power must be high enough to span the distance to the first optical amplifier or the receiver, but not too high to cause nonlinear effects in the fibre or the receiver; (c) the range of the wavelength emitted by the source should not be very broad to a degree that dispersion limits transmission speed; and (d) the source should transfer light efficiently into the transmitting fibre [3, 15, 18, 29].

2.3.2.1 Light Emitting Diodes (LEDs)

A light emitting diode (LED) is a semiconductor p-n junction device that emits incoherent optical radiation when biased in the forward direction. Light Emitting Diodes use p-n junctions to inject electrons and holes into the same region of a semiconductor so they may recombine and emit light by spontaneous emission. LEDs are usually used for lower data rate, shorter distance multimode fibre. They are also not sensitive to temperature and less expensive than a semiconductor laser. On the other hand, semiconductor lasers can be used for higher data rate, longer distance, and multimode and single mode. Moreover, they are very sensitive to temperature and are expensive. The main applications of LEDs are: (a) displays; (b) high-brightness lamps (laptop, cellular phone PDA back light); (c) indicator and traffic lights; (d) solid-state lighting (being developed); (e) opto-isolators/couplers; (f) infrared wireless communication such as for TV remote controls, free space data links; and (g) optical networks and communications.

Light emitting diodes (LEDs) emit light in proportion to the forward current through the diode. LEDs are low voltage devices that have a longer life than incandescent lamps. They are relatively slow devices, suitable for use at speeds of less than 1 Gbps. They exhibit a relatively wide spectrum width. These inexpensive devices are often used in multimode fibre communications. LEDs can be of two types; surface light emitting diodes and edge light emitting diodes.

The receiving end of an optical fibre has a photodiode, which converts the applied light to electrical pulses. According to current technology, the response time of a photodiode is about 1 ns. This limits the operational data rate to 1 Gbps. LEDs that emit red or near infrared light are often used as sources of light for short fibre systems. In general, a diode is made up of two regions, P and N regions that are doped with impurities to give them the desired electrical characteristics. The wavelength emitted by a semiconductor diode depends on its internal energy levels. The band gap between the energy levels depends on the composition of the junction layer of the diode. The usual LED wavelengths for glass fibre are the 820 nm and 85 nm. The near infrared LEDs used with short glass-fibre systems have active layers that are made using gallium arsenide (GaAs) or gallium aluminum arsenide (GaAlAs). The pure GaAs LEDs emit near 930 nm. The aluminum is added in order to decrease the drive current requirements, which helps to increase the lifetime of the diode. Moreover, this also helps to improve the energy gap so light emission occurs at 750 to 900 nm. LEDs can also be made using gallium arsenide phosphide (GaAsP). These types of LEDs emit different wavelengths in the visible red light near the 650 nm. These are used with plastic fibres that are most transparent in the red and transmit poorly at the GaAs wavelengths. Keep in mind that GaAs LEDs have lower performance than GaAlAs LEDs; however, they are less expensive and are good for short and low-speed plastic fibre links [13, 15, 29, 35, 39].

2.3.2.2 Lasers

Laser is the Acronym for Light Amplification by Stimulated Emission of Radiation. It is a device that produces a coherent beam of optical radiation by stimulating electronic, ionic, or molecular transitions to higher energy levels so that when

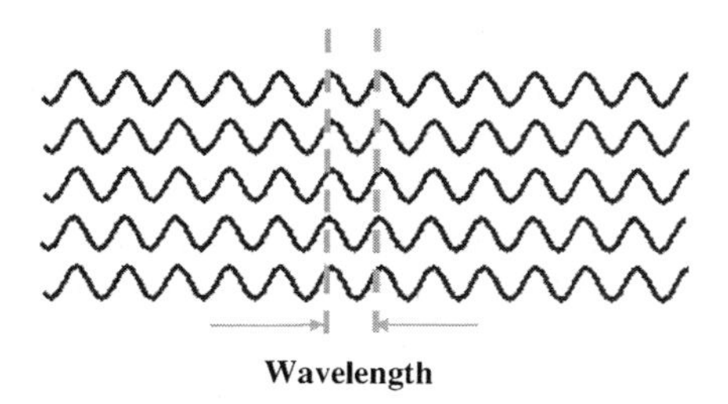

Figure 2.13 Coherent electromagnetic waves.

they return to lower energy levels they emit energy. A laser diode uses a forward biased semiconductor junction as the active medium. It is important to note that laser radiation may be either temporally coherent, spatially coherent, or both. The degree of coherence of laser radiation can exceed 0.88. In a coherent beam of electromagnetic energy, all the waves have the same frequency and phase. Figure 2.13 shows coherent electromagnetic waves with identical phase and similar frequencies [1, 13, 15, 16, 18, 22, 29, 36, 37, 39].

All lasers have an energized substance that can increase the intensity of light passing through it. This is called the amplifying medium or the gain medium, and it can be a solid, a liquid or a gas. Regardless of its physical form, the amplifying medium must contain atoms, molecules or ions, a high proportion of which can store energy that is subsequently released as light [1, 18, 29, 35–45].

In a basic laser, a cavity is designed to internally reflect infrared (IR), visible-light, or ultraviolet (UV) waves so that they strengthen each other. The cavity can contain gases, liquids, or solids. The selection of cavity material determines the wavelength obtained. At each end of the cavity, there is a mirror. One mirror is totally reflective, allowing none of the energy to pass through; the other mirror is moderately reflective, allowing approximately 5% of the energy to pass through. Energy is introduced into the cavity from an external source; this is called pumping. As a result of pumping, an electromagnetic field appears inside the laser cavity at the natural (resonant) frequency of the atoms of the material that fills the cavity. The waves reflect back and forth between the mirrors. The length of the cavity is chosen so that the reflected and re-reflected wavefronts strengthen each other in phase at the natural frequency of the cavity substance. Electromagnetic waves at this resonant frequency come out from the end of the cavity having the partially-reflective mirror. The output may appear as a continuous beam, or as a series of brief, intense pulses.

There are various types of lasers. These are [29, 35–46]:

- *Solid-state lasers*. Solid state lasers have laser material distributed in a solid matrix (such as the Ruby or Neodymium Yttrium-Aluminium garnet "Yag" lasers). The Neodymium-Yag laser emits infrared light at 1,064 nanometres (nm).
- *Gas lasers*. Gas lasers (Helium and Helium-Neon, HeNe, are the most common gas lasers) have a primary output of visible red light. The CO_2 lasers emit energy in the far-infrared, and are used for cutting rigid materials.
- *Excimer lasers*. Excimer lasers (the name is derived from the terms excited and dimers) use reactive gases, such as Chlorine and Fluorine, mixed with inert gases such as Argon, Krypton or Xenon. When electrically stimulated, a pseudo molecule (dimer) is produced. When a laser is used, the dimer generates light in the ultraviolet range.
- *Dye lasers*. Dye lasers use complex organic dyes, such as Rodamine 6G, in liquid solution or suspension as lasing media. They are tunable over a broad range of wavelengths. Tuning range depends on the selected dye—for Rodamine 6G a wavelength of 0.570–0.650 nm can be achieved.
- *Semiconductor lasers*. Semiconductor lasers, sometimes called diode lasers, are electronic devices that are generally very small and use low power. They may be built into larger arrays. There are different semiconductor lasers: (a) homostructure lasers, with a threshold current of (300K) 30000–50000 A/cm2; (b) single heterostructure (300K) 6000–8000 A/cm2; (c) double heterostructure (300K) 500 A/cm2; (d) GRINSCH (Graded-index separate confinement heterostructure), threshold current ~30 mA; (e) mirror type: FP, DFB, DBR; and (f) VCSEL (Vertical Cavity Surface Emitting Laser), threshold current about 1 mA [35–46]. Figure 2.14 shows a simple semiconductor structure of the homostructure type.

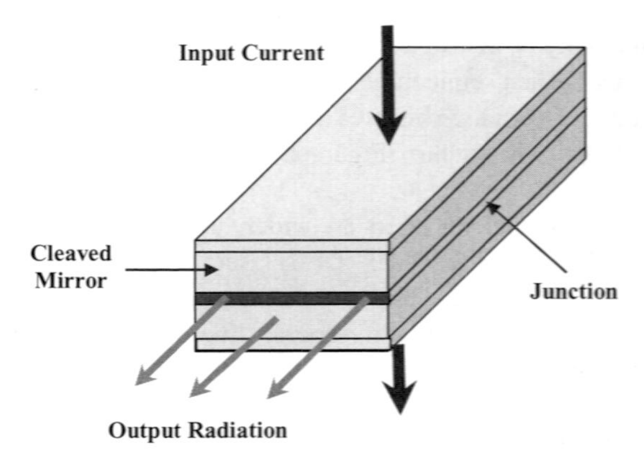

Figure 2.14 A simple semiconductor laser of the homostructure type.

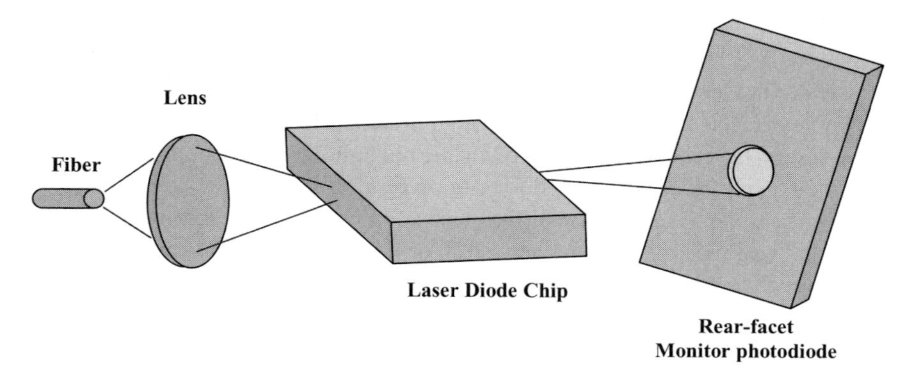

Figure 2.15 Typical laser design.

Semiconductor lasers are excellent for single mode fibre applications due to their outstanding performance. Figure 2.15 shows the general principles of launching laser light into fibre. The laser diode chip emits light in one direction to be focused by the lens onto the fibre and in the other direction onto a photodiode. The photodiode, which is angled to reduce back reflections into the laser cavity, provides a way of monitoring the output of the lasers and providing feedback so that adjustments can be made.

Requirements for lasers include precise wavelength, narrow spectrum width, sufficient power, and control of *chirp* (the change in frequency of a signal over time). Semiconductor lasers satisfy the first three requirements nicely. Chirp, however, can be affected by the techniques used to modulate the signal.

In directly modulated lasers, the modulation of the light to represent the digital data is done internally. With external modulation, the modulation is performed by an external device. When semiconductor lasers are directly modulated, chirp can become a limiting factor at high bit rates (above 10 Gbps). External modulation, on the other hand, helps to limit chirp. Figure 2.16 illustrates the external modulation technique.

Two types of semiconductor lasers are widely used, namely monolithic Fabry-Perot lasers, and distributed feedback (DFB) lasers. The latter type is particularly well suited for DWDM applications, as it emits a nearly monochromatic light. It has a favourable signal-to-noise ratio, superior linearity and can be used for high data rates. The DFB lasers have centre frequencies in the region around 1310 nm, and from 1520 nm to 1565 nm. The latter wavelength range is compatible with EDFAs. There are other types such as the narrow spectrum tunable lasers whose tuning range is limited to approximately 100–200 GHz. Wider spectrum tunable

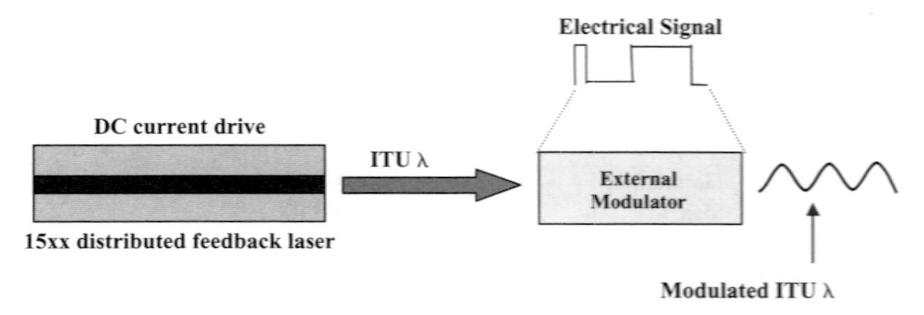

Figure 2.16 External modulation of a laser.

lasers are under development. The latter types will play an important role in the dynamically switched optical networks. There are three requirements for using laser: (a) gain; (b) feedback; and (c) phase match. The main characteristics of lasers are [18, 29, 35, 37–42]:

- *Frequency instability*. This is due to:
  - *Mode hopping*: In present injection lasers, a change in the injection current above a threshold can cause an abrupt jump in the laser frequency.
  - *Mode shift*: This is a change in frequency due to temperature changes.
  - *Wavelength chirp*: This is a variation in the frequency caused by the variation of injection current.
- *Line width*. This is the spectral width of the light generated by the laser. It affects channel spacing and dispersion.
- *Tuning range*. This is the range of wavelengths over which the laser can be operated.
- *Number of longitudinal modes*. This is number of wavelengths that the laser can amplify.
- *Tunability*. This can be continuous tunable (over its tuning range) or discretely tunable (only to selected wavelengths).
- *Tuning time*. This is the time needed for the laser to tune from one wavelength to another.

There are several classes of lasers that include [29, 35–44]:

1. Mechanically tuned lasers where the distance between the two mirrors of an external cavity is changed mechanically,
2. Acoustically and electro-optically tuned lasers where the index of refraction in the external cavity is changed using either acoustic waves or electrical current,

Table 2.1 Operating wavelength ranges of major photodetectors.

Material Type	Wavelength (nm)
Germanium	600–1600
Silicon	400–1100
Gallium Arsenide (GaAs)	400–900
InGaAS	900–1700
InGaAsP	800–1600

3. Laser array where a number of lasers are integrated into a single component, with each laser working at a different frequency
4. Injection-current-tuned lasers that allow wavelength selection with a different diffraction grating. This class has two subclasses: (a) distributed feedback (DFB) where the Bragg grating is placed in the active region; and (b) distributed Bragg reflector where the Bragg grating is placed outside the active region.

2.3.3 PHOTODIODES

These devices detect optical signals and convert them into electrical signals. Their name comes from the fact that they are capable of detecting light. Photodetectors can be made of materials such as germanium, silicon, gallium arsenide, and indium gallium arsenide. In order to produce a photocurrent, photons should have enough energy to raise an electron from the valence band to the conduction band. In other words, their energy should be greater or at least equal to the band gap energy. The photodetector sensitivity tends to drop sharply at the long-wavelength, low energy end of the spectrum. Table 2.1 summarizes the typical operating ranges of typical photodetectors.

There are two types of photodetectors that are widely used: the positive-intrinsic-negative (PIN) photodiode and the avalanche photodiode (APD). The PIN photodiodes work on principles similar to, but in the reverse of, LEDs. That is, light is absorbed rather than emitted, and photons are converted to electrons in a 1:1 relationship. The avalanche photodiodes (APDs) are similar devices to positive-intrinsic-negative (PIN) photodiodes, but provide gain through an amplification process: one photon acting on the device releases many electrons. PIN photodiodes have many advantages, including low cost and reliability, but APDs have higher reception sensitivity and accuracy. However, APDs are more expensive than PIN photodiodes. Moreover, they can have very high current requirements, and they are temperature sensitive.

Table 2.2 Characteristics of typical photodetectors.

Photodetector	Responsivity	Device Current	Rise Time
Pin photodiode (Ge)	0.4 A/W	100 nA	0.1–1 ns
Pin photodiode (Si)	0.5 A/W	1–10 nA	0.1–5 ns
Pin photodiode (InGaAs)	0.8 A/W	0.1–3 nA	0.005–5 ns
Avalanche photodiode (Ge)	Voltage dependent	400 nA (voltage dependent)	03–1 ns
Avalanche photodiode (Si)	10–125 A/W (voltage dependent)	01–2 ns	10–250 nA (voltage dependent)
Avalanche photodiode (InGaAs)	7–9 A/W (voltage independent)	6–160 nA (voltage dependent)	0.1–0.5 ns
Photodarlington (Si)	500 A/W	100 nA	40 μs
Phototransistor (Si)	18 A/W	25 nA	2.5 μs

There is another device called a phototransistor, which both senses and amplifies light generated current. It can only be used for low-cost and low speed applications such as inexpensive sensors, and low-cost, low-speed fibre optic systems. It is worth mentioning here that phototransistor use is limited to systems that operate below the megahertz range. The overwhelming majority of phototransistors are made of silicon [29, 36–40].

A photodarlington is made using two phototransistors, by feeding the output of a phototransistor to the base of a second phototransistor. This configuration increases responsivity of low-cost, low-speed systems.

In direct detection, a photodetector converts the incoming photonic stream into a stream of electrons or current. The current is then amplified and passed through a threshold device. A bit is considered logic 0 or 1 depending on whether the stream is above or below a specific threshold for bit duration. In coherent detection, phase information is used in the encoding and detection of signals. The receiver uses a laser as a local oscillator. The incoming signal is combined with the signal oscillator which generates a signal of the difference frequency. This difference signal, in the microwave range, is amplified and then photodetected [16, 18, 29]. Table 2.2 summarizes the main characteristics of typical photodetectors.

2.3.4 OPTICAL FILTERS

An optical filter is a device that blocks some input light from reaching a particular destination/point. Optical filters are used in WDM systems to separate signals at different wavelengths and route them to different destinations. In other cases,

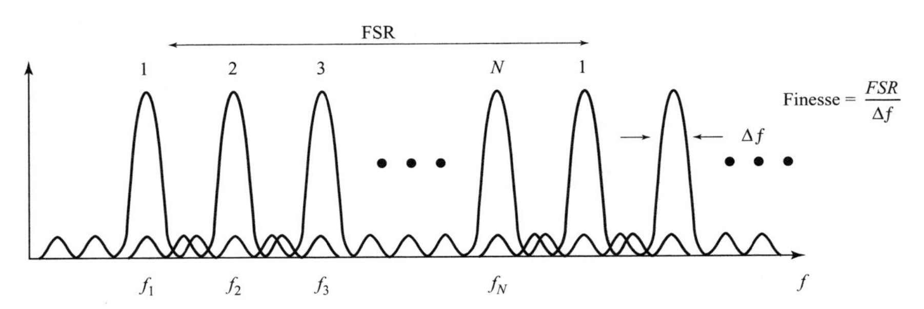

Figure 2.17 Tunable filter characteristics.

optical filters are used to reduce signal intensity, which might, otherwise, overload the receiver to block extraneous signals at other wavelengths or to balance signals transmitted through the same system at different wavelengths.

Some filters selectively block light or, otherwise, transmit only selected wavelengths. Line and band filters can be used in some types of WDM, but more often, they serve to block unwanted wavelength rather than separate signals. The chief function of line and band filters is to restrict light transmission so that only certain wavelengths pass through, blocking other wavelengths, which might generate noise. In general, a WDM filter transmits selected wavelengths and reflects others. The wavelengths that are not transmitted through the filter are usually reflected so that they can go elsewhere in the optical communications system. Clearly, these filters are like mirror shades and one-way mirrors, which reflect more incident light, but transmit enough to see through them [3, 15, 38, 39].

Filters usually have periodic resonant frequencies. The distance between two neighbouring resonant frequencies is often called free spectral range, FSR. Another performance metric of an optical filter is called finesse which is defined as the measure of sharpness of resonance. Finesse is measured as the ratio of FSR to the 3-dB bandwidth of a resonant peak. Figure 2.17 shows the finesse and FSR metrics [1, 15, 22, 29, 36–52].

Optical filters can be cascaded with different FSRs. They also can be classified into two main categories: fixed and tunable filters. The tunable optical filters are efficient in selectively adding or dropping particular wavelength channels from the multi-wavelength network. *tunable filters* can come in different types; these are briefly described below [2, 3, 13, 16, 18, 29, 38, 46]:

- *Etalon*. This is a single cavity formed by two parallel mirrors. By adjusting the distance between mirrors, a single wavelength can be selected to propagate through.

- *Mach-Zehnder (MZ) chain*. This tunable filter splits the incoming wave into two waveguides, and recombines the signals at the outputs. By adjusting the optical path length of one arm, a specific wavelength can be selected to be reflected back. This is due to interference at the combiner.
- *Electro-optic filters*. These filters use crystals with refractive indices that can be changed by electrical currents. Their tuning range is about 16 nm and tuning time is between 1 and 10 μs.
- *Acousto-optic tunable filters (AOTFs)*. In these filters, RF waves are passed through a transducer, which generates acoustical waves in the waveguides. The refractive index changes accordingly and can be constructed into a grating. The latter can select a specific wavelength. This type has a tuning range of about 250 nm and a tuning time of about 10 μs.
- *Liquid-crystal Fabry-Perot filters*. Here, an FP cavity is filled with liquid crystal that has a refractive index, which can be changed by applying an electrical current. These filters have a tuning range of about 30 nm with a tuning time of 0.5 to 10 μs.

The main types of *fixed optical filters* are: (a) grating filters, (b) fibre Bragg grating, and (c) thin-film interference filters [16, 18, 29, 38, 43]. The fibre Bragg grating is of great interest to optical communications as it is used to demultiplex wavelengths. A fibre Bragg grating is a small section of fibre that has been modified by exposition to ultraviolet radiation to create periodic changes in the index of refraction. The result is that light travelling through these refractive index changes is reflected back slightly, but the maximum reflection occurs only at one particular wavelength. The reflected wavelength, known as the Bragg resonance wavelength, depends on the amount of refractive index change that has been applied and also on how distantly spaced these changes are.

The main features of an optical filter are [48, 53]:

- *Insertion loss*. The filter is usually inserted in a network and it should ideally be loss-less.
- *Tuning range (D)*. If the filter is to be tuned over a long-wavelength transmission window such as 1300 nm or 1500 nm, then 25 THz is a fine-tuning range. Gain-flattened amplifiers need wider ranges too.
- *Channel spacing (S)*. In order to avoid a cross-talk degradation, there should be a minimum frequency separation between channels; that is called channel spacing. In general, the cross-talk degradation should be 30 dB less than the desired signal.
- *Maximum number of channels N*. This refers to the maximum number of equally spaced channels. It is defined as the ratio of total tuning range (D) to the channel

spacing (S), that is:

$$N = D/S$$

- *Tuning speed*. This gives a measure of how fast the filter can be reset from one wavelength to another.

There are other features such as: (a) narrow bandwidth and high side-lobe suppression; (b) large dynamic range and fast tuning speed; (c) simple control mechanism for a tunable optical filter; (d) small size and cost effectiveness; (e) stringent requirements on the amplitude response; and (f) linear phase response and constant time delay for optimum bandwidth utilization.

Filters can also be classified into the following classes [29, 48, 53]:

- *Compensating filters*. This class of filters has gradually sloping spectral curves. These filters have been used for various applications. They are used to compensate for wrong lighting such as when indoor film is used outdoors or when outdoor film is used indoors. These filters are of the amber type that lowers colour temperature, and the blue type that raises colour temperature.
- *Neutral-density filters*. Neutral-density filters are uniform (grey) filters that absorb and/or reflect a fraction of the energy incident upon them. The term "neutral" is used here since the absorption and/or reflection characteristics of the filter are constant over a wide wavelength range.
- *Attenuation filters*. These filters are used in order to reduce the intensity of light beam. High-quality attenuation filters are said to have a "flat response". This means that they attenuate all wavelengths of light over their usable spectral range by the same amount. These filters are used over a photosensitive surface when the light signal received is too intense. This would prevent overdriving a photodetector or overexposing photosensitive film. To determine if the photosensitive surface is reacting linearly to the exposure, calibrated attenuation filters are usually used. A light signal being measured by a photodetector is an example of this application. If the photodetector is reacting linearly, the insertion of a 50% attenuation filter in the light beam should cause about 50% decrease in the output electrical signal.
- *Wavelength-selective filters* (*colour filters*). For absorbing filters, the transmittance is an exponential function of thickness [1, 15, 29, 48]. That is:

$$T = t_1 t_2 e^{-\times \alpha_\lambda x}$$

Where:

t_1 = Transmission of front surface of filter
x = Filter thickness in centimetres
α_λ = Filter absorption coefficient (cm–1) at wavelength λ
t_2 = Transmission of back surface of filter

- *Cut-off filters*. If a filter discards the longer wavelengths and transmits the shorter wavelengths, it is called a short-wave-pass filter. However, if it transmits the longer wavelengths and discards the shorter wavelengths, it is called a long-wave-pass filter. In this context, a long-wave-pass filter is called a 'low-frequency-pass filter', and a short-wave-pass filter is called a 'high-frequency-pass filter'.
- *Bandpass filters*. These filters are produced so that they transmit only a very narrow wavelength range. Such filters are often called by several equivalent names such as narrow-pass filters, spike filters, and notch filters. The main application for such filters in electro-optics is the isolation of individual laser lines.
- *Interference filters*. If it is needed to sharp cut off a bandpass, then wavelength selection is used. Spike filters or narrow-pass filters are generally made by accumulating alternating layers with thin coatings of dielectric materials on a glass or quartz window. In order to provide reflection or transmission at the desired wavelengths, selection of the materials and thickness of the coatings is chosen. In general, an interference filter may have as many as 30 layers of coatings. Such multilayer coating techniques are the same as these used to produce high-reflective laser mirrors. Scattering from these surfaces is less than 1%. This means that when these coatings are used as filters, basically all of the light is either transmitted or reflected.

2.3.5 OPTICAL AMPLIFIERS

Due to attenuation, there are limits to how long a fibre segment can propagate a signal with integrity before it has to be regenerated. Before the arrival of optical amplifiers (OAs), repeaters were used to regenerate the signal electronically to compensate for the losses due to the silica fibre as well as the losses caused by other optical components along the line. Electronic regenerators have two main drawbacks: they are expensive and they limit the system's performance since each generator can operate at only one predetermined incoming bit rate, and on one wavelength of a single channel.

Until the advent of the Erbium-doped fibre amplifiers (EDFAs), no practical all-optical amplifier existed. Optical signals were, instead, regenerated electronically and treated in the electronic domain to compensate for losses. The optical amplifier

(OA) has made it possible to amplify all the wavelengths at once and without optical-electrical-optical (OEO) conversion. In addition to being used on optical links, optical amplifiers also can be used to boost signal power after multiplexing or before demultiplexing, both of which can introduce loss into the system.

Since DWDM systems handle information optically rather than electrically, it is important that long-haul applications do not suffer from the effects of dispersion and attenuation. Erbium-doped fibre amplifiers (EDFAs) resolve these problems. EDFAs are silica-based optical fibres that are doped with erbium, which is a rare earth element with appropriate energy levels in its atomic structure for amplifying light at 1550 nm. In order to inject energy into the doped fibre, a 980 nm "pump" laser is used. When a weak signal at 1310 nm or 1550 nm enters the fibre, the light stimulates the rare earth atoms to release their stored energy as additional 1310 nm or 1550 nm light. The process continues as the signal passes down the fibre, repeatedly growing stronger. The EDFA contains several metres of silica glass fibre that have been doped with ions of erbium. When the erbium ions are excited to a metal stable energy state, a population inversion takes place that changes this medium into an active amplifying medium. The amplifier will now accept parallel optical signals at many different wavelengths and amplify them concurrently, regardless of their individual bit rates, modulation formats, or power level [1, 15, 49–53].

By making it possible to carry the large loads that DWDM is capable of transmitting over long distances, the EDFA is considered a key enabling technology. At the same time, it has been a driving force in the development of other network elements and technologies. The spontaneous emissions in the EDFA also add noise to the signal. Figure 2.18 shows a simplified diagram of an EDFA.

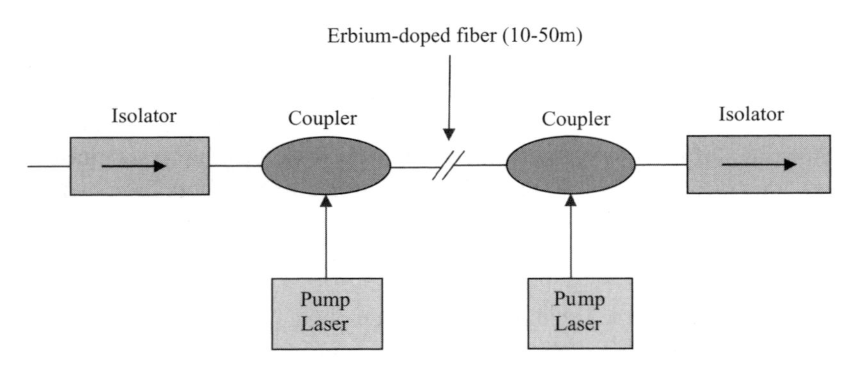

Figure 2.18 A simplified diagram of an erbium-doped fibre amplifier (EDFA).

The key characteristics of optical amplifiers are: gain, gain efficiency, gain bandwidth (flatness), gain saturation, amplified spontaneous emission (ASE), and polarization sensitivity. The gain metric is defined as the ratio of output power of a signal to its input power. EDFAs are normally capable of gains of 30 dB or more and output power of +17 dB or more. The prime parameters that are considered when selecting an EDFA, however, are low noise and flat gain. Gain should be flat because all signals must be amplified uniformly. While the signal gain provided with EDFA technology is inherently wavelength-dependent, it can be corrected with gain flattening filters. Such filters are often built into state-of-the art EDFAs.

The gain efficiency is defined as the measure of the gain as a function of pump power in decibels per milliwatt (dB/mW). The gain bandwidth is the range of frequency over which the amplifier is effective and the gain saturation is the value of output power over which the amplifier is effective. The amplified spontaneous emission (ASE) is considered the dominant source of noise, which arises from the spontaneous emission. Finally, the polarization sensitivity is the dependence of the gain on the polarization of the signal; it is usually measured in dBs.

Low noise is an important requirement since noise, along with signal, is amplified. Due to the fact that this effect is cumulative, and cannot be filtered out, the signal-to-noise ratio is an ultimate limiting factor in the number of amplifiers that can be concatenated and, therefore, the length of a single fibre link. Generally, signals can travel for up to 120 km (74.5 miles) between amplifiers. At longer distances of 600 to 1000 km (372.6 to 621 miles) the signal must be regenerated because the optical amplifier only amplifies the signals and does not perform the 3R functions (reshape, retime, retransmit). EDFAs are available for the C-band and the L-band. It is worth mentioning that the amplification process is independent of the data rate. Due to this advantage, upgrading a system means only changing the launch/receive terminals.

The demand for wider bandwidth is growing at a super-exponential rate. This means that there is a need for more efficient and reliable optical amplifiers. The usable bandwidth of an EDFA is only about 30 nm (1530 nm–1560 nm), but the minimum attenuation is in the range of 1500 nm to 1600 nm. The dual-band fibre amplifier (DBFA) has the potential to solve the usable bandwidth problem. It is divided into two sub-band amplifiers. The DBFA is similar to the EDFA, but it has a wider bandwidth range, from about 1528 nm to 1610 nm. The first range is similar to that of the EDFA and the second is known as extended band fibre amplifier (EBFA). Among the features of the latter are flat gain, slow saturation, and low noise. The EBFA can attain a flat gain over a range of 35 nm. EBFAs have the advantage of reaching a slower saturation keeping the output constant although the input increases [13, 15, 22, 27, 29, 27, 29–53].

2.3.6 PASSIVE STAR COUPLERS

A coupler covers all devices that combine light into or split light out of a fibre. It is a device that can distribute the optical signal (power) from one fibre among two or more fibres. A fibre optic coupler can also combine the optical signal from two or more fibres into a single fibre. Fibre optic couplers attenuate the signal much more than a connector or splice because the input signal is divided among the output ports. For example, with a 1×2 fibre optic coupler, each output is less than one-half the power of the input signal (over a 3 dB loss).

Fibre optic couplers can be either active or passive devices. The difference between them is that a passive coupler redistributes the optical signal without optical-to-electrical conversion while active couplers are electronic devices that split or combine the signal electrically and use fibre optic detectors and sources for input and output. The number of input ports and output ports vary depending on the intended application. Types of fibre optic couplers include optical splitters, optical combiners, X couplers, star couplers, and tree couplers.

A splitter is a coupler that divides the optical signal on one fibre to two or more fibres. The splitting ratio, α, is the fraction of input power that goes to each output. Combiners have the reverse function of splitters, and when turned around can be used as a splitter [26, 27]. Figure 2.19 shows a functional diagram of a passive star coupler.

A passive star coupler is an optical coupler that has a number of input and output ports, used in optical network applications. Note that an optical signal introduced into any input port is distributed to all output ports. Because of the

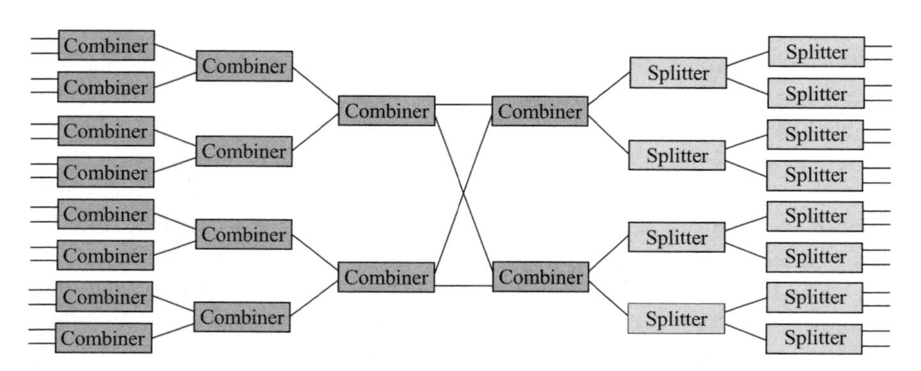

Figure 2.19 A passive star coupler.

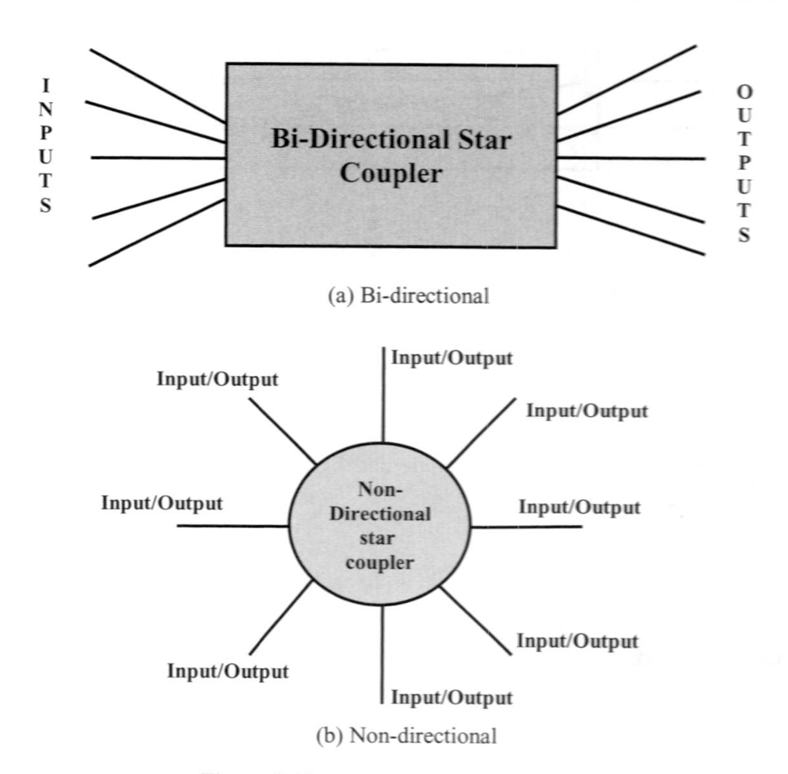

(a) Bi-directional

(b) Non-directional

Figure 2.20 Two types of star coupler.

nature of the construction of a passive star coupler, the number of ports is usually a power of 2.

Star couplers got their name from the geometry used to show their operation in diagrams such as the one shown in Figure 2.20, which depicts two basic types of the star couplers. The first type, see Figure 2.20(a), is directional, mixing signals from all input fibres and distributing them among all outputs. This type is a bi-directional device since it can also transmit light in the opposite direction. The second type, see Figure 2.20(b), is non-directional. Instead it takes inputs from all fibres and distributes them among all fibres, both inputs and outputs [15–19, 22–29].

The passive star coupler (PSC) is a multiport device in which the light coming into any input port is broadcast to every output port. If P_{out} is the output power, P_{in} is the input power introduced into the star coupler by a single node, and N is

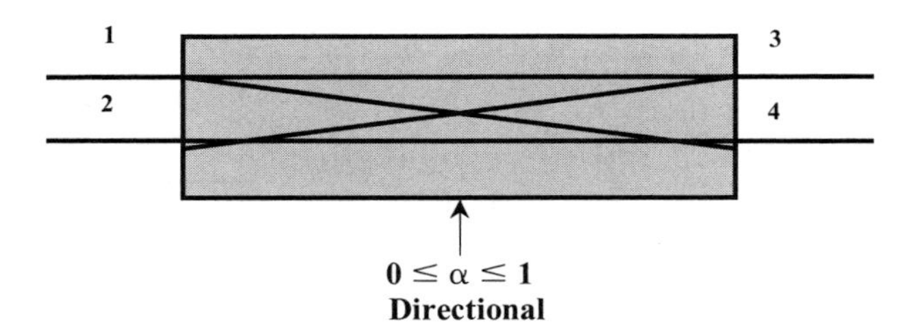

Figure 2.21 A functional diagram of a 2 × 2 directional coupler.

the number of output ports of the star, then:

$$P_{\text{out}} = P_{\text{in}}/\text{N}$$

Passive star coupler is usually implemented by using a combination of splitters, combiners, and couplers.

2.3.7 DIRECTIONAL COUPLERS

In general, a directional coupler consists of two waveguides that are in close proximity, and coupled to each other by evanescent fields. The separation between the waveguides dictates the efficiency of coupling. The closer the waveguides are to each other, the greater the coupling is. In a directional coupler that consists of two completely identical waveguides, as much as 100% of the energy injected into one waveguide can be transferred to the second waveguide after a specific propagation distance L. For lengths less than L, the coupling is less than 100%, whereas for lengths greater than L, energy is coupled back into the original waveguide and the coupling is also less than 100% [15, 16, 18, 22, 23, 26–28].

A 2 × 2 directional coupler is shown in Figure 2.21. It has an optical four port, with ports 1 and 2 designated as input ports and 3 and 4 designated as output ports. Optical power enters the coupler through fibres attached to the input ports, is combined and divided linearly, and leaves via fibres attached to the output ports. If the signals entering each input port originate at distinct optical sources, the action of the coupler can be represented in terms of input signal powers P_1 and P_2 and output powers by P_3 and P_4. These relations can be expressed as below.

$$P_3 = a_{11}P_1 + a_{12}P_2$$
$$P_4 = a_{21}P_1 + a_{22}P_2$$

In the case of ideal symmetric couplers, the power transfer matrix $A = [\alpha_i j]$ is given in the following form:

$$A = \begin{vmatrix} (1-\alpha) & \alpha \\ \alpha & (1-\alpha) \end{vmatrix}$$

where α is a parameter that takes any value between 0 and 1. If the parameter α is fixed, the device is called a static device. However, if α is varied using external electronic or mechanical means, then the device is considered dynamic or controllable.

The majority of directional couplers are of the bi-directional type, which means that they can transmit light in either direction. Generally, directionality or bidirectionality is an advantage in couplers since it sends the signal in the required direction [15–18].

2.3.8 WDM MULTIPLEXERS/DEMULTIPLEXERS

Wavelength Division Multiplexing, WDM, is considered an important technique for utilizing the large available bandwidth in a single mode optical fibre system. The total available bandwidth in a fibre channel is 30 THz. WDM combines a number of wavelengths into the same fibre. The availability of practical wideband optical amplifiers enabled the use of wavelength division multiplexers to multiplex many wavelengths in the same optical fibre. WDM can improve the capabilities of the optical system.

Due to the complex challenges of increasing the rate of information transfer and efficiency, we need cost-effective techniques. One solution is to lay out more fibre; however, laying new fibre will not necessarily enable the optimum utilization of the bandwidth. Another solution could be the use of time division multiplexing (TDM), which increases the capacity of a fibre by slicing time into smaller intervals so that more bits (data) can be transmitted per second. This could be a reasonable solution, but it is not efficient. A better solution is to use wavelength division multiplexing, WDM, which increases the capacity of embedded fibre by first assigning incoming optical signals to specific wavelength (λ) within a selected wavelength band and then multiplexing the resulting signals out onto one fibre. WDM combines multiple optical signals so that they can be amplified as a group and transported over a single fibre in order to increase capacity and efficiency.

In order to show how WDM is superior to TDM, consider a highway analogy where one fibre can be thought of as a multilane highway. Traditional TDM systems use a single lane of this highway and increase capacity by moving faster on this single lane. In optical networking, utilizing WDM is analogous to accessing the unused lanes on the highway (increasing the number of wavelengths on the embedded fibre base) in order to gain access to the huge capacity in the fibre.

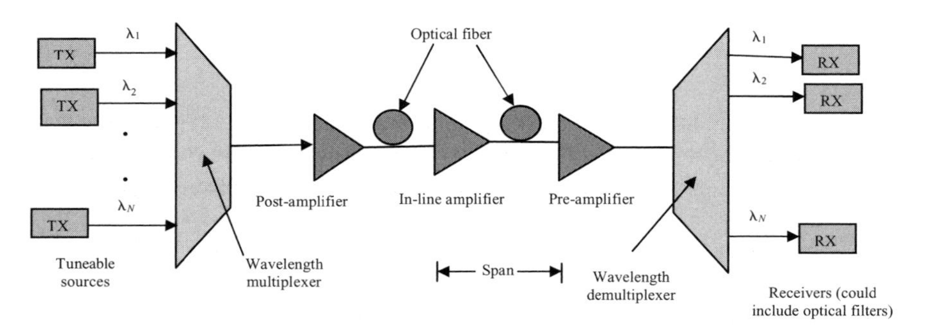

Figure 2.22 A typical waveform division multiplexing, WDM, system.

Moreover, in optical networking this superhighway is blind to the type of traffic that travels on it. Therefore, all types of traffic streams such as ATM, SONET, and IP can be carried efficiently using WDM [3, 4, 7, 9, 12, 49, 50].

In WDM, discrete waveforms form an orthogonal set of carriers that can be separated, routed and switched without interfering with each other. Of course, this holds as long as optical intensity is kept sufficiently low to prevent non-linearity such as Brillourin scattering and four-wave mixing process. Many light beams of different wavelengths can travel along a single fibre carrying different modulated data. Due to the fact that frequency and wavelength are inversely related since $c = \lambda f$ where c is the speed of light, f is the frequency and λ is the wavelength, then we can say that WDM is a form of frequency division multiplexing, FDM. Figure 2.22 shows a typical WDM system.

WDM has the potential to exploit the huge opto-electronic bandwidth mismatch by requiring that each end-user's equipment operate only at electronic rate. Multiple WDM channels from different end-users can be multiplexed on the same fibre. By allowing multiple WDM channels to coexist on a single fibre, one can efficiently utilize the huge fibre bandwidth, with the challenges of designing and developing the appropriate network architectures, and control and management protocols.

The main factors that affect the performance of a WDM system are: (a) fibre chromatic dispersion; (b) nonuniform gain across the desired wavelength range in Erbium-Doped Fibre Amplifiers, EDFAs; (c) scattering processes; (d) nonlinear processes; and (e) reflections from splices and connectors.

WDM started with very wide wavelength spacing. Early systems carried two wavelengths, 1310 and 1550 nm. The invention of the EDFAs has led to the development of techniques to put wavelengths more closely together. Common spacing in traditional improved WDM are 1000, 400, 200, 100, and 50 GHz or 8, 3.2, 1.6, 0.8, and 0.4 nm in the 1550 nm band. In WDM terminology, when channels are spaced at 200 GHz or less, they are often referred to as dense wavelength division multiplexing, DWDM.

DWDM multiplexers and demultiplexers can handle closely spaced optical wavelengths. These designs require narrow passbands, usually 0.4 nm wide, steep roll-off to reject adjacent channels, and stable operation over increased temperature. Recently, multiplexers have gained versatility, moving beyond the wideband wavelengths and into densely packed wavelengths that can be integrated into a multiple high frequency [9, 15, 22–29, 39, 40]

DWDM systems send signals from several sources over a single fibre, therefore, they must include some means to combine the incoming signals. This can be done with a multiplexer, which takes optical wavelengths from multiple fibres and joins them into one beam. At the receiving end, the system must be able to separate out the components of the light so that they can be discreetly detected. This separation function is performed by demultiplexers where the received beam is separated into its wavelength components. The multiplexer also couples these wavelength components to individual fibres. Demultiplexing must be done before the light is detected, because photodetectors are inherently broadband devices that cannot selectively detect a single wavelength. In a unidirectional system, a multiplexer at the sending end and a demultiplexer at the receiving end are needed. Two systems would be required at each end for bidirectional communication, and two separate fibres would be needed. In a bidirectional system, there is a multiplexer/ demultiplexer at each end and communication is performed over a single fibre, with different wavelengths used for each direction [9, 15, 22–29, 39, 40–53].

Multiplexers and demultiplexers can be either passive or active in design. Passive designs are based on prisms, diffraction gratings, or filters. On the other hand, active designs combine passive devices with tunable filters. The primary challenges in the design of these devices are minimizing the cross-talk and maximizing channel separation. Cross-talk is a measure that is used to measure how well the channels are separated, while channel separation is used to measure the ability to distinguish each wavelength.

A simple form of multiplexing or demultiplexing of light can be done using a prism. A parallel beam of polychromatic light impinges on a prism surface; each component wavelength is refracted differently. This is the "rainbow" effect. In the output light, each wavelength is separated from the next by an angle. A lens then focuses each wavelength to the point where it needs to enter a fibre. The same components can be used in reverse to multiplex different wavelengths onto one fibre. Figure 2.23 illustrates a prism refraction demultiplexing process.

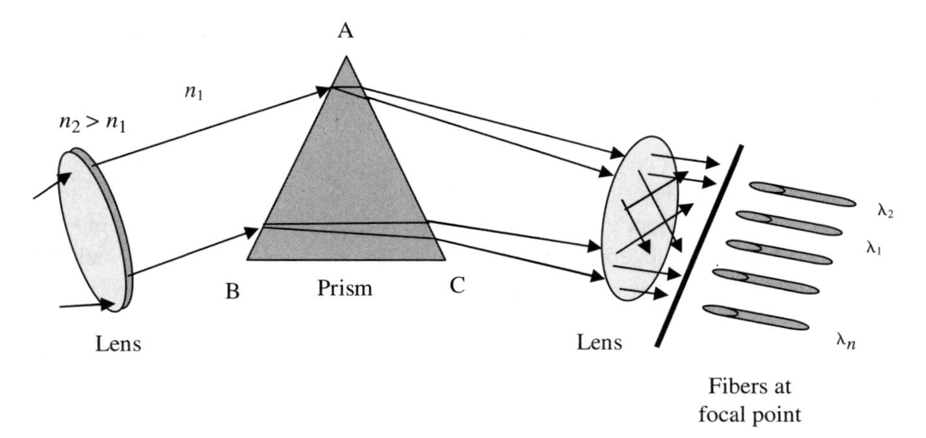

Figure 2.23 A prism refraction demultiplexing process.

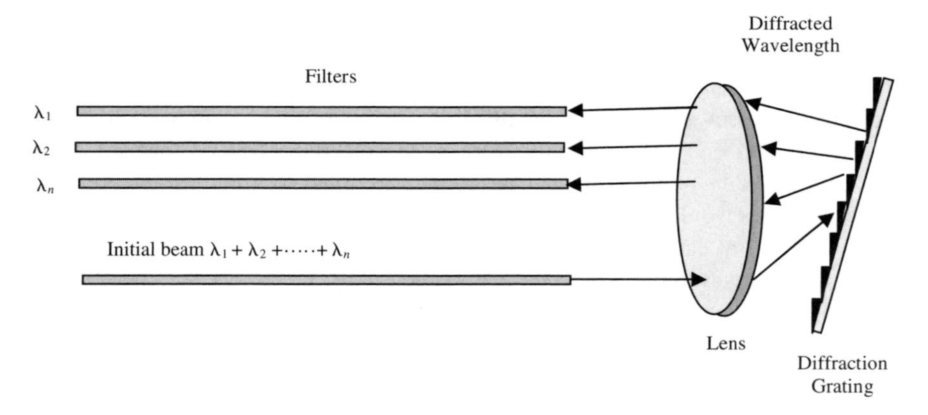

Figure 2.24 A waveguide grating diffraction process.

Another possible technique is based on the principles of diffraction and of optical interference. When a polychromatic light source is applied on a diffraction grating, each wavelength is diffracted at a different angle and therefore to a different point in space. By using a lens, these wavelengths can be focused onto individual fibres; see Figure 2.24.

Arrayed waveguide gratings (AWGs) are also based on diffraction principles. An AWG device, sometimes called an optical waveguide router or waveguide grating

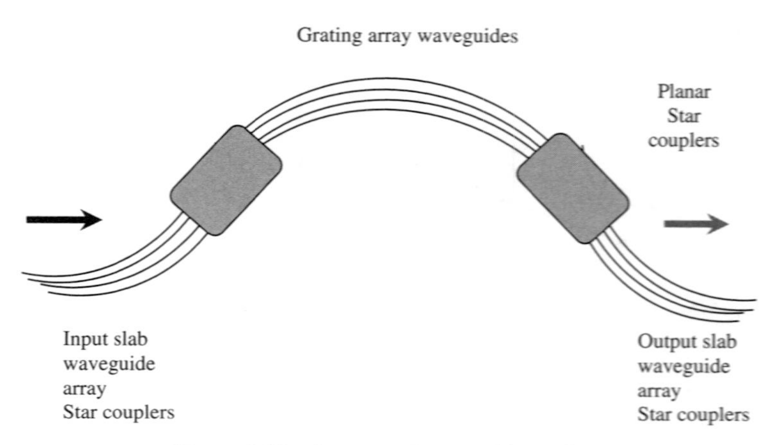

Figure 2.25 An arrayed waveguide grating.

router, consists of an array of curved-channel waveguides with a fixed difference in the path length between adjacent channels. The waveguides are connected to cavities at the input and output. When the light enters the input cavity, it is diffracted and enters the waveguide array. The optical length difference of each waveguide introduces phase delays in the output cavity, where an array of fibres is coupled. The process results in different wavelengths having maximal interference at different locations, which correspond to the output ports. Figure 2.25 shows an arrayed waveguide grating.

There is also a third technique, which uses interference filters in devices, called *thin film filters* or *multilayer interference filters*. By placing filters, consisting of thin films, in the optical path, wavelengths can be demultiplexed. The property of each filter is that it transmits one wavelength while reflecting others. By cascading these devices, many wavelengths can be demultiplexed, see Figure 2.26.

Of these techniques, the AWG and thin film interference filters are the most popular. Filters offer good stability and isolation between channels at reasonable cost, but with a high insertion loss. The AWGs are polarization-dependent, which can be compensated for, and they exhibit a flat spectral response and low insertion loss. One potential disadvantage for AWGs is that they are temperature sensitive, which may limit their use for specific environments. Their chief advantage is that they can be designed to perform multiplexing and demultiplexing operations simultaneously. Moreover, the AWGs are better for large channel counts, whereas the use of cascaded thin film filters is impractical [15, 24–27, 38–52].

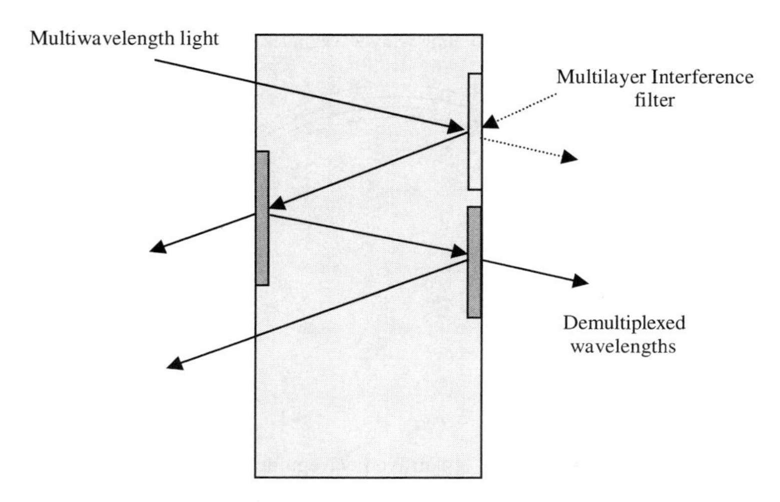

Figure 2.26 Multilayer interference filters.

In a WDM system, there is an area in which multiple wavelengths exist between multiplexing and demultiplexing points. It is often desirable to remove or insert one or more wavelengths at some point along a length. This can be done using an optical add/drop multiplexer (OADM). Instead of combining or separating all wavelengths, the OADM can remove some while passing others on. They are considered key enabling components toward achieving the all-optical networks goal.

OADMs are similar in many respects to the SONET ADM, except that only optical wavelengths are added and dropped, and no conversion of the signal from optical to electrical takes place. Figure 2.27 shows a functional diagram of the add-drop process. This example includes both pre- and post-amplification, which may or may not be present in an OADM, depending upon its design [15, 29–31].

There are two general types of OADMs: (a) a fixed device that is physically configured to drop specific predetermined wavelengths while adding others; and (b) a second generation type that is reconfigurable and capable of dynamically selecting which wavelengths to be added/dropped. Thin-film filters have emerged as a preferred technology for OADMs in most current metropolitan DWDM systems. This is due to their low cost and good stability. For the promising second generation of OADMs, other technologies, such as tunable fibre gratings and circulators, are expected to be distinguished [15, 24–31, 38–53].

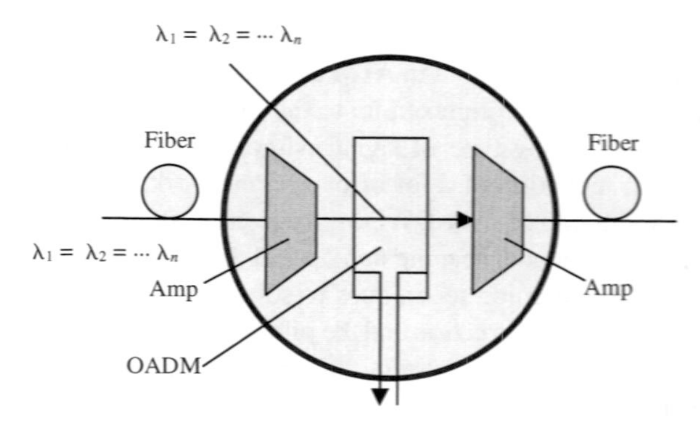

Figure 2.27 Functional diagram of the optical add and drop multiplexers (OADMs).

2.3.9 APPLICATIONS OF DWDM

DWDM has a great potential to be used in many applications. It has already been proven to be well suited for several vital applications such as long-reach high-speed parallel interfacing. For example, certain computer applications require that computer centres be interconnected with multiple high-speed channels that have capacity and availability requirements, as well as interlink delay restrictions. If we need to connect two mainframe computers over a long distance, and the high rate of information being exchanged between the two computers requires multiple parallel OC-N connections, then bits being transmitted at any instant in these lines have a tight skew-delay requirement. In other words, they need to arrive at the far end within a very short deterministic time, multiplexed to a higher SONET rate, and then the randomness introduced by pointer processing in the SONET terminal would introduce sufficiently high nondeterministic delay between channels. This makes it impossible to meet this skew-delay constraint. However, if the OC-Ns are wave division multiplexed onto the same fibre, the signals are guarded to traverse the same physical path and, therefore, they have identical transit and processing delays. Thus, the skew-delay requirement is met. DWDM optical transport benefits all delay-sensitive applications such as real time voice and video that are also called constant bit rate traffic (CBR).

ATM over DWDM has become popular recently due to the increasing demand on bandwidth. Telecommunications service providers are faced with huge investments in order to fulfil capacity demands. Moreover, the demand for Quality of Service (QoS) has increased recently. There seems to be a general move towards providing QoS, while still maintaining the same capacity. ATM over DWDM solves the bandwidth and QoS issues in a cost-effective manner. In DWDM networks, if there is a carrier that operates both in ATM and SONET networks there is no need for the ATM signal to be multiplexed up to the SONET rate. This is because the optical layer can carry any type of signal without any additional multiplexing. This results in the reduction of a lot of overlay network. While there are many advantages in running ATM over DWDM, there are certain issues that should be addressed. These are channel spacing and optical attenuation. Therefore, we need good wavelength conditioning techniques to solve this problem. The techniques used are the forward error correction and the pilot light schemes. By using the latter technique network management systems are able to ensure connectivity, signal on each channel and also identify faults [22–30, 39–52].

A good solution is to take IP directly over DWDM, which would bring about scalability and cost-effectiveness. Currently, there exist commercial products that actually implement IP over DWDM such as Monterey Networks (bought by Cisco in August 1999) that has its own Monterey 20000 Series Wavelength Router. It is claimed that by using their product, service providers can traffic-engineer and rapidly scale up survivable mesh optical cores without using intermediate ATM switches or increasing legacy SONET multiplexers and cross-connects. This means that we are completely eliminating ATM and SONET layers from the networks. The real assessment is whether it would be possible to create an end-to-end optical Internet operating in the range between OC-3 and OC-48 and build systems around an optical Internet infrastructure. With the development of erbium-doped fibre amplifiers most systems that use IP over DWDM using SONET frames have removed the SONET multiplexers. GTS Carrier Service launched a few years ago the first high capacity transport platform in Europe that uses IP over DWDM technology. Moreover, major carriers such as AT&T, Sprint, and others have all started to realize the huge economic potential of IP over DWDM and there is no longer any scepticism about the viability and reliability of this technology [24–30, 36–53].

2.3.10 WAVELENGTH ROUTERS

Any optical network consists of wavelength routers and end nodes that are connected by links in pairs. The wavelength-routing switches or routing nodes are interconnected by optical fibres. Despite the fact that each link can support many signals, it is required that the signals be of distinct wavelengths.

Wavelength routers are found in the literature with various names such as optical switches, optical cross-connects, wavelength switches and wavelength cross-connects. They typically inspect every packet so these packets can be forwarded in the correct direction. Coupling Internet protocol (IP) routers with wavelength-selective optical cross-connects makes it possible to support existing Internet infrastructure in a wavelength division multiplexing (WDM) optical network. Since optical wavelength routing is transparent to IP, it is possible to achieve a very high throughput and low delay when packets are made to bypass the IP forwarding process by being routed directly through the optical cross-connect. This scheme is called *packet over wavelength* (*POW*) [38–41].

WDM links have already been deployed and it is highly advantageous to use these links to interconnect the routers that comprise the global Internet. The main advantage of the POW architecture is to switch as much traffic as possible directly by means of wavelength routers since IP forwarding is relatively expensive. POW has some common features such as *label switching*. The latter is used when an IP router includes a switching fabric that can be used to bypass IP forwarding. Due to the fact that switching speeds are much greater than forwarding speeds, one way is to place as large a fraction of packets as possible on the switched path, leaving as small a fraction of packets as possible on the forwarded path. In order to accomplish this, we need some good intelligence in the switch-router. The router must have software that recognizes that a flow of packets can be passed through the switching fabric. There is a need for a signalling protocol that assists in notifying switches that the recognized flow should be carried over a switched path rather than a routed path. Finally, a hop-by-hop sequence of switches carries the flow of packets from one router to another [7, 13, 17, 19, 25, 26, 38, 39].

An optical network with scalability characteristics can be constructed by taking several WDM links, and then connecting them with wavelength-selective switching subsystems. The path of the signal through the network is individually determined by the wavelength of the signal and the port through which it comes into the network. There are two main types of wavelength switching: (a) one of which dynamically switches signals from one path to the other by changing the WDM routing in the network; and (b) the other type is basically wavelength conversion, in which the information on a signal is transferred from an optical carrier at one wavelength to another. In such cases, the same wavelength can be reused in some other port of the network as long as both lightpaths do not use it on the same fibre. Due to the fact that such a spatial reuse of wavelengths is supported by wavelength routing networks, they are much more scalable. Another characteristic, which enables these networks to cross long distances, is that the energy put in the lightpath is not split to irrelevant destinations. The issue of routing and assigning wavelengths to lightpaths in a network is a complex and necessary one. Intelligent schemes are needed to ensure that this function is performed using a minimum number of wavelengths.

A router transmits signals on the same wavelength on which they are received. An all-optical wavelength-routed network carries data across from one access station to the other without any Optical/Electronic (O/E) conversions. The main categories of Wavelength Switches are [14, 15, 37, 38]:

- *Non-reconfigurable (or static) switches.* In these types of switches, each input port and each wavelength transmit onto a fixed set of output ports at the same wavelength. These cannot be modified once the switch is built. Networks that contain only such switches are called non-reconfigurable networks.
- *Wavelength-independent reconfigurable switches.* These types of switches have an input-output pattern that can be changed dynamically. However, the input-output pattern is independent of the wavelength of the signal.
- *Wavelength-selective reconfigurable switches.* These switches combine the features of the first two categories.

The above three types of wavelength switches were also addressed in subsection 1.5.3 of Chapter 1; refer to Figures 1.17, 1.18 and 1.19 respectively. (Additionally, a diagram of an example wavelength switch of the third type is provided later in subsection 2.3.12 on wavelength converters.) Another classification of optical switches considers the following two types: optical-electrical-optical (O-E-O) switches and the all-optical switches.

The all-optical switches are made possible due to the progress in a number of technologies. Such technologies allow managing and switching photonic signals without the need to convert them to electronic signals. Indeed, only a couple of technologies are ready to make the move from laboratory environment to the real network environment. The primary technology for developing an economically viable, scalable all-optical switch is 3D MEMs (Micro Electro Mechanical Systems). MEMs technology currently provides the best opportunity of providing an all-optical switch matrix, which can scale to the size needed to support a global communication network's node with multiple fibres, each carrying hundreds of wavelengths.

An optical switch adds manageability to a dense wave division multiplexing (DWDM) node that could potentially grow to hundreds of channels. It has the potential of managing those light signals without the need to convert them to electrical and then back to optical. Such switches are very appealing, especially to those carriers that operate large offices where over 75% of the traffic is expected to pass through the office on its way to locations around the world. The 3D MEM devices use control mechanisms to tilt mirrors in 3 directions; hence the name 3D MEM was given. This high level of control can direct light to a higher number of ports with negligible impact on the insertion loss. This is considered the key to supporting thousands of ports with a single stage device. The 3D MEM-based optical switches are expected to be produced in various sizes ranging from 256 × 256 to 1000 × 1000 bi-directional port devices. Moreover, some early research and development outcomes seem to indicate that 8000 × 8000 ports will be practical within the near future.

The network management functions, which are considered an important part of any network, are available today using optical switches that have an electronic-based switching matrix. These intelligent optical switches address the need for high bandwidth management while continuing the tradition of providing easy fault location and performance monitoring information required for monitoring and reporting on the status and reliability of the network. An intelligent optical switch using an electronic fabric provides bandwidth grooming, which is not available in an all-optical switch.

An intelligent optical-electrical-optical (OEO) switch can support a new class of high bandwidth services. This is an incremental step in the operations and maintenance of a new service class that is not troublesome to a carrier's normal mode of operations. It addresses the need to manage a larger portion of bandwidth. By utilizing an electronic-based fabric, the intelligent OEO switch is able to overcome the network impairments that currently limit the use of an all-optical switch in a dynamic mesh architecture. An OEO switch combines the latest generation hardware with sophisticated software to better accommodate the data centric requirements of a dynamic optical network. The inherent 3-regeneration functions allow the intelligent optical switch to be deployed in different network environments. The intelligent optical switch therefore promotes the use of mesh that is more bandwidth efficient and supports a flexible set of bandwidth intensive services. The electronics used in an intelligent optical switch also allow it to make use of the well-accepted evolution of the intelligent optical switch including the support of evolving standards such as generalized multi-protocol label switching (GMPLS). The latter is an emerging standard, which is based on the well-accepted data oriented MPLS (multiprotocol label switching) standard. MPLS is a standardized suite of commercially available data protocols that handles routing in a data network [3, 9, 12, 13, 16, 18, 19, 22–24, 26, 27, 38–51].

2.3.11 COMBINERS/SPLITTERS

A combiner is defined as a passive device that combines optical power carried by two input fibres into a single output fibre. On the other hand, a splitter is defined as a passive device that splits the optical power carried by a single input fibre into

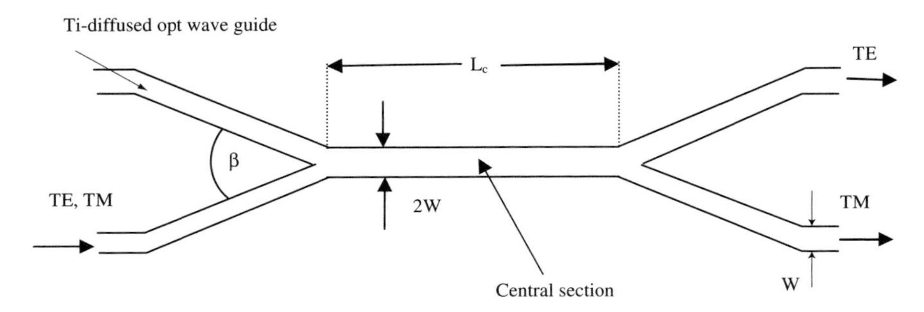

Figure 2.28 A splitter model.

two output fibres. Clearly, the function of a splitter is the reverse that of a combiner [26, 38, 52].

Integrated optical polarization splitters are considered the fundamental elements of the integrated acousto-optic circuits. Basically, they are either used to separate or to recombine the polarization components TE and TM. In order to achieve a wavelength independent operation, a passive directional coupler structure is used. The operation principle of the polarization splitter is based on two-mode interference: symmetric and asymmetric modes, which are guided for both polarizations. The incoming wave stimulates these symmetric and asymmetric modes with equal power, which then propagate at different speeds. The ratio of the power splitting at the end of the structure is found by the relative accumulated phase difference of these modes. In order to construct a polarization splitter, the structure has to be dimensioned such that the phase difference results in either a coupling to the cross-state output for TE and bar-state output to TM or vice versa (see Figure 2.28 and [3, 29, 37, 38]).

In order to design a directional coupler with almost ideal splitting performance, the central section and the branching angle have to be optimized. One way is to choose a central section with a width of twice the width of the incoming optical waveguides. Optimization of the central section length, Lc, and the branching angle, β, can be carried out.

2.3.12 WAVELENGTH CONVERTERS

The basic function of the wavelength converter is to convert an input wavelength to possibly different output wavelengths within the operational bandwidth of DWDM systems in order to improve their overall efficiency. Therefore, the reuse factor is increased. Wavelength converters are one of the important building blocks of any

DWDM system as they enable the reuse of wavelengths in the system. This process is needed in order to increase the overall system bandwidth and for wavelength routing.

There are four possible forms of waveform conversion: full conversion, limited conversion, fixed conversion and sparse wavelength conversion. In the full conversion type, any wavelength shifting is possible and therefore channels can be converted regardless of their wavelengths. In the limited conversion type, wavelength shifting is restricted so that not all combinations of channels may be connected. In the fixed conversion type, restricted form of limited conversion that each node has, each channel may be connected to exactly one predetermined channel on all other links. Finally, in the sparse wavelength conversion, networks are comprised of a mix of nodes having full and no wavelength conversion.

An ideal converter has the following characteristics [15–22]:

- transparent to bit rates and signal formats;
- ability to convert to both short and longer wavelength;
- has fast set-up time;
- has low chirp output and high signal-to-ratio ratio;
- simple to implement;
- polarization insensitivity;
- straightforward implementation.

As multiwavelength fibre optic systems proliferate, interest is growing in converting signals from one wavelength to another. The simplest method to convert wavelengths today is opto-electronically where the input signal is converted to electronic form, and then is used to modulate a transmitter operating at the desired wavelength. This process is common these days wherever wavelengths must be converted, but like electro-optical repeaters, it is cumbersome and inefficient. In order to convert the wavelength, we need a complete receiver-transmitter set. It would seem better to convert wavelengths by purely optical means, but that is difficult in practice. However, this is changing as we start to see new developments in this area. New schemes for wavelength conversion in the optical domain have been developed recently. One approach is to use a process like four-wave mixing. In this case, we combine the input signal with light at another wavelength to generate a different wavelength. A second approach is to use light at one wavelength to control a semiconductor laser operating at another wavelength. The input light changes the population of current carriers in the laser, modulating its output. Although simple in concept, it turns out to be difficult in practice. A third scheme is to build an optically controlled gate, which is essentially a modulator controlled by the input of light rather than by a voltage signal. It directly modulates a laser output or controls

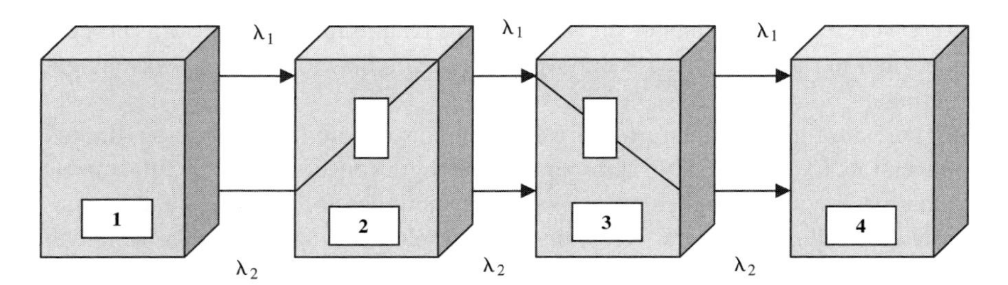

Figure 2.29 Illustration of the use of converters.

an external modulator that modulates another laser. Developers have demonstrated several variations on this approach, some of them quite complex. Among these are the schemes that are based on non-linear wave mixing effects (e.g., four-wave mixing (FWM) and difference frequency generation (DFG) [15–19, 29].

In Figure 2.29, let us assume that there exists a path of wavelength λ_1 between nodes 1 and 2, a path of λ_2 between nodes 2 and 3 and a path of λ_1 between nodes 3 and 4. It is not possible to establish a light path from node 1 to node 4 for wavelength λ_2 since the wavelength has already been used between nodes 2 and 3. One solution to this is to use wavelength converters at nodes 2 and 3 to convert the wavelength to λ_1, which is an unused wavelength in this part of the network. Functionally, such a network is similar to a circuit switched network. Figure 2.29 shows an illustration of a switch that contains dedicated wavelength converters, while Figure 2.30 depicts an illustration of a switch that has a dedicated wavelength converter.

2.4 SUMMARY

This chapter presents the main enabling technologies for multiwavelength optical local area networks (LANs). The chapter sheds some light on components used to implement such a multiwavelength optical LAN system. Among the components that are covered are: light-emitting diodes, photodiodes, optical filters, directional couplers, passive star couplers, WDM multiplexers/demultiplexers, wavelength routers, combiners/splitters, wavelength converters, and DWDM.

The capacity of fibre is assumed to be limited by the restrictions and shortcomings of optical transceivers, amplifiers and cross-connects. Such limitations affect the maximum available spectrum, wavelength/waveband spacing, and maximum bit rate per channel. Trade-offs between optical and electronic methods of

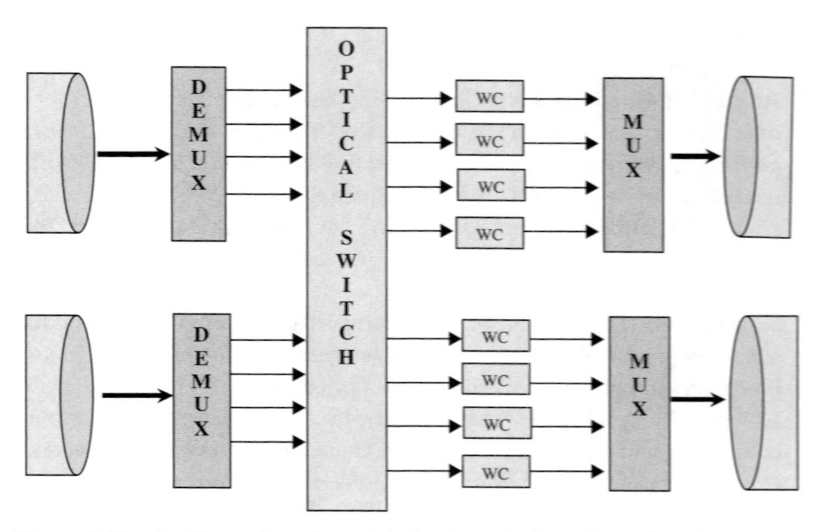

Figure 2.30 An illustration of a switch that has a dedicated wavelength converter.

implementing connectivity and routing are proposed in the literature. We reviewed some of these approaches as they apply to individual components. The goal is to reach an optimum design that provides a reasonable cost-performance ratio. After about a decade of intensive research and development, photonic and optoelectronic technology is now undergoing amazing achievements. We present in this chapter the essential features and characteristics of the main building blocks for the implementation of multiwavelength optical LANs. The goal is to communicate an understanding of the fundamental functions of these components together with some notion about their performance and structures. Of course, it is not easy to give in one chapter all the details, but we believe we have presented a reasonable coverage of the state-of-the art of current photonic and optoelectronic technology as it applies to multiwavelength optical LANs [1–53].

REFERENCES

[1] Papadimitiou GI, Obaidat MS and Pomportsis AS, "Advances in optical networking", *Wiley International Journal of Communication Systems*, vol. 15, no. 2–3, pp. 101–113, March–April 2002.

[2] Veeraraghavan M, Karri R, Moors T, Karol M and Gobbler R, "Architectures and protocols that enable new applications on optical networks", *IEEE Communications Magazine*, vol. 39, no. 3, pp. 118–127, March 2001.

[3] Green P, "Progress in optical networking", *IEEE Communications Magazine*, vol. 39, no.1, pp. 54–61, January 2001.

[4] Ahmed CB, Boudriga N and Obaidat MS, "Supporting adaptive QoS for multiple classes of services in DWDM Networks", in *Proceedings of 2001 International Conference on Parallel Processing Workshops, ICPP-01/Workshop of Optical Networks 2001*, Valencia, Spain, pp. 283–288, September 2001.

[5] Ahmed CM, Boudriga N and Obaidat MS, "Adaptive QoS schemes in DWDM networks", in *Proceedings of the IEEE ICC 2002*, New York, pp. 2881–2885, April/May 2002.

[6] Papadimitriou GI, Obaidat MS and Pomportsis AS (eds), "Special issue on advances in optical networks", *Wiley International Journal of Communication Systems*, vol. 15, no. 2–3, March–April 2002.

[7] Ahmed CB, Boudriga N and Obaidat MS, "QoS routing with wavelength conversion and connection admission connection in DWDM networks", in *Proceedings of IEEE International Conference on Computer Networks and Mobile Computing ICCNMC2002*, Beijing, China, pp. 61–66, October 2001.

[8] Papadimitriou GI, Pomportsis AS and Obaidat MS, "Adaptive bandwidth allocation schemes for lightwave LANs with asymmetric traffic", in *Proceedings of IEEE International Performance, Computing and Communications Conference IPCC'01*, Phoenix, AZ, pp. 45–50, April 2001.

[9] Senior JM, Handley MR and Leeson MS, "Developments in wavelength division multiple access networking", *IEEE Communications Magazine*, vol. 36, no. 12, pp. 28–36, December 1998.

[10] Ramaswami R and Segall A, "Distributed network control for optical networks", *IEEE/ACM Transactions on Networking*, vol. 5, no. 6, pp. 936–943, December 1997.

[11] Zang H, Sahasrabuddhe L, Jue JP, Ramamurthy S and Muckerjee B, "Connection management for wavelength-routed WDM networks", in *Proceedings of IEEE GLOBECOM'99, Rio de Janeiro, Brazil*, pp. 1428–1432, December 1997.

[12] Brackett CA, "Dense wavelength division multiplexing networks: principles and applications", *IEEE Journal on Selected Areas in Communications*, vol. 8, no. 6, pp. 948–964, August 1990.

[13] Green PE, "Optical networking update", *IEEE Journal on Selected Areas in Communications*, vol. 8, no. 5, pp. 764–779, June 1996.

[14] Chatterjee S and Pawlowski S, "All optical networks", *ACM Journal of Communications*, vol. 32, no. 36, pp. 75–83, June 1999.

[15] Mukherjee B, *Optical Communication Networks*, McGraw-Hill, New York, 1997.

[16] Ramaswami R and Sivarajan KN, *Optical Networks: A practical perspective*, Morgan Kaufmann, Second edition, 2001.

[17] Stern TE and Bala K, *Multiwavelength Optical Networks: A Layered Approach*, Addison-Wesley, Reading, MA, 2000.

[18] Green PE, *Fibre Optic Networks*, Prentice Hall, Englewood Cliffs. NJ. 1993.

[19] Borella A and Cancellieri G, *Wavelength Division Multiple Access Optical Networks*, Artech House, 1998.

[20] Papadimitriou GI, Obaidat MS and Pomportsis AS, "Adaptive protocols for optical LANs with bursty and correlated traffic", *Wiley International Journal of Communication Systems*, vol. 15, no. 2–3, pp. 115–125, March–April 2002.

[21] Guo A, Henry M and Salamo GJ, "Fixing multiple waveguides induced by photorefractive solitons: directional couplers and beam splitters", *Optics Letters*, vol. 26, no. 16, pp. 1274–1276, August 15, 2001.

[22] Mukherjee B, "WDM optical networks: progress and challenges", *IEEE Journal on Selected Areas in Communications*, vol. 18, no. 10, pp. 1810–1824, October 2000.

[23] Yao S, Dixit S and Mukherjee B, "Advances in photonic packet switching: An overview", *IEEE Communications Magazine*, vol. 38, pp. 84–94, February 2000.

[24] Knoke M and Hartmann HL, "Fast optimum routing and wavelength assignment for WDM ring transport networks", in *Proceedings IEEE International Conference on Communication ICC'02*, New York City, vol. 5, pp. 2740–2744, 2002.

[25] Riziotis C and Zervas MN, "Performance comparison of Bragg grating-based optical add-drop multiplexers in WDM transmission systems", in *Proceedings IEE Circuits and Systems*, vol. 149, no. 3, pp. 179–186, June 2002.

[26] Sivalingam KM and Subramaniam S, *Optical WDM Networks Principles and Practice*", Kluwer Academic Publishers, Dordrecht, 2000.

[27] Goralski W, *Optical Networking and WDM*, McGraw-Hill, New York, 2001.

[28] Ramaswami R and Sivarajan KN, *Optical Networks: A Practical Perspective*, Academic Press, New York, 2002.

[29] Hecht J, *Understanding Fibre Optics*, Third Edition, Prentice Hall, Englewood Cliffs, NJ. 1998.

[30] Ramaswami R and Sasaki G, "Multiwavelength optical networks with limited wavelength conversion", *IEEE/ACM Transactions on Networking*, vol. 6, no. 6, pp. 744–754, December 1998.

[31] Tanenbaum A, *Computer Networks*, fourth edition, Prentice Hall, Englewood Cliffs, NJ. 2003.

[32] Stallings W, *Data and Computer Communications*, sixth edition, Prentice Hall, Englewood Cliffs, NJ. 2000.

[33] Stallings W, *ISDN and Broadband ISDN with Frame Relay and ATM*, fourth edition, Prentice Hall, Englewood Cliffs, NJ. 1999.

[34] McDysan D and Spohn D, *ATM Theory and Applications*, McGraw-Hill, New York, 1999.

[35] Chang-Hsnain CJ, "Tunable VCSEL", *IEEE Journal on Selected Topics in Quantum Electronics*, vol. 6, no. 6, pp. 978–987, November/December 2000.

[36] http://www.lightreading.com/document.asp

[37] http://www.iec.org/tutorials/dwdm

[38] http://www.lucent.com/solutions/core_optical.html

[39] http://www.oplink.com/Products/DWDM

[40] http://ieeexplore.ieee.org/iel4/45/16132/00747242.pdf

[41] http://www3.50megs.com/jhsu/tech-dwdm.html

[42] http://www.alcatel.com/backbone/bckb2b.htm

[43] http://www.techguide.com/comm/sec_html/dwave.shtml

[44] http://www.ciena.com/products

[45] http://www.cisco.com/warp/public/779/servpro/solutions/optical/dwdm.html

[46] http://www.cis.ohio-state.edu/~jain/cis788–99/dwdmW

[47] http://www.howstuffworks.com/laser.htm

[48] http://abhijit.8m.com/optical.htm.

[49] http://www.fibre-optics.info/articles/dwdm.htm

[50] http://www.fibre-optics.info/articles/dwdm.htm

[51] http://www.comsoc.org/livepubs/surveys/public/2q99issue/pdf/Yates.pdf

[52] http://www.tpub.com/neets/tm/108–11.htm

[53] http://www.centrovision.com/tech2.htm

WWW RESOURCES

[1] http://www.howstuffworks.com/led1.htm

[2] http://oemagazine.com/fromTheMagazine/aug01/pdf/tutorial.pdf.

[3] http://www.comsoc.org/livepubs/surveys/public/2q9

[4] http://www.fibre-optics.info/articles/dwdm.htm

[5] http://www.eng.nus.edu.sg/EResnews/0601/rd/rd_8.html.

[6] http://networks.cs.ucdavis.edu/~mukherje/book/ch01/node4.html

[7] http://searchnetworking.techtarget.com/sDefinition/0,,sid7_gci214538,00.html

[8] http://currentissue.telephonyonline.com/ar/telecom_wavelength_routers_bumped
[9] http://studytour.herkimer.net/hightech2000/Pres/WR%2015900.pdf
[10] http://cc.uoregon.edu/cnews/summer2000/fibre.html
[11] http://certcities.com/editorial/features/story.asp?EditorialsID=25
[12] http://www.allbookstores.com/browse/TEC011000:7
[13] http://www.3m.com/market/telecom/enterprise/volition/
[14] http://www.corning.com/opticalfibre/products_services/
[15] http://www.tpub.com/neets/tm/108–11.htm
[16] http://msdn.microsoft.com/vstudio/techinfo/articles/upgrade/Csharpintro.asp
[17] http://www.comsoc.org/livepubs/surveys/public/2q99issue/pdf/Yates.pdf

3

Medium Access Control Protocols

The common strategy implemented in the network and adopted by all nodes in order to communicate with each other comprises the so-called *media-access control (MAC) protocol*. In a WDM broadcast-and-select network the above mentioned 'media' are in fact the various optical wavelengths (channels) available to the network nodes for communication. Media-access control protocols are simply methods or ways of accessing the available media and correspond to the Data Link Layer of the OSI model; more precisely, they are protocols of the MAC layer (sublayer of the Data Link Layer).

Since the time when WDM broadcast-and-select networks were introduced, the subject of how nodes should access the various channels has received a great deal of attention resulting in a vast amount of literature proposing and studying MAC protocols. Most protocols, almost right from the start, assume tunability on at least one side at each station (transmitters or receivers) and, in most cases, they also consider that the tunability range of these devices includes all available channels, so that there is full connectivity between every node-pair in just one hop. In other words, *single-hop* WDM broadcast-and-select systems are considered and the same applies for the protocols examined throughout this chapter too.

Clearly, the main objective of a protocol is to provide the way (or set of rules) according to which stations have to regulate and perform their *broadcast-and-select*

Multiwavelength Optical LANs, G.I. Papadimitriou, P.A. Tsimoulas, M.S. Obaidat and A.S. Pomportsis.
 ISBN 0-470-85108-2.

operations. In simple terms a good protocol is essential to coordinate transmissions between various stations in the network and also, when necessary, to determine how a node should select among various transmissions destined to it at the same time. Assuming tunability of transmitters, it is obvious that two or more nodes might transmit on the same channel at the same time. When this happens we say that a *channel collision* (or just a collision) occurs, resulting in the corruption of information and the need for retransmission; obviously, this would degrade network performance. Moreover, in the case of a single tunable receiver per node, when two or more stations transmit (on different wavelengths) messages destined to the same station at the same time, we say that a *receiver collision* occurs. It may be the case that before a node finishes receiving on a certain wavelength, another message destined to it on another wavelength arrives. The latter message would have to be ignored and transmitted by its source again later, unless some kind of priority was defined in the framework of the MAC protocol, which determined that the new incoming message is more important. If two or more messages arrived at a destination simultaneously,[1] a kind of a selection would have to be made, e.g. selection of the wavelength with the smallest number, priority-based selection or just random selection. A MAC protocol usually aims to *prevent channel (and possibly receiver) collisions from occurring* and *specify how selections should be made* when necessary.

Before proceeding to outline the basic features of certain interesting protocols it would be helpful to discuss some fundamental parameters differentiating protocols:

- *Pre-transmission coordination or pre-allocation based.* On the basis of this parameter, the most basic differentiation of MAC protocols is made. There are protocols which require using (at least) one separate wavelength as a *control channel* for coordination between nodes before actual transmission of data. These protocols are generally called pre-transmission coordination-based protocols. When no control channel is used, the protocols are typically based on some form of pre-allocation of wavelengths to transmitters or receivers; thus we have the pre-allocation-based protocols. Accordingly, broadcast-and-select systems employing pre-transmission coordination-based protocols are sometimes characterized as *WPC* (*With Pre-transmission Coordination*), while the ones employing pre-allocation based schemes are termed *WOPC* (*Without Pre-transmission Coordination*). As an exception to the above general definitions, however, note that a comparatively small set of pretransmission coordination-based protocols do not entail any separate control channel, but assume control packets are transmitted over the same channels used for data.

- *'Tell-and-go' or reservation-based WPC protocols*. When at least one separate control channel is used (for pretransmission coordination), there are protocols specifying that a node should first transmit a control packet on the control channel just to inform the potential destinations and then (immediately after that) transmit the data. These protocols are said to have the *tell-and-go* feature and their main advantage is the low access delays.[2] On the other hand, there are protocols according to which a node transmits the control packet and has to wait for a round-trip delay from the star until it receives it back successfully. Then it may be allowed to transmit or not. This results in higher network throughput, but access delays get higher too. It should be noted that sometimes in the literature the term '*reservation protocols*' implies all pre-transmission coordination-based protocols, even though the ones with the tell-and-go feature mentioned before are not actually reservation-based in the strict sense. Among purely reservation-based protocols, there are some schemes appropriate for circuit switched traffic or traffic with long holding times. These might determine that a successful data packet transmission on a channel actually *reserves* this channel for the corresponding source node for a somewhat long period during which the node may transmit a long message comprising many data packets.
- *Number of data (and control if present) channels*. Protocols may assume that the overall number of channels is smaller than the number of nodes; or alternatively, that the number of data channels equals the number of nodes. When pretransmission coordination exists, one could distinguish between using just one or more control channels. In the former case and supposing that the number of nodes is high, some amount of electronic bottleneck would probably be introduced as each node would constantly have to process control information coming from all nodes and being useless most of the time for an individual node. Processing power can be saved if multiple control channels are used, but this comes at the price of higher implementation costs.
- *Scheduling transmissions*. A significant part of the literature on pretransmission coordination-based MAC protocols concerns schemes that involve *scheduling* of packet transmissions and receptions in a way that (usually) both channel and receiver collisions are avoided and the network throughput gets considerably improved. To achieve this, they, first of all, abandon the tell-and-go feature, so that all nodes have enough time to gather the necessary feedback (control information) from the network before deciding the appropriate schedule each time. Throughput performance improvement, however, is achieved at the cost of more complexity, higher access delays and additional implementation cost, the exact form of which depends on the specific scheduling scheme. If the scheduling algorithm is distributed, which is quite a frequent case, all nodes of the network

have to maintain some global information and update it every now and again, according to information obtained by the shared control channel(s). A characteristic form of global information is a *traffic demand matrix* $D = [d_{ij}]$, where d_{ij} represents the number of data packets[3] at node i that are destined for node j. Other types of global status information include e.g. the tables *receiver available time* (*RAT*) and *channel available time* (*CAT*), with $RAT[j] = t$ meaning that node j will be idle after t time slots and $CAT[c] = t$ implying that channel c will be available after t time slots [12]. All nodes run the same distributed scheduling algorithm based on the same (global) data to decide how transmissions and receptions should be made. In centralized scheduling schemes (e.g. [29–31]) it is assumed that the hub is no longer passive, but except for the optical passive star coupler it also includes a scheduler. Typically this scheduler accepts requests for transmissions from nodes, schedules them and gives the necessary directions back to nodes. The latter could possibly be done by means of two control channels, one for accepting transmission requests and one for sending back scheduling decisions (assignments). Scheduling algorithms will be further discussed in the introduction to pretransmission coordination-based schemes without receiver collisions, in Subsection 3.3.2.

- *Synchronization*. When nodes are synchronized according to a common clock and data channels are time-slotted, the protocol is said to be *synchronous* (or else, of course, *asynchronous*).
- *Hardware equipment per node* (*node architecture*). This relates to a discussion in the introductory Chapter 1 about the number and types of transceivers per node. Concerning the type, tunability is assumed on at least one side, either that of transmitters or receivers.
- *Active operations at the hub*. Typically for broadcast-and-select networks the case of a centralized passive hub (with just a passive star coupler) is considered. However, several interesting protocols that have been developed recently assume that the hub has some additional complexity and the responsibility to perform certain active operations.
- *Adaptation to network feedback*. An interesting part of the above mentioned protocols incorporating active operations at the hub are *adaptive* in that their operation is based on network feedback information. Protocols entailing network architectures which assume the hub remains passive could be adaptive too. In both cases, by adapting their operation according to the current state of the network, these protocols typically achieve a higher performance than several other schemes. Such protocols will be studied in Chapter 4.
- *Tuning (or switching) latencies of transceivers*. This parameter refers to the time the tunable transmitter(s) (and/or receiver(s)) need for tuning to the appropriate

channel each time, before actual transmission (reception). Tuning latencies may be assumed to be negligible or non-negligible by the various protocols.

- *Tuning range*. Usually taken to be wide enough to 'host' all available channels of the network, but this may not be the case in some schemes.
- *Packet length*. May be assumed to be fixed (and, for example, equal to one data slot, if the protocol is synchronous) or of variable-length.
- *Network throughput performance*. The use of each protocol brings about a certain level of network throughput performance, i.e. determines how efficiently the available bandwidth is used.
- *Policy regarding collisions*. There are schemes eliminating both channel and receiver collisions, avoiding only one collision type or allowing both. The latter two cases usually imply simpler schemes (at the cost of lower throughputs in general), which additionally have to specify what has to be done in case a collision occurs.

In this chapter we are going to describe the main features of several noteworthy MAC protocols, which are chosen from the enormous set of protocols that have been proposed for WDM broadcast-and-select networks throughout the years. The proposed schemes vary greatly, extending from some early (from about the late 1980s) proposals that basically involve straightforward extensions of older single channel-oriented schemes to the multichannel WDM case, up to more complicated recent protocols whose operation is based on fairly intricate scheduling algorithms.

The structure of this chapter is based on the classification of MAC protocols depicted in Figure 3.1. The two major categories of protocols are the pre-allocation

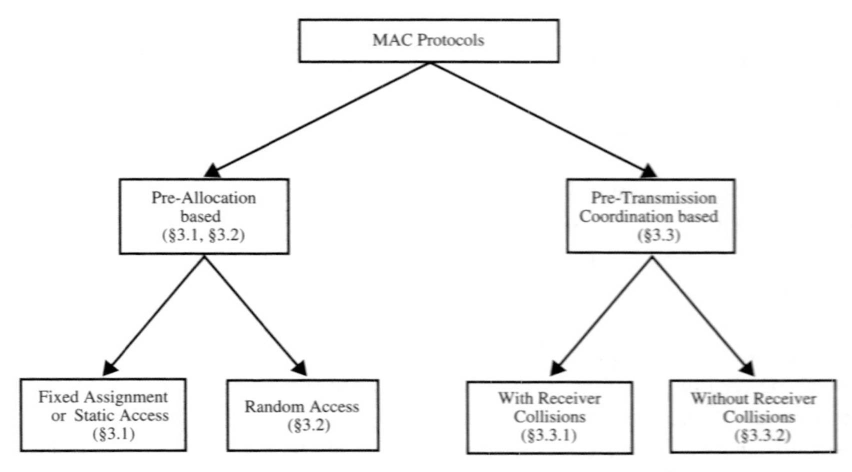

Figure 3.1 Classification of MAC protocols according to the structure of this chapter.

based and the pre-transmission coordination based, as was already mentioned. The former category is examined in Sections 3.1 and 3.2, while the latter is discussed in Section 3.3.

In pre-allocation schemes the available wavelengths are used only as data channels and no wavelength serves as a shared control channel for pre-transmission coordination purposes. The wavelengths are pre-assigned to transmitters or receivers in a fixed manner, and each wavelength comprises the so-called *home channel* of the node it is pre-allocated to. If we assume that N and W denote the number of network nodes and wavelengths respectively, two quite common home channel allocation schemes are the following [37]:

- *interleaved allocation*: $\lambda_i = i \text{ MOD } W$;
- *neighbour allocation*: $\lambda_i = \left\lfloor \frac{i}{\lceil N/W \rceil} \right\rfloor$.

In the above home channel allocation schemes, node i is assigned channel λ_i as its home channel, where $0 \leq i \leq N - 1$ and $0 \leq \lambda_i \leq W - 1$. Thus, assuming for example TT-FR node architecture, it is obvious that a source node can determine the home channel of a destination node through the destination node number (i), the number of nodes (N) and the number of wavelengths (W).

Since we consider single-hop systems, each node must be capable of communicating with any other node of the network in just one hop. Thus, assuming the number of nodes N is not too small to make a solution of N fixed-tuned transmitters (or receivers) per node practical, it follows that either each node's transmitter(s) or receiver(s) are required to have tuning capabilities over all wavelengths in the system. More specifically, it suffices that each node has either one tunable transmitter or one tunable receiver, as long as it can tune over the whole range of wavelengths; this assumption is a commonplace in many MAC protocols. An alternative would be to equip each node with a small number of tunable transmitters (or equally receivers) each one with limited tuning capability, which, however, can cover all wavelengths in collaboration, so that single-hop communication is always possible in the system. Using one tunable transmitter with limited tuning capability and a small number of fixed-tuned receivers per node would be another solution (e.g. in [15]). However, the simplest configuration remains either an FT-TR or a TT-FR system, where the tunable device can tune over the entire spectrum range.

In the first approach, i.e. FT-TR, receivers are required to tune to the appropriate home channel of a node transmitting to them (channels are pre-allocated to transmitters). The problem here is how nodes should be notified of an oncoming packet transmission destined to them so as to tune their receivers appropriately. Obviously, some kind of coordination should exist, which is quite difficult in the absence of a control channel. One rather simplistic scheme in [14] determines that each source node continuously transmits on its home channel, while the transmitted data is collected by some idle receiver. Another pre-allocation based FT-TR approach would require that each receiver tunes to all wavelengths in a round-robin fashion, so that within a tuning period it always tunes to each channel for some time.[4] It can be noticed that no sophisticated coordination can exist, which makes the second approach, i.e. TT-FR systems, rather more attractive when pre-allocation schemes are implemented. In such systems channels are pre-assigned to the receivers of nodes and each receiver is fixed-tuned to a home channel. In this way, the decision of where to transmit (choice of a destination) made by a source node actually entails the tuning decision of its tunable device (transmitter in this case) to the home channel of the destination, which is known in advance; thus, no coordination, as in the case of FT-TR, is required here. Besides being simpler and inherently eliminating the coordination problem that exists in the presence of tunable receivers, TT-FR systems also have the advantage of costing less in most cases, since tunable transmitters (TTs) are usually cheaper than tunable receivers (TRs). For this reason, we shall mostly have to deal with pre-allocation protocols that assume tunable transmitters and fixed-tuned receivers in this chapter.

The category of pre-allocation based protocols can be further divided into fixed-assignment (or *static access*) *protocols* and *random access protocols*, which will be examined in Sections 3.1 and 3.2 respectively (see Figure 3.1). Fixed-assignment schemes predetermine the exact time instants the nodes are allowed to access the shared channels. Hence, they are also called static access protocols. Channel collisions are avoided, since the prearrangement is done in a way that prevents two or more stations from transmitting on the same channel at the same time. This reminds us of time-division multiplexing presented in the first chapter (Section 1.2). On the other hand, random access protocols allow nodes to access the shared channels in a random way. This typically leads to collisions, whose frequency depends on the offered traffic each time. Note that in both cases of pre-allocation based protocols, if the number of wavelengths equals the number of nodes in the system, *optical self-routing* [4] is achieved, that is to say each node has its own home channel. If the wavelengths are less than the nodes, the same home channel may be shared by two or more nodes, which implies *partial self-routing*. Note also that in both Sections 3.1 and 3.2 we mainly concentrate on protocols for non-experimental WDM optical LANs of the star topology. However, we also make brief reference to a couple of MAC protocols designed for the folded bus topology (see Figure 1.29) or have been used in experimental systems in practice.

The other major category of MAC protocols includes pre-transmission coordination-based schemes and it is studied in Section 3.3 (see Figure 3.1). When we presented the fundamental parameters of the various MAC protocols in the beginning of this chapter, we saw that there are pre-transmission coordination-based protocols that have the 'tell-and-go' feature and others that are reservation-based. However, as we have noted before, it is quite customary in the literature (e.g. [4], [37]) to use the term 'reservation protocols' for all pre-transmission coordination-based schemes. These schemes differ from pre-allocation ones in that they designate at least one wavelength (which is the most frequent case, but may be two or more) to be used as a control channel that coordinates access to the remaining data channels. Thus, it is often the case that nodes are additionally equipped with a transmitter and a receiver, both fixed-tuned on the control channel. As far as data channels are concerned, nodes may use tunable devices on both sides or at least on one side (transmitting or receiving); the necessary coordination in case tunable receivers are used for data is naturally provided here by making use of the control channel(s).

A control packet typically contains the destination node and the desired data channel for transmission. In 'tell-and-go' schemes, a control packet notifies all other nodes, which monitor the common control channel, about the data packet transmission that will follow immediately. On the other hand, in purely reservation protocols it attempts to reserve access on the desired data channel. In the special case of scheduling protocols, the control information is normally used to form the necessary status information-input to the scheduling algorithm, which may also be viewed as reservation for access. As we can note, in pre-transmission coordination schemes, unlike pre-allocation ones, data transfer actually consists of two phases: a reservation (or just notification in 'tell-and-go' protocols) and a data transmission phase. Channel collisions may be allowed to occur in both phases (equivalently both in control and data channels), only in one or in none of them, depending on the protocols used. Note that the required MAC protocols for the control and data channels are typically different; sometimes the symbolism X/Y is used to denote that protocol X is used for the control channel and protocol Y for the data channels. However in most cases, e.g. in scheduling schemes, the data channel protocol is more complex and therefore will mostly draw our attention. As far as receiver collisions are concerned (if we assume tunable receivers at nodes for data reception), there are pre-transmission protocols that allow receiver collisions and others that eliminate them. As Figure 3.1 shows, we shall deal with the former type in Subsection 3.3.1 and with the latter in Subsection 3.3.2.

Comparing the two major protocol categories, we see that pre-allocation-based protocols are generally much simpler in operation and more easily implemented. Moreover, they do not have to waste any wavelengths for control purposes and they can operate efficiently with simple node configurations, e.g. with nodes having only one tunable transmitter and one fixed-tuned receiver (TT-FR). Note that the latter has been the main objective for the development of such protocols in order to reduce implementation costs. Pre-transmission-based schemes, on the other hand, are often capable of utilizing the available bandwidth more efficiently, i.e. lead the network to higher performance levels. They are also the preferred choice when a large and time-varying number of network nodes is considered, since they offer more flexibility than pre-allocation schemes. Thus, a vast number of researchers have been focusing on pre-transmission coordination-based MAC protocols throughout the years, making the relative literature really huge. But before proceeding to study noteworthy samples from this literature (Section 3.3), in the next two sections we (respectively) explore some interesting fixed assignment (static access, Section 3.1) and random access (Section 3.2) pre-allocation based protocols.

3.1 FIXED-ASSIGNMENT PROTOCOLS

In this section some of the most typical fixed-assignment protocols will be presented. We begin with I-TDMA.

3.1.1 I-TDMA

The most characteristic example of a fixed-assignment MAC protocol is *I-TDMA* [4, 37] i.e. *Time Division Multiple Access* protocol, where 'I' stands for *interleaved home channel allocation*, which was described before. The network model considered is a WDM Broadcast-and-Select optical network of the star topology with N nodes and W wavelengths. Each node has one tunable transmitter capable of tuning rapidly (i.e. tuning time assumed zero) over the whole range of the W wavelengths and one receiver fixed tuned to its home channel; that is to say, the assumed node architecture is TT-FR. Packets are taken to be of fixed-size and time is slotted so that packet transmission lasts exactly one slot and is synchronized to begin at the start of each time slot.

I-TDMA is actually an extension of the well-known Time Division Multiplexing technique (proposed for a single channel) over a multichannel WDM environment. Thus, time is divided into cycles and each node of the network is pre-assigned exactly one slot per cycle to transmit on every channel. Each source node is assumed to have a single queue for buffering the arriving packets and transmitting the packet at the top of the queue (after the necessary tuning to the destination's home channel) when it is assigned to do so by the protocol. If the number of nodes is equal to the number of wavelengths ($N = W$), cycle length is considered to be $N - 1$ slots and it follows that each node is essentially pre-assigned one slot per cycle to transmit

Table 3.1 Sources/destinations allocation map for I-TDMA and I-TDMA* for an example network with $N = W = 5$ and zero switching latency ($\alpha = 0$).

Sources	Destinations						
0	1	2	3	4	1	2	...
1	2	3	4	0	2	3	...
2	3	4	0	1	3	4	...
3	4	0	1	2	4	0	...
4	0	1	2	3	0	1	...
Time slots:	1	2	3	4	1	2	...
Cycles:		← 1st cycle →				← 2nd cycle →	

to every possible destination (except for itself). Moreover, as we explained before, self-routing is achieved for $N = W$. An example sources/destinations *allocation map* for a network with five nodes and five wavelengths applying the I-TDMA protocol[5] is illustrated in Table 3.1. In this example it is assumed that the *switching latency* α is zero. As we can see, the cycle length is equal to four slots.

On the other hand, in the more general case where there are more nodes than wavelengths in the network ($N > W$), home channels should be shared by some destinations and the cycle length is N slots. Each node has exactly one chance per cycle to transmit on each possible home channel, while it is idle for $N - W$ slots; partial self-routing applies in this case. As an example, let us consider a network with seven nodes and five available wavelengths. The channels/sources allocation map for this network is depicted in Table 3.2(a), where it is obvious that the cycle length is equal to seven time slots. From the sources/channels allocation map shown in Table 3.2(b), it becomes clear that each node is assigned to transmit on $W = 5$ slots and remains idle for exactly $N - W = 2$ slots per cycle, in accordance to what we said before. The same assumption about zero switching latency applies here as well.

The performance of I-TDMA is generally high under heavy traffic conditions, while it gets significantly worse in the opposite case; when there are not many new generated packets for transmission and the number of nodes, which determines the cycle length, is quite large, one source node that is ready may have to wait for quite a long time before its turn to transmit comes. In other words, access delays become considerably high under low load conditions. This explains why I-TDMA performs poorly in this case.

Moreover, there is another factor contributing to the performance degradation of I-TDMA protocol, which relates to the assumption of one queue (only) per

Table 3.2 I-TDMA and I-TDMA* for an example network with $N = 7$, $W = 5$ and zero switching latency ($\alpha = 0$). (a) Channels/sources allocation map. (b) Sources/channels allocation map.

Channels	Sources								
0	0	1	2	3	4	5	6	0	1	...
1	1	2	3	4	5	6	0	1	2	...
2	2	3	4	5	6	0	1	2	3	...
3	3	4	5	6	0	1	2	3	4	...
4	4	5	6	0	1	2	3	4	5	...
Time slots:	1	2	3	4	5	6	7	1	2	...
Cycles:			← 1st cycle →					← 2nd cycle →		

(a)

Sources	Channels									
0	0	X	X	4	3	2	1	0	X	...
1	1	0	X	X	4	3	2	1	0	...
2	2	1	0	X	X	4	3	2	1	...
3	3	2	1	0	X	X	4	3	2	...
4	4	3	2	1	0	X	X	4	3	...
5	X	4	3	2	1	0	X	X	4	...
6	X	X	4	3	2	1	0	X	X	...
Time slots:	1	2	3	4	5	6	7	1	2	...
Cycles:			← 1st cycle →					← 2nd cycle →		

(b)

transmitter. More specifically, using just one queue per transmitter for buffering all generated packets independently of destinations, causes severe head-of-line effects as explained in [24]. This further decreases network performance quite considerably and has led to the development of an extension of I-TDMA protocol described next. More details about I-TDMA protocol along with a detailed performance analysis can be found for example in [37].

*3.1.2 I-TDMA**

In an effort to eliminate the head-of-line effects of I-TDMA, [4, 36] proposed an extension of it, denoted as I-TDMA*, which assumes that the number of queues of each transmitter is equal to the number of channels W. In fact, each queue corresponds to a different channel and buffers only the arriving packets that are about to be transmitted on this channel.

The protocol remains channel collision-free on account of the allocation policy of channels to source nodes used. In fact, allocation of channels to sources is similar to I-TDMA when zero switching latency ($\alpha = 0$) is considered. Thus, the example allocations presented before for both networks, the one with equal number of nodes and wavelengths (five) and the other with more nodes (seven) than wavelengths (five), apply in this case too. These are shown in Tables 3.1 and 3.2 respectively.

The performance of I-TDMA* protocol is shown in [36] to be quite notably improved as compared to that of I-TDMA, due to the addition of more queues per transmitter. The highest performance levels are reached here under high traffic conditions (again) and when zero switching latency is assumed. It is worth noting that performance tends to degrade as switching latency (α) tends to increase. The impact of switching latency on the performance of this protocol is studied in [36]. Here, we shall briefly describe how the allocation policies of channels to sources are proposed to be modified when switching latency is non-negligible, which is quite a realistic assumption.

In fact, there are three variations of I-TDMA*, which are proposed for different values of (actually relations between) switching latency α, number of nodes N and number of wavelengths W [36]. Before describing them briefly, let us denote by T the time slots that are required for a node to complete a packet transmission on a certain channel. Thus, generally we have $T = \alpha + 1$, since α time slots are needed for tuning the transmitter to the channel and one slot for actual packet transmission.[6]

The first variation of I-TDMA* is called *Scheme* 0 in [36] and is also denoted by *I-TDMA*$_0^*$. This is applied when $WT \leq N$. As in the case of negligible switching latency, the length of each cycle is equal to N time slots for $W < N$ and $N - 1$ slots for $W = N$, if we assume a source node transmits to all nodes except for itself. The scheme tries to overlap switching latency with cycle synchronization. According to the allocation scheme of this protocol, source node i (nodes are numbered 0, 1, 2, ..., $N - 1$) is initially assigned to transmit on channel 0 (channels are numbered 0, 1, 2, ..., $W - 1$) at time slot $(i+1)$ MOD N. After that channel 1 is assigned to this node T time slots later, so that it has just enough time (equal to α slots) to tune on this channel and transmit immediately, since $T = \alpha + 1$, and so on, each subsequent channel up to channel $W - 1$ is allocated to node i after T time slots. This allocation scheme determines also that node i will have to remain idle for $N - WT$ slots before it is assigned again to transmit on channel 0.

The allocation procedure becomes clearer if we consider an example network with nine nodes ($N = 9$), four wavelengths ($W = 4$) and switching latency equal

Table 3.3 I-TDMA$_0^*$ for an example network with $N = 9$, $W = 4$ and switching latency $\alpha = 1$ (thus $T = 2$). Obviously the case $W < N$ and $WT \leq N$ is considered. (a) Channels/sources allocation map. (b) Sources/channels allocation map.

Channels	Sources											
0	0	1	2	3	4	5	6	7	8	0	1	...
1	7	8	0	1	2	3	4	5	6	7	8	...
2	5	6	7	8	0	1	2	3	4	5	6	...
3	3	4	5	6	7	8	0	1	2	3	4	...
Time slots:	1	2	3	4	5	6	7	8	9	1	2	...
Cycles:				← 1st cycle →						← 2nd cycle →		

(a)

Sources	Channels											
0	0	X	1	X	2	X	3	X	X	0	X	...
1	X	0	X	1	X	2	X	3	X	X	0	...
2	X	X	0	X	1	X	2	X	3	X	X	...
3	3	X	X	0	X	1	X	2	X	3	X	...
4	X	3	X	X	0	X	1	X	2	X	3	...
5	2	X	3	X	X	0	X	1	X	2	X	...
6	X	2	X	3	X	X	0	X	1	X	2	...
7	1	X	2	X	3	X	X	0	X	1	X	...
8	X	1	X	2	X	3	X	X	0	X	1	...
Time slots:	1	2	3	4	5	6	7	8	9	1	2	...
Cycles:				← 1st cycle →						← 2nd cycle →		

(b)

to one slot ($\alpha = 1$). Notice that since $T = \alpha + 1 = 2$, we are indeed in the case $WT \leq N$ and also we have $W < N$, that is to say the cycle length has to be equal to nine slots according to the discussion above. The channels/sources and sources/channels allocation maps for this network applying the I-TDMA$_0^*$ protocol are shown in Table 3.3(a) and Table 3.3(b) respectively. If we consider for example source node 0, the protocol's allocation policy determines it is assigned to transmit on channel 0 at time slot 1, on channel 1 at time slot $T + 1 = 3$, on channel 2 at slot $2T + 1 = 5$ and on channel 3 at slot $3T + 1 = 7$. Thus, it always has just enough time to tune to the appropriate channel before transmitting its packets. Also before channel 0 is allocated again to it, it stays idle for $N - WT = 9 - 4 \cdot 2 = 1$ slot, as

Table 3.4 Channels/sources allocation map of I-TDMA$_1^*$ for an example network with $N = 9$, $W = 4$ and switching latency $\alpha = 2$ (implying $T = 3$). Thus, case $W < N$ and $WT > N$ is considered.

Channels	Sources														
0	0	1	2	3	4	5	6	7	8	X	X	X	0	1	...
1	X	X	X	0	1	2	3	4	5	6	7	8	X	X	...
2	6	7	8	X	X	X	0	1	2	3	4	5	6	7	...
3	3	4	5	6	7	8	X	X	X	0	1	2	3	4	...
Time slots:	1	2	3	4	5	6	7	8	9	10	11	12	1	2	...
Cycles:						← 1st cycle →							← 2nd cycle →		

is obvious from the tables (that is to say it is allocated again channel 0 after $T + 1$ slots and not just T). The allocation of channels to the remaining source nodes is done analogously. One idle slot for all source nodes is always present before the (re-)allocation of channel 0 to them.

The two remaining variations of I-TDMA* are applied when $WT > N$. The first one is called *Scheme 1* in [36] and is also denoted by *I-TDMA*$_1^*$. This is a straightforward extension of *I-TDMA*$_0^*$, which considers a longer cycle length equal to *WT* time slots, maintaining the interleaving of *I-TDMA*$_0^*$. This is exactly the time required for each source to tune and transmit on all channels[7] available in the system. For example in Table 3.4 we see the channels/sources allocation map decided by I-TDMA$_1^*$ for a network with nine nodes ($N = 9$), four wavelengths ($W = 4$) and switching latency equal to two time slots ($\alpha = 2$, which implies T = 3). According to the scheme, a cycle lasts exactly $WT = 12$ slots. It is quite obvious that the scheme is an extension of the previous one, where, for example, source 0 is given permission to transmit on channel 0 at time slot 1, on channel 1 at time slot $1 + T = 4$ and so on for all channels and nodes.

The third variation of I-TDMA* is applied when $WT > N$ as well. It is called *Scheme 2* in [36] and is also denoted by *I-TDMA*$_2^*$. This scheme aims at retaining a shorter cycle length as compared to I-TDMA$_1^*$ so as to perform better under light traffic conditions, since shorter cycle implies shorter access delays. On the other hand, its disadvantage is that it has to abandon interleaving. The length of each cycle is retained equal to N time slots as in the first variant. There is a fixed-assignment of channels to sources as in the case of zero switching latency (see for example Table 3.2(a) for the case $N > W$). However, a source node may not be able to transmit on a certain channel during a time slot in which it is allocated this channel,

Table 3.5 Channels/sources allocation map for I-TDMA$_2^*$ applied by an example network with $N = 4$, $W = 3$ and $\alpha = 1$ (thus $T = 2$). Obviously, we have $WT > N$ and $N > W$.

Channels	Sources											
0	0	1	2	3	0	1	2	3	0	1	2	...
1	1	2	3	0	1	2	3	0	1	2	3	...
2	2	3	0	1	2	3	0	1	2	3	0	...
Time slots:	1	2	3	4	1	2	3	4	1	2	3	...
Cycles:	← 1st cycle →				← 2nd cycle →				← 3rd cycle →			

even if it has a packet buffered in the corresponding queue; this is the case when it does not have enough time to tune its transmitter to this channel, because switching latency is non-negligible here. In this instance this slot should be left unused and the source node must wait for a chance later. Generally, the important difference of this scheme compared to the previous ones is that it is impossible for a source node to transmit on all channels within just a cycle, even if it has packets in all of its W queues. This is simply because it does not have enough time to tune to all of them and transmit, since cycle length is N slots and $WT > N$.

Next we describe generally how the sources decide to tune to the channels and then we present a simple example. Let us denote by $c[i, j]$ the channel that is allocated to node i at time slot j. Thus, $c[i, j] = \lambda$ means that node i is assigned to transmit on channel λ at time slot j, where $0 \leq i \leq N - 1$, $1 \leq \mathrm{j} \leq N$ and $0 \leq \lambda \leq W - 1$. Also, $c[i, j] = null$ quite obviously means that no channel is allocated to node i at time slot j. Now assume that source node i does transmit on $c[i, j]$ at slot j. The protocol determines that after transmitting on channel $c[i, j]$, node i begins tuning its transmitter to channel $\mathrm{c}[i, 1 + (j + \alpha) \text{ MOD } N]$, if $\mathrm{c}[i, 1 + (j + \alpha) \text{ MOD } N] \neq null$ and if the node's queue for this channel is not empty. The time slots between slot j and slot $1 + (j + \alpha) \text{ MOD } N$ are not used (even if the node had some channel allocated to it) to allow enough time, i.e. at least equal to α slots, for the node to tune its transmitter. Note that if either the relevant queue was empty or $\mathrm{c}[i, 1 + (j + \alpha) \text{ MOD } N] = null$, the node waits for the next slot and checks again (apparently with $j = j + 1$).

The channels/sources allocation map for I-TDMA$_2^*$ when applied by an example network with four nodes ($N = 4$), three channels ($W = 3$) and switching latency equal to one time slot ($\alpha = 1$ implying $T = 2$ slots) is shown in Table 3.5. We notice that it the same as the case switching latency is taken to be negligible (zero) and

$N > W$. Let us see for example, how source node 1 decides to tune its transmitter according to the general scheme described in the previous paragraph. Initially we have c[1, 1] = 1 and let us assume that node 1 transmits a packet on channel 1. Since c[i, $1 + (j + \alpha)$ MOD N] = c[1, 1 + (1 + 1) MOD 4]= c[1, 3] = null, it has to wait for the next slot ($j = j + 1 = 2$) to check again. Notice that even though c[1, 2] = 0, node 1 does not have enough time to tune to channel 0 and cannot transmit any packet that may be possibly waiting in its queue for channel 0; hence, this slot has to be lost for node 1 this time. During time slot 2, node 1 checks and sees that c[i, $1 + (j + \alpha)$ MOD N] = c[1, 1 + (2 + 1) MOD 4] = c[1, 4] = 2. Consequently, if it has at least one packet buffered for channel 2, it decides to start tuning its transmitter to this channel; tuning will actually take place during slot 3 and the node will be ready to transmit at slot 4 as it should. If it had no packet for this channel, it would just have to wait for the next slot to check again.

The scheme described so far could be unfair. Notice for example that if source node 0 always had packets waiting for channel 0 and channel 2, it would never be able to transmit on channel 1 even if it had packets for it as well (see Table 3.5). This would be because it would always decide to tune and transmit successively on these two channels (channel 0 and channel 2) and ignore channel 1. Generally [36] suggests that if there is a positive integer x such that $x(\alpha + 1) = N$, an extra slot should be skipped when cycle boundaries are crossed, so that a situation where the same set of channels is visited during each cycle is avoided and the protocol remains fair. More details about I-TDMA* and the three variants, along with further remarks regarding system throughput and a detailed performance analysis, may be found in [36] and [4].

3.1.3 R-TDMA

R-TDMA (*Random—Time Division Multiple Access*) [15] is another noteworthy variant of the general TDMA idea. This protocol quite realistically defines that the tunable transmitter of each node has limited tuning capability and thus a node can only transmit over a subset of the W available channels in the system. Concerning the receiving part of a node's optical component equipment, each node i has r_i fixed-tuned receivers, where generally $r_i \neq r_j$ for different nodes ($i \neq j$). Hence, the overall broadcast-and-select WDM star network can be characterized as $TT - FR^{r_i}$ (where r_i is not fixed as explained above). The presence of (possibly) more than one fixed-tuned receivers at a node, apparently allows it to receive *simultaneously* on the respective wavelengths. Time is slotted across all channels and synchronization is provided on the basis of packet transmission time, i.e. the slot duration is equal to a packet transmission.

The broadcast-and-select WDM system of the protocol is taken to be single-hop, which is provided by the following assumptions:

1. Each channel is included in the transmitter's tunability range of at least one node in the system, i.e. there is at least one node that can transmit on it.
2. At least one node in the system has a receiver fixed-tuned to a channel, for all channels, i.e. each channel can be received at least by one node.
3. For every possible pair of source-destination nodes, there is at least one channel on which they can communicate,[8] i.e. the source can transmit on and the destination can receive from. Let us call this a *communication channel* for this source-destination pair.

Before the beginning of each time slot, every node runs the same distributed algorithm which determines the nodes that will be scheduled for transmission during the following slot. The same scheduling decisions are therefore made by all stations and collisions are avoided as we will see. Before presenting the algorithm, let us define the following:

- N: Number of nodes in the system.
- W: Number of available channels (wavelengths) in the system.
- C: Set of channels for which no node has been scheduled to transmit on yet. Initially, when the algorithm starts running for an oncoming slot, the set contains all channels i.e. $C = \{1, 2, \ldots, W\}$.
- U_j: Set of nodes, which are not scheduled for transmission yet (on any channel) and are capable of transmitting on channel j. Initially, this set contains all nodes of the system that have channel j in the tuning range of their transmitter.

The following algorithm determines the nodes that will transmit on each channel [15]:

1. Initialize the sets $C$and U_j as described before.
2. Choose randomly one free channel k (without any node scheduled to transmit on it yet), i.e. choose one $k \in C$.
3. Choose randomly one node i that is not scheduled to transmit on any channel yet and is able to transmit on channel k, i.e. choose one $i \in U_k$.
4. Schedule node i to transmit on channel k.
5. Remove node i from all sets of unscheduled nodes: For $j = 1$ to W do $U_j = U_j - \{i\}$.
6. Remove node k from the set of free channels: $C = C - \{k\}$.
7. If $C \neq \emptyset$ go to step 2.

Notice that this algorithm is distributed; therefore, for an identical assignment of channels to nodes by all stations running the algorithm, the same channels and nodes should always be chosen in steps 2 and 3 respectively. This can be achieved, according to [15], if the system's nodes have the *same random number generator* and the *same seed*. The algorithm ensures that each channel is assigned exactly one transmitting node per slot and therefore at most W simultaneous transmissions occur per slot. As is quite clear from the above algorithm, these transmissions never collide, making R-TDMA a collision-free protocol, like all fixed-assignment protocols described so far. When a node is scheduled to transmit on a certain channel, it may choose from its buffer any of the packets that are destined for nodes capable of receiving on this channel (have a receiver fixed-tuned on it).

The R-TDMA protocol's performance is studied in [15] and compared to a slotted Aloha protocol of the same network model presented later in Section 3.2. Generally the protocol performs well under moderate to high traffic loads, which is quite typical of fixed-assignment schemes. Its maximum performance is achieved when the number of fixed-tuned receivers per node is equal to the available wavelengths, which, however, is not always feasible to implement. Thus, there is always a trade-off between performance and number of fixed-tuned receivers used per node.

Lastly, it should be pointed out that even though the assignment of channels to nodes presupposes certain random choices[9] (implying it is not fixed and repeated identical over time), the assignment is still *predetermined*, since it is decided before the beginning of each slot. Thus, the protocol cannot be characterized as random access, which implies dynamic access of channels by nodes, and naturally falls into the class of fixed-assignment or static access schemes.

3.1.4 TDMA AND VARIANTS BASED ON TT-TR NODE ARCHITECTURES

The TDMA protocols described so far were based on the TT-FR node architecture, which implied fixed (and repeated over time) allocation of home channels (each pre-assigned to one or more destination nodes) to source nodes. The important advantages of this architecture are simplicity and low implementation cost. However, some earlier schemes suggest using one tunable transmitter and one tunable receiver per node i.e. are based on the TT-TR node architecture. The first one described here is very similar to the previously presented protocols. Two variants of it are also briefly presented. These variants are indeed fixed-assignment protocols, but lose one key feature of the majority of these schemes as a trade-off of shortening access delays; namely, they do not avoid channel or receiver collisions.

Table 3.6 Channels/node-pairs allocation map for the TDMA applied to an example network with $N = 3$ and $W = 2$. This is an exemplar allocation map for this network.

Channels	Node-pairs				
0	(1,2)	(1,3)	(2,1)	(1,2)	...
1	(2,3)	(3,1)	(3,2)	(2,3)	...
Time slots:	1	2	3	1	...
Cycles:		← 1st cycle →		← 2nd cycle →	

Hence, they could be more appropriately categorized as *partial fixed-assignment protocols* as in [32].

In [10] the *TDMA* protocol is presented for networks whose nodes have one tunable transmitter and one tunable receiver (TT-TR) capable of tuning over the whole range of available wavelengths in negligible time. The actual difference with the previous schemes is that here we have fixed allocation of channels to node-pairs and not sources, since both transmitters and receivers are tunable. When a node-pair denoted as (i, j) is assigned a certain channel at some time slot, it follows that source node i has permission to transmit to node j on this channel and during this specific time slot. This is better understood if we consider a simple example of a network with three nodes ($N = 3$) and two wavelengths ($W = 2$) [32]. A possible allocation map of channels to node-pairs for this network is shown in Table 3.6. Of course, for communication between two nodes it is required that both tune appropriately to the same channel, which is assumed to be done rapidly. In [10] the allocation map is generalized for an arbitrary number of nodes and channels.

The first variant of this scheme is the so-called *Destination Allocation* (*DA*) protocol. According to this, during a time slot more that one node-pairs are given permission to communicate per channel, with the limitation that they should determine the same node as destination; hence the name Destination Allocation. Note that the same node is not included as a source in two or more different channels at the same time slot. Thus, the total number of node-pairs that can communicate at a time slot increases from W to N, which makes access delays shorter and is quite beneficial for performance, especially under light traffic conditions. As a trade-off, the eventuality of channel collisions is not avoided in this protocol, which is more of a characteristic of random access schemes, as was discussed in the introduction of this chapter. Hence, DA protocol is classified as a partial fixed-assignment protocol.

Table 3.7 Channels/node-pairs allocation map for the Destination Allocation (DA) protocol applied to an example network with $N = 4$ and $W = 2$. This is just one possible allocation map for this network.

Channels	Node-pairs						
0	(1,2) (3,2)	(1,4)	(1,3)	(4,2)	(1,2) (3,2)	(1,4)	...
1	(2,3) (4,3)	(2,1) (3,1)	(4,1)	(3,4) (2,4)	(2,3) (4,3)	(2,1) (3,1)	...
Time slots:	1	2	3	4	1	2	...
Cycles:		← 1st cycle →				← 2nd cycle →	

The allocation of channels to node-pairs determined by this scheme can be better understood by an example. Let us consider a network with four nodes ($N = 4$) and two channels ($W = 2$) applying the DA protocol. An exemplar allocation is shown in Table 3.7. Notice, for example, that at time slot 1 four (N) node-pairs are given permission to communicate, which makes access delays for the corresponding source nodes shorter, but introduces at the same time the possibility of channel collisions. If, for instance, both source nodes 1 and 3 were ready to transmit on channel 0 during slot 1, there would be a channel collision.

The second variant of TDMA is the so-called *Source Allocation* (*SA*) protocol, which is essentially the opposite of DA. According to SA, during a time slot, up to $N - 1$ node-pairs are allowed to communicate over the same channel. The restriction for these pairs is that they should determine the same source node; hence the name Source Allocation. Thus, the total number of node-pairs that may be allowed to communicate during a single time slot rises to $W \cdot (N - 1)$. Channel collisions cannot occur since only one source is allowed to transmit on a certain channel each slot, but the eventuality of receiver collisions is introduced, because more than one source nodes may choose to transmit to the same destination over different channels simultaneously.

For example, let us consider the same network as above with four nodes ($N = 4$) and two channels ($W = 2$), applying the SA protocol this time. One possible allocation is shown in Table 3.8. Notice, for example, that at time slot 1 three ($N - 1$) node-pairs are given permission to communicate over each of the two available channels, all determining the same source and in total including all possible destinations (except for the source itself). Thus, if for example at time slot 2 both source nodes 3 and 4 were willing to transmit to destination node 1

Table 3.8 Channels/node-pairs allocation map for the Source Allocation (SA) protocol applied to an example network with $N = 4$ and $W = 2$. This is just one possible allocation map for this network.

Channels	Node-pairs			
	(1,2)	(3,1)	(1,2)	
0	(1,3)	(3,2)	(1,3)	...
	(1,4)	(3,4)	(1,4)	
	(2,1)	(4,1)	(2,1)	
1	(2,3)	(4,2)	(2,3)	...
	(2,4)	(4,3)	(2,4)	
Time slots:	1	2	1	...
Cycles:	← 1st cycle →		← 2nd cycle →	

over channels 0 and 1, respectively, there would be a receiver collision, since the destination would not know where to tune; in the best case it could only accept one transmission and ignore the other. Allowing receiver collisions is a significant drawback that quite severely degrades the performance of this protocol. Note that like in DA, other allocations would also be applicable.

As far as performance is concerned, the TDMA scheme is still performing poorly under low traffic conditions, such as the interleaved home channel allocation versions for TT-FR architectures. The two variants considered here, i.e. DA and SA, are trying to cope with this deficiency, but introduce another, by allowing either channel or receiver collisions to occur, respectively.

3.1.5 SPECIAL CASES OF FIXED-ASSIGNMENT PROTOCOLS

We consider schemes that are either proposed for another topology besides the commonly assumed (by so many proposals) star or that have been applied in practice for optical WDM network testbeds, as special cases of fixed-assignment protocols.

In this part two fixed-assignment protocols for the *folded unidirectional bus* topology are firstly described in brief. We have already referred to this topology in Section 1.6 and presented it schematically in Figure 1.29. The first protocol is *AMTRAC* [9]. The title of the reference straightforwardly explains how the name was formed. The network model of AMTRAC assumes each node is equipped with one tunable transmitter and one receiver fixed-tuned to the home channel of the node; that is to say, the system can be characterized as TT-FR,

Table 3.9 Channels/sources allocation map for the AMTRAC protocol applied to an example bus network with $N = 3$ and $W = 2$. Cycle length is equal to four slots, as expected ($2(N - 1)$).

Channels	Sources						
0	0	1	2	X	0	1	...
1	1	2	X	0	1	2	...
Time slots:	1	2	3	4	1	2	...
Cycles:		← 1st cycle →				← 2nd cycle →	

but, of course, it differs from the systems described so far in the physical topology. Transmitter tuning times are assumed to be negligible and the allocation of home channels to nodes follows the interleaved allocation policy, as before. Let us denote again by N and W the number of nodes and channels available in the network, respectively. Time is slotted and synchronization across all channels is provided on the basis of propagation delay Δ between adjacent nodes, i.e. the length of each time slot is equal to Δ. The protocol is based on time-division multi-channel access, that ensures each node is given permission to transmit exactly once per cycle on each channel, quite like the TDMA schemes for star. Thus, AMTRAC assumes fixed (and repeated over time) allocation of channels to source nodes; however, cycle length here is taken to be quite long, equal to $2(N - 1)$ time slots. Moreover, there is another important difference from previous TDMA schemes. Since time is slotted on the basis of Δ, when a node is assigned a time slot to transmit on a certain channel, it does not necessarily follow that this channel will be free of uncompleted transmissions at this point. Therefore, the protocol determines that each node willing to transmit on a channel has to check whether this is free exactly one time slot before the actual slot pre-assigned to it. If the channel happens to be free, the node proceeds on transmitting at its pre-assigned slot (the next one) or otherwise it waits for a chance in the subsequent cycle.

The allocation of channels to sources for an example bus network with three nodes ($N = 3$) and two channels ($W = 2$) applying the AMTRAC protocol is depicted in Table 3.9. Notice that for each channel one slot per cycle is not assigned to any source node. Generally the number of wasted (not assigned) slots for each channel per cycle is obviously equal to $2(N - 1) - N = N - 2$, since N slots are allocated to sources.

The second protocol for the same bus topology is *B-TDMA*, which essentially is an extension of AMTRAC. The same assumptions concerning the node architecture (TT-FR) and tuning latencies of transmitters (negligible) hold in this scheme as well. However, synchronization across all channels for B-TDMA is provided on the basis of packet transmission time, i.e. one time slot is equal to the time required for packet transmission. The length of each cycle is equal to $N + 1$ time slots if N is an integral multiple of $(1 + \Delta)$ or N otherwise. Channels are allocated to source nodes in the typical TDMA way.

Last of all, a scheme applied in practice in two optical LAN network testbeds, which may also be considered as a special case of fixed-assignment protocols, is described briefly. The testbeds applying the protocol in question are *Rainbow I* [19] and its follow-up *Rainbow II* [17] developed by IBM. These networks consider FT-TR node architectures, while they have 32 nodes and equal number of channels. As was mentioned in the introduction, the FT-TR architecture introduces some difficulties when pre-allocation protocols are considered. That is to say, there is need for some coordination between sources and destinations that would ensure the latter are notified about the channel they have to tune before transmission begins, which happens to be quite tricky given the absence of a control channel. In order to provide a solution for this, these testbeds apply a protocol determining that receivers have to perform a circular search among home channels (pre-assigned to transmitters) in a round-robin fashion, looking for a possible connection request by a transmitter. Note that nodes make full-duplex circuit switched connections. According to the *polling protocol* used, a node sends a 'connection-request' message for a predetermined time on its home channel, which somehow determines the desired node it wants to connect to. It also tunes each receiver to the home channel of the desired destination, so that it is able to receive a 'connection-accept' message back from it. After waiting for the 'connection-accept' message for a predetermined amount of time and assuming it receives this, it sends back a 'connection-confirm' message implying that the exchange of actual data may begin. Quite obviously, the protocol results in considerably long connection set-up times making it rather inappropriate for packet switched traffic. Notice that the polling protocol used in Rainbow I and Rainbow II determines that receivers have time-multiplexed access to the available channels in order to search for connection requests, which in a way makes this approach complementary of having channels pre-allocated for data packet reception with time-multiplexed access of transmitters to each channel, as in I-TDMA [37]. That is why this protocol can be classified as fixed-assignment. For additional details on these testbeds the reader may refer to [17, 19].

3.2 RANDOM ACCESS PROTOCOLS

As we saw in the introduction to this chapter, random access protocols generally do not predetermine the way nodes access the available channels. In a slightly more confining definition, they are schemes that allow each node to access a channel at any time. Of course, schemes that allow stations access channels randomly (with no predetermined order), but not exactly at any time, could be viewed as random access too. Anyway these general observations will become clearer later. The most characteristic example of a random access scheme for WDM optical local area networks of the star topology is described first.

3.2.1 I-SA

I-SA [37] is a typical random access MAC protocol for broadcast-and-select networks. The initials stand for *Slotted Aloha* (*SA*) with *Interleaved* (*I*) home channel allocation. It is essentially an extension of the original slotted Aloha (presented along with pure Aloha as early as 1970 for a single-channel network [1]) over a multi-channel environment. According to Aloha, each ready station transmits its packet immediately, while slotted Aloha considers synchronization based on the maximum size of a packet and determines that a ready node starts transmitting its packet immediately at the start of the slot that follows packet generation.

I-SA extends this simple idea over a multi-wavelength environment. The nodes of the network are assumed to be equipped with one tunable transmitter and one receiver fixed-tuned to the appropriate home channel decided by the interleaved allocation policy described before (TT-FR system). Based on this known allocation policy, a source node tunes each transmitter to the home channel of the desired destination and actually transmits the packet on this channel right in the next slot after packet generation. Generated packets destined to any node are buffered in a single queue and the top packet is chosen to be transmitted each time. Time is slotted across all channels on data channel boundaries.

Note that in [13] this approach is named $SA^{(4)}$; an alternative considered in this study assumes a packet transmission lasts L minislots and synchronization is provided on the basis of these minislots; this implies that a ready station transmits at the next minislot following packet generation. This alternative scheme is denoted by $SA^{(3)}$ in [13] and shown to have poorer performance than $SA^{(4)}$. (That is why we concentrate on $SA^{(4)}$ only.)

It is quite clear to see that if one channel is busy and some other node decides to transmit on it as well or if two nodes start transmitting on the same channel simultaneously, a channel collision will occur. According to the protocol, packets have to be retransmitted in such an instance. The question is how transmitting stations are informed about collision or success of a packet transmission. A straightforward way to do this is to predetermine an amount of time a source node has to wait for an *acknowledgement* sent back to it by the destination in the event of successful reception. As we see in the next paragraph, in I-SA this timeout period is just included on a single packet slot. Of course, if no such acknowledgement is received within the predetermined period owing to a collided packet, the source node understands that its transmission was not successful. In this case, I-SA determines that a source should retransmit the packet in the following slots with some back-off probability denoted by p_r, until it gets an acknowledgement back from the destination.

Turning to the actual way of acknowledging transmitted packets, we already mentioned that in I-SA the waiting time for acknowledgements is included in a packet slot. Thus, a packet slot actually comprises a packet transmission subslot and an acknowledgement subslot. The latter has to be long enough to include packet header decoding, CRC verification, tuning to the source's home channel and transmitting the actual acknowledgement (which is really small in size) [35]. Therefore, each source node is aware of the result of its transmission by the end of the packet slot, interpreting the absence of acknowledgement as collision and need to retransmit. Note that when the number of nodes equals the number of channels, acknowledgements never collide; otherwise, a time-multiplexing approach could be used which, however, adds to the overall complexity of the protocol. This acknowledgement scheme used by I-SA is called *Extended Subslot Scheme* (*ESS*) [35]. Adoption of ESS by I-SA makes it really sensitive to propagation delay, switching latency and protocol processing overhead. An alternative acknowledgement scheme has also been proposed in [35], which is used by the next protocol.

As far as performance is concerned, it should be noted that a network applying the I-SA protocol may suffer from a great number of collisions under heavy traffic conditions. This comprises the weakness of slotted Aloha schemes and would obviously degrade performance of I-SA severely. An analytic performance study of I-SA along with comparisons to the performance of other protocols, like I-TDMA and I-TDMA*, are included e.g. in [4, 37].

*3.2.2 I-SA**

An improved version of I-SA, denoted as *I-SA**, has been proposed in [36]. The improved version is not only denoted in the same way as I-TDMA*, the improved version of I-TDMA (i.e. using an asterisk), but it is also analogous in assuming W separate queues per node—one per channel. The assumption of multiple queues per node copes with the head-of-line effects present in I-SA and contributes to the overall performance improvement introduced by I-SA*.

I-SA* also differs in the acknowledgement scheme it adopts; namely, it assumes *explicit acknowledgements* [35]. This means that the acknowledgement subslot (second part of the packet slot) of the ESS scheme is here eliminated. The length of a packet slot is therefore shorter than I-SA (equal to the packet transmission time only) and acknowledgements are transmitted explicitly like packets. Hence, unsuccessful packet transmissions may be caused either by packet or acknowledgement collisions. Since acknowledgements are no longer expected within the same packet slot (like in ESS), a *timeout* period has to be defined by the protocol during which a source node waits for acknowledgment and appropriately figures out whether its packet transmission was successful or not. Note that when there are more nodes than channels in the system, stations send acknowledgments upon successful reception of packets destined *only* to them and not other nodes with the same home channel.

It is worth observing that the scheme described thus far implies a kind of transmitter contention among generated packets and acknowledgements that have to be sent on the same channel at the same time. In [36] the I-SA* protocol intelligently deals with this situation by allowing *acknowledgements piggy-backed* onto the data packet that has arrived for transmission on the same channel. Acknowledgements that are explicitly transmitted on empty slots like packets (without piggybacking, i.e. in the absence of transmitter contention with any data packet) are called *forced acknowledgements*. The acknowledgement scheme adopted by I-SA* is especially more appropriate than ESS when there are much more nodes than channels in the network. For example, it may allow that just one acknowledgment is sent to nodes sharing the same home channel. Another interesting feature of the protocol is that it provides stations with the ability to *transmit a window of packets* after tuning to a channel. This may significantly reduce switching latency per packet, but, on the other hand, a large window size may lead to more collisions. That is why the size of the allowed window of transmitted packets must be carefully selected.

More details about I-SA* and its acknowledgement scheme may be found in [36] and [35], respectively, along with extended performance analyses. Performance of I-SA* has been shown to be much better that I-SA for the following reasons:

- It eliminates the head-of-line effect of I-SA by using multiple (W) queues per node corresponding to the W available channels (one queue per channel).
- It assumes a shorter packet slot equal to the transmission of a data packet only, i.e. it eliminates the extension of the packet slot to include an acknowledgement

subslot (as in ESS scheme). This makes the protocol less sensitive than I-SA to propagation delay, switching latency and protocol processing overhead.

- It allows transmission of a window of packets on a channel, which also reduces the impact of propagation delay and switching latency on the protocol's performance.

3.2.3 *INTERESTING VARIANTS OF THE MULTICHANNEL SLOTTED ALOHA*

The multichannel slotted Aloha schemes discussed so far, i.e. I-SA and I-SA*, assumed a TT-FR system, where the tunable transmitter of each node is capable of tuning over all available wavelengths. Some interesting variants that differ from I-SA (and I-SA*), e.g. in transceiver equipment per node or tunability range of tunable components, are discussed in this subsection.

First of all, a generalization of multichannel slotted Aloha would be to consider that each ready station transmits on the home channel of the destination with probability p and suspends transmission with the remaining probability $1 - p$. Apparently, for I-SA and I-SA* $p = 1$. The retransmission probability p_r (in case of a collision) might differ from p or not. Under heavy traffic conditions a value of p not very close to one might be preferred, in order to avoid an excessively high number of collisions and bring about a more stable network operation.

Furthermore, one remarkable extension of the slotted Aloha protocols described so far would be to assume each node is equipped with one tunable transmitter and one tunable receiver (TT-TR). This would make the system much more flexible, since the tunable receiver could also be used for sensing collisions over the transmission channel. As a result, the acknowledgement schemes that contribute to the degradation of the protocol performance would not be required in this architecture. However, the system would evidently become more complex and expensive to implement as a trade-off.

Another noteworthy variant of slotted Aloha defines that the tunable transmitter of each node has limited tuning capability and thus a node can only transmit over a subset of the W available channels in the system [15]. This assumption is much closer to reality than other schemes assuming infinite tuning capability for transceivers. Notice that this protocol is suggested in the same study as the previously described R-TDMA scheme and thus considers the same network model. Specifically, it assumes a $TT - FR^{r_i}$ system, implying that each node i has one tunable transmitter and an array of r_i fixed-tuned receivers. Transmitters have limited tuning capability (not over all available wavelengths W), but the system is still single-hop in exactly the same way as was explained for R-TDMA before.

The operation of the slotted Aloha protocol is quite simple. When node i gets ready to transmit to another node,[10] it transmits with probability p_i on the communication channel that exists for this source-destination pair (according to assumption number three in Subsection 3.1.3 about R-TDMA). If there are more than one communication channels, it *randomly* chooses one and transmits on it. As in previous slotted Aloha schemes, channel collisions may occur when two or more nodes transmit on the same channel simultaneously and collided packets should persistently be retransmitted in subsequent slots until success of transmission.

The performance of this slotted Aloha scheme is analyzed in detail and compared to R-TDMA protocol which assumes the same network model [15]. As was expected, the slotted Aloha protocol has superior performance under low traffic conditions and, like R-TDMA, achieves maximum performance when W fixed-tuned receivers are used at each node, which, however, has a high implementation cost.

3.2.4 MULTICHANNEL CSMA AND CSMA/CD

Besides slotted Aloha, another classic example of a random access protocol for single-channel networks is the *CSMA* protocol, where the initials stand for *Carrier Sense Multiple Access*. As was the case for slotted Aloha, it was attempted to extend this scheme over a multichannel environment quite early [27]. The same happened with an interesting extension of CSMA called *CSMA/CD*, where 'CD' stands for *Collision Detection*.

According to CSMA, a station wishing to transmit first senses the channel for carrier. If it finds the channel is idle at the moment, it begins transmission; otherwise, it keeps on sensing until the channel becomes free. If two or more stations happen to sense an idle channel at about the same time, they will all decide to transmit and a collision will definitely occur. On the other hand, CSMA/CD improves CSMA in the following way. During transmission, a node keeps on sensing the channel for (possible) collisions with other transmitted packets. If a collision is detected, further transmission is aborted and the other nodes are notified about the fact until the channel becomes idle again. Retransmission may be decided by each source node involved in a collision after a (different) random amount of time.

The extension of the above two single-channel random access schemes in a multichannel environment, such as a broadcast-and-select WDM optical LAN which is our main focus, entails some problems. Specifically, in order for a node to be able to sense every channel (and keeping in mind that sensing implies receiving optical signal back from the star coupler), it surely has to be equipped with a tunable receiver as well. Since we are talking about pre-allocation based protocols where no control channel-based coordination exists, channels have to be pre-assigned to receivers (for example) in a fixed manner. Assuming we do not want transmission to depend on simultaneous desired reception (and vice versa) for a node, the system has to be CC-TT-TR FR, where the fixed-tuned receiver is always tuned to the home channel of the node and the tunable one senses a channel before transmission; in the case of CSMA/CD the tunable receiver also detects collisions on the channel just used by a certain node for transmission. Thus, the implementation cost for both schemes is quite considerable. Note that this would not be of great importance if there was an analogous network performance improvement as a trade-off. However, in realistic high-speed optical WDM local area networks, the ratio of the channel propagation delay to packet transmission time is increased, resulting in a poor performance of the above protocols. Hence, if we combine a considerably high implementation cost with a poor performance, it follows that CSMA and CSMA/CD may not be a practical solution for the high-speed optical LANs considered in this book.

3.2.5 SPECIAL CASES OF RANDOM ACCESS PROTOCOLS

As special cases of random access, in this part we consider two schemes that are proposed for the folded unidirectional bus topology (see Figure 1.29). These schemes are not strictly random access in exactly the same sense as the slotted Aloha-based protocols.[11] In simple terms, nodes are not assumed to transmit at any time (immediately) in the framework of these protocols, but they still access channels randomly i.e. without a fixed predetermined order.

First, a sample protocol for the folded unidirectional bus topology (Figure 1.29) that could be characterized as random access is presented in brief. The so-called *Fairnet* [2] protocol assumes each node has one tunable transmitter with infinite tuning capability and one fixed-tuned receiver (TT-FR architecture for the nodes). Allocation of home channels to receivers is done according to the interleaved allocation policy. The protocol determines that the head node, which is the first from the left and is assumed to have special skills, generates data slots of fixed size across all channels. Ready station i, after tuning to home channel c of the desired destination, checks if the BUSY bit of a slot on this channel is set or not. If it is not set, it transmits with probability p_{ic} and sets the BUSY bit for this slot. According to Fairnet, transmission probabilities p_{ic}, $i = 1, \ldots, N$ and $c = 1, \ldots, W$, can be determined appropriately based on *a priori* traffic distribution. More details about this scheme may be found in [2].

The second scheme is *D-Net/C* which is a generalization of the single-channel oriented D-Net [44] protocol, also proposed for the folded bus topology. Note that this topology (Figure 1.29) comprises two parts—the *outbound bus* for transmission (upper bus) and the *inbound bus* for reception (lower bus). According to D-Net/C, a generator at the left end of the bus launches a locomotive on an outbound bus channel when it senses an *end-of-train* (*EOT*) on the corresponding inbound channel. A ready node waits for the locomotive on the right outbound bus channel and transmits immediately. At the same time, the node senses for upstream transmission which would imply a collision. If collision is detected, transmission is aborted and the nodes have to wait for an *end-of-carrier* (*EOC*) on the same outbound channel before attempting to transmit again.

Short descriptions of the above two protocols and performance comparison to protocols for other topologies (star and double bus) may be found in [39].

3.3 PRETRANSMISSION COORDINATION-BASED PROTOCOLS

In this section we present protocols that are based on pretransmission coordination. As we explained in the introduction to this chapter, these protocols make use of one (or more) control channel(s) for coordination between the various nodes of the network. It is recalled (from Figure 3.1) that in Subsection 3.3.1 we deal with schemes that do not avoid receiver collisions, while in the following Subsection 3.3.2 we focus on protocols that prevent the occurrence of receiver collisions. The protocols outlined are only worth noting samples taken from a really vast relevant literature.

3.3.1 PROTOCOLS WITH RECEIVER COLLISIONS

In this subsection we present the main characteristics of several MAC protocols for WDM broadcast-and-select star networks, which assume at least one control channel for coordination and do not prevent the occurrence of receiver collisions. The described schemes include (among others) some of the earliest pre-transmission coordination protocols, which were really innovative for their time (about the late 1980s). Most of them will be denoted as X/Y; this symbolism was already introduced in the beginning of this chapter and implies that protocol X is used for the control channel and protocol Y for the data channels.

3.3.1.1 Aloha/Aloha

This scheme was originally proposed in [16] and deliberately lets overall throughput be low, which is claimed to be tolerable on account of the really huge available bandwidth of single-mode fibre, in an effort to achieve its main goal of minimizing access delays. Indeed, the (unslotted or pure) Aloha scheme used in both

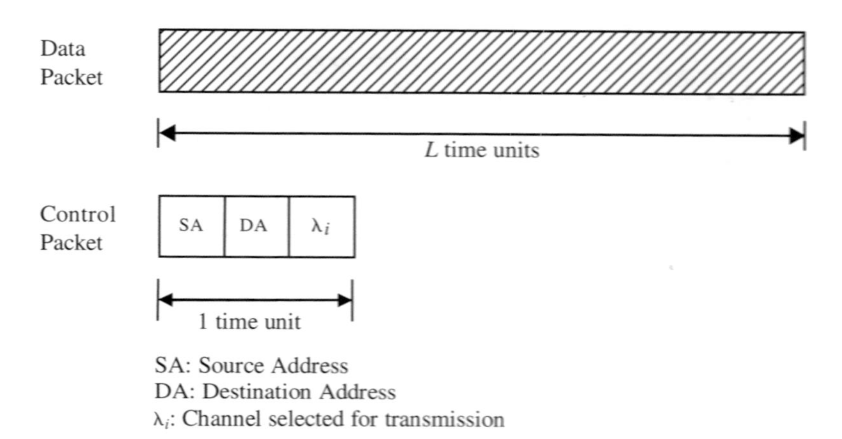

Figure 3.2 Data packet and control packet with its contents (source address, destination address and selected channel). Data packet is L time units and control packet is one time unit long. An example with $L = 3$ is illustrated.

control and data channels minimizes access delays by considering transmission of a control packet at any time (regardless of what other nodes do) and immediate transmission of the data packet afterwards; therefore it is clearly a tell-and-go protocol. The destinations and transmission channels are chosen at random, which adds to the overall simplicity. However, the protocol obviously allows both control and data channel collisions and, as we will see, receiver collisions are not avoided either.

The network model assumes a WDM broadcast-and-select star network with N nodes that are generally more than the W available wavelengths.[12] Each node is equipped with one tunable transmitter and one tunable receiver, both capable of tuning over all available wavelengths in negligible time. Thus, the system can be characterized as CC-TT-TR, which is a common feature of most schemes in [16]. It is also common among schemes of this study that the duration of a control packet is one time unit, while a data packet lasts L time units. A control packet contains the source address, destination address and channel number on which the corresponding data packet transmission is desired to take place right after the control packet (see Figure 3.2).

Figure 3.3 shows example transmissions on the control channel λ_0 and on one data channel λ_i according to the Aloha/Aloha protocol. Transmission on the control channel starts at time instant t_0 and lasts exactly one time unit, after which the data packet is transmitted on the desired channel λ_i. This data packet along with its

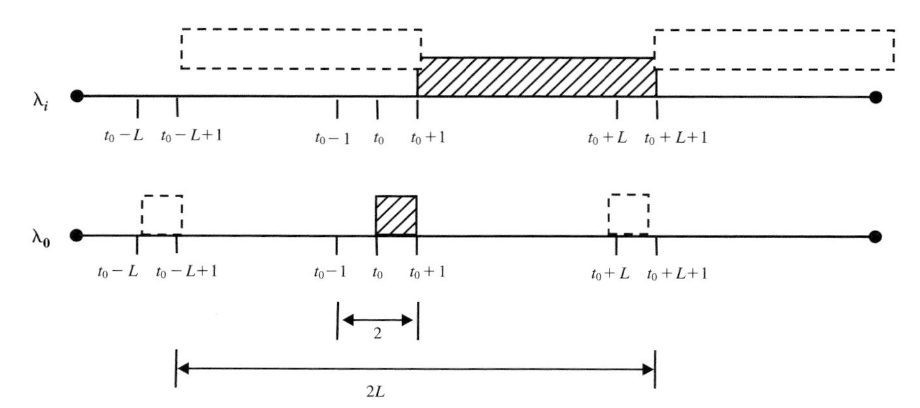

Figure 3.3 Aloha/Aloha protocol. Transmissions on the control channel λ_0 and on one data channel λ_i are shown. The vulnerable periods on which collisions may happen on both channels are also depicted. For the control channel the vulnerable period is (almost) 2 time units, while for the data channel (almost) $2L$ time units. The dashed rectangles show two hypothetical packet transmissions that imply data packet collision; the two extremes are shown, i.e. data packet transmissions starting right after time instant $t_0 - L + 1$ and right before time instant $t_0 + L + 1$.

corresponding control packet that went before, are depicted with diagonal lines inside. It is quite easy to note that the vulnerable period for the control channel, i.e. the period in which a collision might occur, is (almost) two time units long. We say 'almost', since, according to the Aloha protocol, if a control packet is transmitted by some other node just after time instant $t_0 - 1$ or just before time instant $t_0 + 1$, a collision occurs, but this is not the case if the corresponding time instants are *exactly* $t_0 - 1$ or $t_0 + 1$, respectively; hence, the vulnerable period is $(t_0 - 1, t_0 + 1)$[13] for the control channel. The rectangles with dashed-line edges show the two extreme cases of data packet collision. They correspond to control packet transmission right after time instant $t_0 - L$ or right before time instant $t_0 + L$, which respectively imply a data packet transmission beginning right after time instant $t_0 - L + 1$ or right before $t_0 + L + 1$. As shown in Figure 3.3, both extreme cases result in a data channel collision; of course, all data packet transmissions in between would have the same consequence. Hence, transmission of any data packet that begins within the interval $(t_0 - L + 1, t_0 + L + 1)$ would result in collision with the initial data packet, making the duration of the overall vulnerable period for data channel collision (almost) equal to $2L$. Note that this implies control packet transmission starting within the interval $(t_0 - L, t_0 + L)$, as shown in Figure 3.3; if we consider only the case of successful control packet transmissions preceding a data packet transmission, the data channel's vulnerable period becomes $2L - 2$ (as in estimations and analysis of [16]).

Turning to the other type of collisions (not explained in Figure 3.3) making this scheme fall into the class of pre-transmission coordination-based protocols with receiver collisions, it is quite easy to see how these might occur. Specifically in the case of Figure 3.3, a receiver collision would occur, for example, if there was a control packet transmission by another source node starting within the interval ($t_0 - L$, $t_0 + L$), and designating the same node as destination and a wavelength $\lambda_j \neq \lambda_i$ for transmission. Generally, a receiver collision might also occur if a control packet on λ_0 is intended for a certain node, which happens to receive on some data channel at the time. The latter is due to the relatively simple implementation (CC-TT-TR) and could be avoided, if for example each node also used two fixed-tuned transceivers dedicated for use on the control channel (CC-FTTT-FRTR). The main argument for ignoring receiver collisions in [16] (and other related studies) is that their impact could possibly be low for large population systems and, also, that they might be taken care of by higher-layer protocols, if needed. More details about Aloha/Aloha, especially concerning its (low) performance, can be found in [16].

3.3.1.2 Slotted Aloha/Slotted Aloha Protocols

A straightforward extension of the previously examined scheme is to use slotted Aloha (on both control and data channels), which has been shown to have improved performance as compared to the unslotted Aloha scheme. The *slotted Aloha/Aloha* protocol was initially proposed in [16], considering the same network model and time slotting as the Aloha/Aloha protocol. According to slotted Aloha/Aloha, time is slotted on the control channel and one time slot is equal in length to the transmission of one control packet. A data packet is taken to be L times the length of a control packet, therefore lasting exactly L time slots. The system falls into the same class of CC-TT-TR systems. The protocol determines that when a node generates a packet for transmission, it transmits a control packet on channel λ_0 right at the next time slot (after packet generation) and proceeds on transmitting the actual data packet immediately after that (occupying time equal to the following L time slots). Notice that in [16] the unslotted Aloha protocol was considered for the data channels, which is quite the same as using slotted Aloha, especially for $L = 1$. Thus, the *slotted Aloha/slotted Aloha* protocol will be considered here regardless of the value of L, since, besides being quite similar in operation, it has been extended in several ways by related work (e.g. [28, 42]). In this scheme time is slotted across all data channels too, with one data slot lasting exactly L control slots. The control and data packet transmissions are as described before for

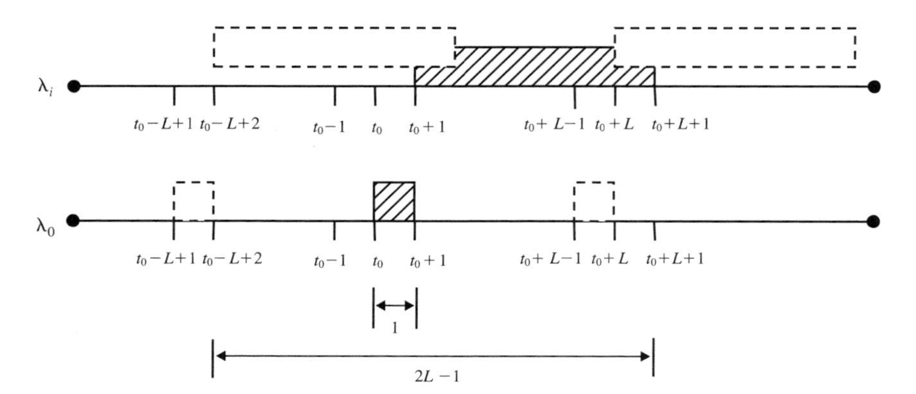

Figure 3.4 Slotted Aloha/slotted Aloha protocol. Transmissions on the control channel λ_0 and on one data channel λ_i are shown. The vulnerable periods on which collisions may happen on both channels are also depicted. For the control channel the vulnerable period is 1 control slot long, while for the data channel $2L - 1$ control slots. The dashed rectangles show two hypothetical packet transmissions that imply data packet collision; the two extremes are shown, i.e. data packet transmissions beginning at control slots $[t_0 - L + 2, t_0 - L + 3)$ and $[t_0 + L, t_0 + L + 1)$.

slotted Aloha/Aloha, implying that successive data slots overlap (unlike several extensions examined later in this part).

Figure 3.4 illustrates example transmissions on the control channel λ_0 and on one data channel λ_i according to the slotted Aloha/slotted Aloha protocol. Transmission on the control channel starts at time instant t_0 and lasts exactly one control slot, while transmission of the data packet begins right in the following control slot on the desired data channel (assumed to be λ_i) and lasts L control slots as stated before. This data packet and the corresponding control packet are depicted with diagonal lines inside. It is quite easy to note that the vulnerable period for the control channel is now exactly one time slot long, i.e. the vulnerable period is $[t_0, t_0 + 1)$. The rectangles with dashed-line edges show the two extreme cases of data packet collision. They correspond to control packet transmission at time slot $[t_0 - L + 1, t_0 - L + 2)$ or at time slot $[t_0 + L - 1, t_0 + L)$, which respectively imply a data packet transmission beginning at control slots $[t_0 - L + 2, t_0 - L + 3)$ or $[t_0 + L, t_0 + L + 1)$. Obviously, these extreme data packet transmissions will respectively occupy data slots $[t_0 - L + 2, t_0 + 2)$ and $[t_0 + L, t_0 + 2L)$. As shown in Figure 3.4, both extreme cases result in a data channel collision. Of course, all data packet transmissions in between would have the same consequence. Hence, the data channel's vulnerable period is the interval $[t_0 - L + 2, t_0 + L + 1)$; observe that in order to estimate this we include all the *control slots* at which transmission of a data packet that will collide with the initial one begins. Equivalently, we could consider the transmission interval of the corresponding control packets, namely $[t_0 - L + 1, t_0 + L)$. If we consider only the case of successful control packet transmissions preceding a data packet transmission, we have to subtract from $2L - 1$ the length of the control channel vulnerable period which is equal to one; thus the data channel's vulnerable period becomes $2L - 2$ (as in estimations and analysis of [16]).

Receiver collisions are not avoided in this scheme either. Concerning the example of Figure 3.4, a control packet transmitted by any other node in control slots included in the interval $[t_0 - L + 1, t_0 + L)$, and designating the same node as destination and a wavelength $\lambda_j \neq \lambda_i$ for transmission, would imply a receiver collision. In the simple CC-TT-TR system considered, a receiver collision would also occur, if a control packet on λ_0 was intended for a certain node which happened to receive on some data channel at the time. The latter receiver collision case could be avoided e.g. in a CC-FTTT-FRTR system with dedicated transceivers for the control channel. The same arguments as in Aloha/Aloha for ignoring receiver collisions apply here as well, according to [16].

Several interesting extensions and modifications to the described basic slotted Aloha/slotted Aloha protocol have been reported. For example, [42] proposes six slotted Aloha/slotted Aloha protocols, which mainly differ in the way of slotting the control and data channels; this has a significant impact on the behaviour of the overall scheme. The proposed schemes are referred as 'cases' (one to six) of slotted Aloha/slotted Aloha in [42] and will be denoted here as $SA/SA^{(1)} - SA/SA^{(6)}$. All these protocols assume exactly the same WDM optical broadcast-and-select network with N nodes, W wavelengths and CC-TT-TR node architecture; the latter can be extended to CC-TT-FRTR, if we want each node to monitor the control channel constantly (by means of the FR), which helps to achieve full synchronization [42]. Protocols $\mathrm{SA/SA}^{(1)} - \mathrm{SA/SA}^{(6)}$ are briefly described next.

According to $\mathrm{SA/SA}^{(1)}$, synchronization across the control and data channels is provided on the basis of the same time reference, which is called a 'cycle' and is shown in Figure 3.5. Within a single cycle, control packets may be transmitted only on one of the W control minislots, while there is also one data slot for a data packet transmission immediately after the Wth control minislot. This means that there are no control minislots available on the control channel after the Wth control minislot and this time (equal to L minislots) is wasted for the control channel within a cycle. Analogously, the first W minislots are wasted for the data channels, which may only transfer one data packet in the last L minislots (comprising one data slot). The overall length of the cycle is therefore $W + L$ minislots. The scheme determines that the W control minislots of each cycle are assigned in a fixed manner to the W data channels of the system, i.e. minislot number one to data channel λ_1 and so on. Thus, if, for example, a node has generated a data packet for channel λ_3,

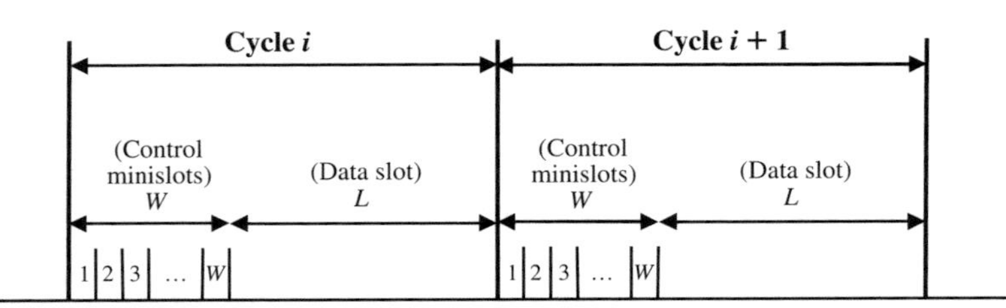

Figure 3.5 The synchronization (common cycle used) across control and data channels for SA/SA$^{(1)}$. Cycle length is $W + L$ minislots (or time units). Only the first W minislots per cycle are used in the control channel and only the last L minislots (one data slot) per cycle are used in the data channels. Cycles i and $i + 1$ are shown as an example.

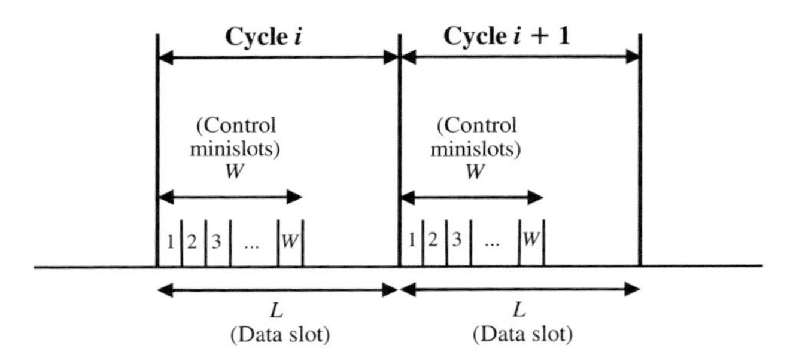

Figure 3.6 The synchronization (common cycle used) across control and data channels for SA/SA$^{(2)}$. Cycle length is L minislots (or time units). Only the first W minislots per cycle are used in the control channel, while all of the L minislots (one data slot) per cycle are used in the data channels. Cycles i and $i + 1$ are shown as an example.

it first transmits a control packet on the third minislot of cycle i[14]and then it transmits the data packet on channel λ_3 during the data slot of the same cycle i. As a result of this fixed pre-assignment of minislots to data channels, if a control packet is transmitted successfully in a control minislot (third control minislot in our example), the corresponding data packet will also be transmitted successfully (on channel λ_3 in our example) in the data slot of the same cycle for sure (Figure 3.5).

The SA/SA$^{(2)}$ protocol of the same article [42] extends the previous scheme by shortening the cycle length to L minislots in order to avoid the previously described waste of minislots (Figure 3.6). The Wcontrol minislots are still pre-assigned to data channels in the same fixed manner. The difference is that if for example a node transmits a control packet during cycle i, it transmits the corresponding data packet in cycle $i + 1$. Success of control packet transmission guarantees success of the corresponding data packet transmission as before.

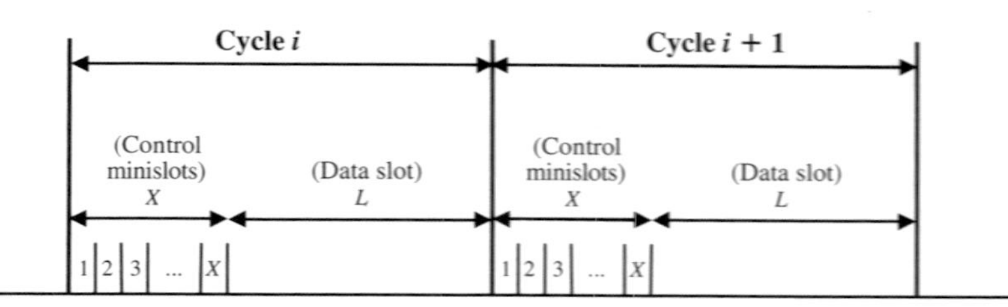

Figure 3.7 The synchronization (common cycle used) across control and data channels for SA/SA$^{(3)}$. Cycle length is $X + L$ minislots (or time units). Only the first X minislots per cycle are used in the control channel and only the last L minislots (one data slot) per cycle are used in the data channels. Cycles i and $i + 1$ are shown as an example.

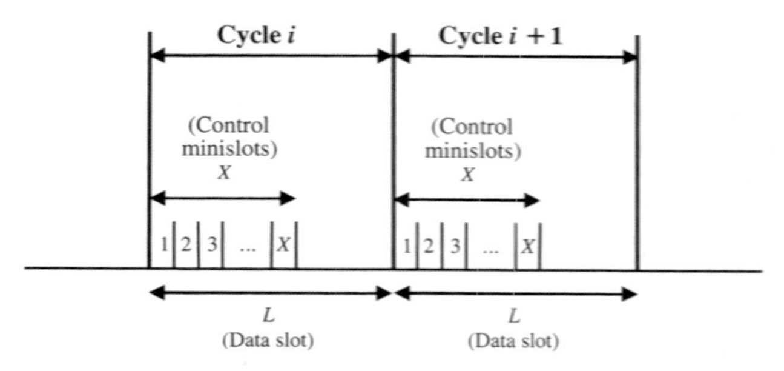

Figure 3.8 The synchronization (common cycle used) across control and data channels for SA/SA$^{(4)}$. Cycle length is L minislots (or time units). Only the first X minislots per cycle are used in the control channel, while all of the L minislots (one data slot) per cycle are used in the data channels. Cycles i and $i + 1$ are shown as an example.

Turning to SA/SA$^{(3)}$ protocol, synchronization is similar to the first protocol but with two differences (see Figure 3.7). First, the number of control minislots is now X making the cycle length equal to $X + L$ and second, there is no fixed pre-assignment of minislots to data channels. For the same example considered in SA/SA$^{(1)}$ protocol, the node could transmit the control packet in one of the Xminislots of cycle i chosen at random and then transmit the data packet on channel λ_3 during the data slot of the same cycle. Data packet collision may occur even after a successful control packet transmission.

The SA/SA$^{(4)}$ protocol extends the previous scheme in the same way the second protocol extends the first one. Thus, each cycle is shorter (L minislots long and not $L + X$) and the X minislots of the control channel are not pre-assigned to data channels too (see Figure 3.8). A node chooses at random one of the X minislots for the control packet and then it transmits the data packet on the desired channel

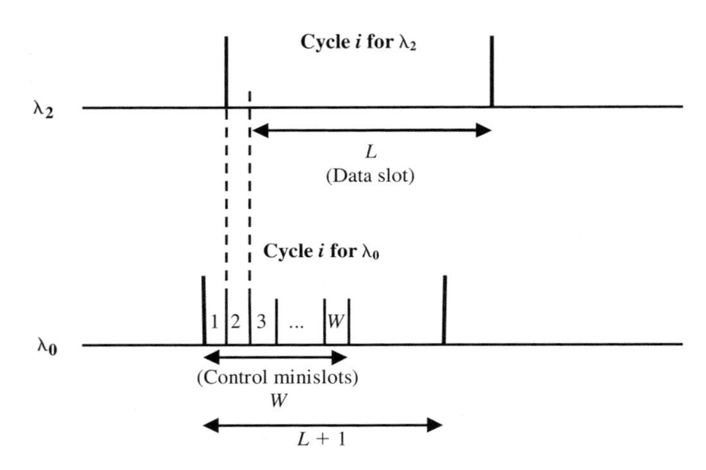

Figure 3.9 Time slotting of the control channel λ_0 and data channel λ_2 for SA/SA$^{(6)}$ protocol. Cycle i for each channel is shown. Obviously, cycles are asynchronous.

in the next cycle. The fifth protocol (SA/SA$^{(5)}$) is simply a special case of this scheme with $X = L$.

Finally, the SA/SA$^{(6)}$ protocol is somewhat different. It assumes *asynchronous cycles* for the control and data channels, which, however, have the same length, equal to $L + 1$ minislots. The control channel is divided into W minislots and some wasted space after the Wth minislot up to $L + 1$. The cycle for the ith data channel begins concurrently with the ith control minislot and contains one unused minislot followed by one data slot (equal to L minislots, thus, in total cycle length is $L + 1$ minislots). The control channel cycle along with the cycle of data channel λ_2 (as an example) are shown in Figure 3.9. If a node wished to transmit on channel λ_2, it would just have to transmit a control packet in minislot number two of the control channel and, immediately after that, send its data packet on λ_2 during the data slot of the corresponding cycle for this data channel. One unique feature of this scheme is that retransmissions are assumed to be done always on the same channel that was chosen in the initial attempt.

All of the extensions of the basic slotted Aloha/slotted Aloha schemes do not prevent receiver collisions from occurring. In the analysis of [42] receiver collisions are ignored for the same reasons that were previously mentioned for Aloha/Aloha in Section 3.3.1.1. More details on the protocols along with a performance comparison with each other and the basic slotted Aloha/slotted Aloha scheme, may be found in the same study.

3.3.1.3 Slotted Aloha/Delayed Slotted Aloha Protocols

All of the slotted Aloha/slotted Aloha protocols presented before have the tell-and-go feature, which has been mentioned in the introduction of this chapter. Thus, each ready station first *tells* (announces to) the other nodes the destination and channel of its transmission, and, *immediately* after that, it *goes* on to transmit the data packet. Obviously, the transmission of the data packet is done blindly, i.e. independently of the success of the corresponding control packet transmission. If the control packet experiences a collision, the following data packet transmission is done in vain and probably might just degrade channel throughput by colliding with another data packet. Of course, the good point with tell-and-go schemes is that access delays are very short.

In an effort to improve the throughput performance of the basic slotted Aloha/slotted Aloha protocol of [16], [28] suggested that a source node should first wait to see whether the control packet transmission was successful, before proceeding to transmit the corresponding data packet. That is to say, the data packet is not transmitted immediately, but it experiences a delay equal to the round-trip propagation delay (time needed for the control packet to reach the star coupler and arrive back to the source, according to the broadcast-and-select nature of the network). If the source figures out that the control packet has experienced a collision, it simply does not transmit the corresponding data packet and attempts to retransmit the control packet in line with the retransmission strategy of the protocol (e.g. after a random time interval).

As an example of this scheme, called slotted Aloha/delayed slotted Aloha due to the delay introduced to the data channels as described before, let us refer to the transmission of Figure 3.4. Let us also denote the round-trip propagation delay by α. The control packet is transmitted in time slot $[t_0, t_0 + 1)$, therefore the result of this transmission is known to the source node by time instant $t_0 + \alpha + 1$. Hence, in the event of a successful control packet, the corresponding data packet is transmitted in data slot $[t_0 + \alpha + 1, t_0 + L + \alpha + 1)$,[15] instead of data slot $[t_0 + 1, t_0 + L + 1)$ of the tell-and-go scheme transmission shown in Figure 3.4.

This straightforward modification of the basic slotted Aloha/slotted Aloha protocol, also mentioned as the 'wait-and-see' modification, implies better utilization of each channel's bandwidth. On the other hand, access delays get longer as a trade-off, since the protocol is no longer tell-and-go.

The idea of waiting to see the result of the control packet transmission [28], was also adopted by [42] to propose the corresponding improved versions, *SA/delayed SA*$^{(1)}$ to *SA/delayed SA*$^{(6)}$, of the six tell-and-go protocols (*SA/SA*$^{(1)}$ to *SA/SA*$^{(6)}$) that were presented before as extensions of the basic slotted Aloha/slotted Aloha scheme. The results taken from the analysis show that the performance of

SA/delayed $SA^{(3)}$, SA/delayed $SA^{(4)}$ and SA/delayed $SA^{(5)}$ is improved as compared to the corresponding tell-and-go protocols. Furthermore, the delay versus offered load performance characteristic of SA/delayed $SA^{(2)}$ is better than that of the corresponding tell-and-go scheme [42].

It should be noted that all of the described slotted Aloha-based schemes that adopt the wait-and-see modification do not avoid receiver collisions either. This is easy to see, if someone notices that no special mechanism (as compared to the previous protocols) is applied here to take care of receiver collisions. The impact of receiver collisions on the performance of slotted Aloha/delayed Aloha protocol has been studied for finite population systems and it was shown that this is significant for relatively small numbers of nodes, as expected (since fewer nodes imply higher probability of same destination selection by two or more transmitters at about the same time). The appropriate number of data channels W, as a function of data slot length L, for optimally dimensioning the available bandwidth of the network, has also been determined. For more information on the last two issues the reader may refer to [32].

3.3.1.4 Aloha/CSMA and Slotted Aloha/CSMA

In this part two additional pre-transmission coordination-based protocols with receiver collisions are presented, namely *Aloha/CSMA* and *slotted Aloha/CSMA* [16]. The two schemes are very similar in operation. Both assume that the CSMA protocol is used on the data channels, while the unslotted or slotted Aloha scheme is applied, respectively, on the control channel. They assume the same broadcast-and-select WDM optical local area network as Aloha/Aloha and basic slotted Aloha/slotted Aloha protocols (presented in the same paper [16]). Thus, the network has a star topology, N nodes and $W + 1$ wavelengths, $\lambda_0, \lambda_1, \ldots, \lambda_W$, where λ_0 is used as a control channel.

According to the protocols, when a node generates at least one packet for transmission, it senses the data channels in a round-robin fashion until it finds a free channel, as the CSMA protocol determines. After that, the node transmits a control packet on channel λ_0 as determined by Aloha or slotted Aloha, while it jams the data channel (found to be free) at the same time for a period equal to a control packet transmission.[16] This way all other nodes that are ready to transmit during this period, find that the specific data channel is not free and keep on sensing other data channels according to CSMA. Immediately after the control packet time, the node transmits the corresponding data packet.

Notice that simultaneous control packet transmission and jamming of the data channel actually require an additional transmitter per node, making the node architecture for these two protocols CC-FTTT-TR; the FT is assumed to be fixed-tuned to the control channel. The authors in [16] propose an alternative operation of the protocol that may still work in a CC-TT-TR system. Specifically, it is suggested that the sensing and jamming period is longer than a control packet transmission, which ensures that other ready nodes are still deterred from using the same data channel that was found to be free by a certain station.

The remarks about the degraded performance of multichannel CSMA in realistic high-speed optical LANs (Subsection 3.2.4) apply in the case of the two protocols described here as well. Also observe that, like the previously examined pre-transmission coordination-based schemes, receiver collisions are not avoided by Aloha/CSMA and slotted Aloha/CSMA either.

3.3.1.5 Aloha/Slotted-CSMA

The *Aloha/slotted-CSMA* protocol [34] was proposed as an improvement of the original Aloha/CSMA scheme of [16] presented before.

It assumes the same network architecture with N nodes and $W + 1$ wavelengths, where W wavelengths are used as data channels and the additional one is the control channel for pre-transmission coordination. Each node is assumed to have one tunable transmitter and one tunable receiver, capable of tuning over all available wavelengths.

The name slotted-CSMA for the protocol applied on the data channels is due to the fact that carrier sensing of data channels is done at the beginning of *CSMA slots*; the latter are assumed to be of length equal to the propagation delay. An idle node (not transmitting or receiving) that has generated a packet for transmission senses data channels—one at a time—at the beginning of each CSMA slot [34]. When it finds a free channel, it sends a control packet on the control channel according to the unslotted Aloha protocol. One notable improvement of the scheme is that no jamming of the data channel is required, but the node tunes its receiver to the control channel in order to figure out the result of the control packet transmission. This tuning is done if the node is not requested to receive from some other station in the mean time. Due to the broadcast-and-select nature of the network, the node determines whether its control packet was transmitted successfully after a round-trip propagation delay. If it did, it transmits the actual data packet immediately on the data channel that was found to be idle before. If there was a control channel collision, the user repeats the process from the beginning after some random time. The idea of this improvement to the original Aloha/CSMA protocol is obviously parallel to the improvement introduced by slotted Aloha/delayed slotted Aloha to the basic slotted Aloha/slotted Aloha scheme.

The performance of Aloha/slotted-CSMA protocol is analyzed and compared to the one of the original Aloha/CSMA [34]. It is found that, especially under heavy traffic conditions and for large values of L,[17] there is a considerable performance improvement in comparison to Aloha/CSMA. Another plus of the proposed protocol is that, since no simultaneous jamming of the data channel is required, it can easily be implemented with less costly transceiver equipment per node (CC-TT-TR).

More details on this scheme may be found in [34]. Note that the drawback of general performance degradation on account of using carrier sensing in practical high-speed optical LANs still holds. Moreover, the occurrence of receiver collisions is not avoided by Aloha/slotted-CSMA protocol either.

3.3.1.6 CSMA/Aloha and CSMA/Slotted Aloha

These two protocols make use of Aloha (or slotted Aloha) and CSMA schemes just as the previous two in 3.3.1.4, but apply them in the opposite way, i.e. CSMA on the control channel and Aloha (or slotted Aloha) on the data channels. Both *CSMA/Aloha* and *CSMA/slotted Aloha* have been proposed for exactly the same network model described before for the previous two protocols [16].

Each ready node first senses the control channel until it finds it free and then transmits the control packet (CSMA). The data channel specified in the appropriate field of the control packet is chosen at random. Transmission of the actual data packet is done on the chosen data channel immediately after the control packet transmission, either in the Aloha or slotted Aloha way. Control and data channel collisions may obviously occur and receiver collisions are not avoided either. Performance degradation owing to the use of CSMA in a high-speed optical LAN environment is the main weakness of the schemes, as explained before.

3.3.1.7 CSMA/W-Server Switch

The *CSMA/W-server switch* protocol[18] has been proposed in the same study [16] and thus assumes the same network model and node architecture (CC-TT-TR) as the previous schemes.

This protocol is quite innovative in determining that each idle (not receiving or transmitting) station has to monitor the control channel for L consecutive time units, which is the duration of a data packet transmission. Monitoring implies examination of the control packets transmitted during this time interval, which implies that each idle node obtains all the necessary information to know which destinations and channels will be idle next. Let us consider a specific idle node that performs this monitoring for L time units. Assuming a packet has been generated by this node, the node will know whether the desired destination is idle and also whether there is any free channel for transmission. If this is the case, the source node attempts to send a control packet to the destination in the CSMA way (by first sensing the control channel). The control packet specifies a channel that was found free by the prior control channel monitoring. Assuming the control channel is sensed idle (as CSMA requires), the control packet is transmitted and the data packet is sent immediately after that on the appropriate data channel.

Observe that this protocol is data channel collision free. If the data channels are seen as *servers* [16], it can be said that they act as a W-server switch. When all data channels are found busy (by the monitoring of the control channel), the protocol determines that the ready node should not attempt to transmit. This can be seen as if a W-server switch is blocked and turns away this node; hence the name of the protocol used on data channels.

The CSMA/W-server switch protocol does not prevent control channel and receiver collisions from occurring, similar to the previous pre-transmission coordination-based schemes. The use of CSMA also degrades its performance in realistic high-speed optical LANs as described earlier.

3.3.1.8 Slotted Aloha/W-Server Switch

The *Slotted Aloha/W-Server switch* protocol [28] was proposed as an improvement of the previous scheme. Thus, it assumes the same network model and node architecture. It tries to avoid the performance degradation due to carrier sensing, by indicating that the slotted Aloha scheme should be used on the control channel (instead of CSMA). The W-Server switch protocol is still used on data channels (W is the number of available data channels).

As before, the protocol determines that an idle node, which is ready to transmit (has generated a new packet), has to monitor the control channel for L consecutive control slots, in order to acquire all the necessary information about idle channels is the system. If no idle channels are found, the set of channels (servers) acts as a W-server switch that is blocked, i.e. does not allow any other node to transmit (get service). If at least one data channel is found to be idle, the node in question transmits a control packet right in the following control slot (after the L control slots) according to the slotted Aloha protocol. The control packet specifies that data packet transmission will be done on the data channel found to be idle (or a randomly selected among idle channels, if they are found to be two or more). The node transmits the actual data packet only if the corresponding control packet was transmitted successfully, as in slotted Aloha/delayed slotted Aloha schemes. Note that in the original study [28] immediate (at the end of the control slot) feedback about the result of the transmission is assumed. Hence, the actual data packet is transmitted immediately after successful control packet transmission. If either the W-server switch is blocked (no idle data channels found) or the transmitted control packet experiences a collision, the control channel monitoring process may be repeated after a random number of time slots.

Just like the CSMA/W-Server switch, the slotted Aloha/W-Server switch protocol does not incorporate any special mechanism in order to eliminate receiver collisions.

3.3.1.9 R-Aloha Protocols

In this section two *reservation protocols* are presented. These are denoted here by *R-Aloha*$^{(1)}$ and *R-Aloha*$^{(2)}$, where 'R' stands for reservation. They combine the idea of reserving data channels for a relatively long period (longer than a single data packet transmission) with the slotted Aloha technique adopting at the same time the slotting of channels present in the two previously presented schemes [42]. Remember from the introduction to this chapter that all pre-transmission coordination protocols are sometimes characterized as reservation protocols in the literature, which may not be the case in the strict sense (e.g. for the tell-and-go protocols). The protocols presented here, however, are purely reservation based. The idea of allowing a node to reserve a certain data channel for quite a long period, i.e. obtain the exclusive right to transmit on this channel, could be applied to circuit switched traffic or traffic with long holding times.[19]

The R-Aloha$^{(1)}$ protocol resembles the SA/delayed SA$^{(1)}$ scheme described earlier in Section 3.3.1.3, both in synchronization across control and data channels and the fact that a data packet is transmitted only if the corresponding control packet was successful. The same time reference (cycle) is used in the control and data channels as explained for SA/SA$^{(1)}$ protocol and shown in Figure 3.5. The additional feature of R-Aloha$^{(1)}$ is that successful transmission on the control and on one data channel by a node essentially implies that this node makes a reservation of the specific data channel for the following data slots, until it finishes transmitting its 'long message'. Note that according to the slotting of control and data channels and the fixed pre-assignment of control minislots to data channels, in order for a node to reserve data channel λ_i, it suffices to transmit a control packet successfully in control minislot i of a cycle; this is because the latter ensures successful transmission of the corresponding data packet within the same cycle.

Reservation of data channels is done as follows. Assume a node wishes to transmit on a specific data channel and succeeds in transmitting a control packet in the corresponding minislot of a cycle. This means that it also transmits successfully the first data packet of its long message within the same cycle. In order to make the reservation, the node transmits a jam signal in the same minislot of all subsequent cycles, until it finishes transmitting the entire (multi-data packet) message. This guarantees reservation (exclusive use) of the data channel in question, since any other control packet transmitted in the same minislot will collide with the jam signal; hence the corresponding nodes will not proceed with transmitting the actual data packets. After the session is finished, all other nodes will have a chance to reserve this channel by attempting to transmit a control packet in the same (contention-based) slotted Aloha fashion.

The second reservation protocol, R-Aloha$^{(2)}$, resembles SA/delayed SA$^{(6)}$ in assuming asynchronous cycles for the control and data channels as described before and also in adopting the idea of transmitting a data packet only after successful transmission of the corresponding control packet. The slotting of the control and data channels is described in Section 3.3.1.2 and an example is shown in Figure 3.9.

In order to describe how reservations are made, assume that a node transmits successfully a control packet in the ith control minislot of one cycle of the control channel. This implies that the node reserves data channel λ_i. Notice that the previously mentioned minislot coincides (in time) with the beginning (first minislot) of a certain cycle of data channel λ_i. The first data packet will be transmitted successfully on the same cycle of λ_i right in the following L minislots (data slot). According to the protocol, the node has to transmit a jam signal in the ith minislot of subsequent cycles of the control channel, for as long as it needs exclusive use of this data channel. This guarantees reservation as explained before for R-Aloha$^{(1)}$. It should be noted that R-Aloha$^{(2)}$ performs slightly better than R-Aloha$^{(1)}$, as becomes evident after the necessary analysis [42].

Observe that jamming of the control channel is not done concurrently with data transmission; hence, CC-TT-TR node architecture suffices to implement both protocols. Notice also that both protocols reserve channels for a source node and *not destinations*. This means that, although some bulk data transfer may be taking place from a source node to a specific destination on a reserved channel, all other nodes are not discouraged in any way from deciding to transmit to the same destination on some other data channel. Thus, receiver collisions are not avoided by these schemes; this is ignored by the analysis in [42], but might lead to considerable performance degradation in practice, especially when relatively small population systems are considered.

3.3.1.10 DT-WDMA

The *Dynamic Time—Wavelength Division Multiple Access* protocol [5] is another remarkable protocol, which has received much attention from the research community and has motivated the development of several extensions.

It assumes a star-coupled broadcast-and-select WDM optical LAN with equal number of nodes and data channels, i.e. $N = W$, if we follow the previous notation. The total number of available wavelengths in the system is actually $W + 1$, where the additional wavelength is used as a control channel; hence, the protocol is pre-transmission coordination based like all the schemes in Section 3.3. This protocol entails a somewhat more costly implementation than schemes described thus far, since each node is assumed to have two fixed-tuned transmitters, one fixed-tuned receiver and one tunable receiver; therefore, DT-WDMA assumes a CC-FT^2-FRTR system. The higher implementation cost, however, as we will see, is balanced by a considerable performance improvement in comparison to most of the previous protocols.

The transceivers of nodes are used as follows. The first fixed-tuned transmitter (FT) and the fixed-tuned receiver (FR) of each node are always tuned to the control channel, while the remaining devices (second FT and TR) are used for the data channels. Observe that, since the number of data channels is equal to the number of nodes, each node is actually pre-assigned a home channel for transmission; thus, its second fixed-tuned transmitter (FT) is always tuned to the home channel. The node transmits all data packets (independent of destination) on this channel. Let us denote the data channels as $\lambda_1, \lambda_2, \ldots, \lambda_N$ and assume that they are pre-assigned to nodes 1, 2, ..., N, respectively, i.e. one transmitter of node i is fixed-tuned to data channel λ_i, for $i = 1, \ldots, N$. Apparently, the protocol is data channel collision-free.

As far as synchronization is concerned, the control and data channels have the same time reference, which we call 'cycle' to keep uniformity with previously presented schemes and is shown in Figure 3.10. Each cycle is divided into N control slots on the control channel, pre-assigned to the N nodes in a TDM manner. Let us assume that control slot j of each cycle is pre-assigned to node j, $j = 1, \ldots, N$. This means that each node has one control slot for exclusive use per cycle and control channel collisions are avoided. The entire cycle consists of one data slot on each data channel, as shown in Figure 3.10.

According to DT-WDMA, a node which has generated a new data packet transmits a control packet on the control channel in its pre-assigned slot. It transmits the actual data packet on its home channel during the following cycle. For example, let us assume that before the beginning of cycle i of Figure 3.10, node k already has a data packet for node m, where $k, m \in \{1, \ldots, N\}$ and, of course, $k \neq m$. Considering the assumptions mentioned before, the protocol determines that node k has to transmit a control packet in the kth control slot of cycle i and the actual data packet on its home wavelength λ_k during cycle $i + 1$. Consequently, DT-WDMA also has the tell-and-go property.

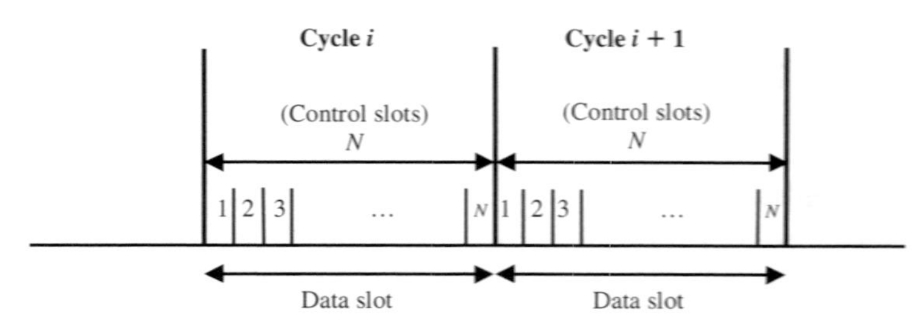

Figure 3.10 The synchronization (common cycle used) across control and data channels for DT-WDMA. Cycle length is $N(=W)$ control slots or one data slot. Each control slot per cycle is pre-assigned to a node in a fixed manner. Cycles i and $i+1$ are shown for example. Transmission of a control packet by a node in its pre-assigned minislot of cycle i implies transmission of the actual data packet (on the home channel of the node) during cycle $i+1$.

It should be noted here, that it suffices for the control packet to contain *only* the desired destination address m, with no source address and data channel. After a round-trip propagation delay, node m will have received the control packet from the kth control slot designating it as a destination. Node m will therefore know (just by the control slot number) that the source is node k and the data channel will be λ_k. Hence, it tunes its receiver to this wavelength and receives the data packet which comes next (after the same round-trip propagation delay).[20] As mentioned before and is also obvious from the example, the protocol is both control and data channel collision free.

Observe that in the previous example, we did not consider the case where node m receives at least one more control packet designating it as a destination *within the same cycle*. This implies that two or more source nodes want to transmit on the same destination (m) at the same time. The DT-WDMA protocol does not actually include a receiver collision avoidance mechanism, but it has the innovative feature (as compared to the previous schemes) determining that one of the sources will be selected and its transmission will be fine. The selection of one specific source (equivalently one data packet) and rejection of the rest are performed according to an algorithm (e.g. priority-based) that is known by all nodes in the system. Thus, the source nodes that were rejected will know that they have to retransmit in subsequent cycles.

For more details and the relevant performance analysis of DT-WDMA, the reader may refer to [5]. It should be noted here that the performance of DT-WDMA is considerably improved as compared to most of the schemes described so far. On the other hand, besides being quite costly in implementation, the scalability of the DT-WDMA protocol is at least questionable, as a consequence of the requirement that the number of data channels in a network applying this protocol must be equal to the number of nodes.

3.3.1.11 Quadro Mechanism

The *Quadro mechanism* was introduced in [8] as an extension to the DT-WDMA protocol. It assumes the same network model and node architecture as DT-WDMA, which were described before. It is more of a hardware enhancement to DT-WDMA than a new protocol as we will see.

The motivation behind the development of this mechanism was to deal with the considerable reduction of network throughput owing to receiver collisions. DT-WDMA determines that if two or more sources transmit a data packet to the same destination at the same time, one of the transmitted data packets is received successfully and the other(s) are lost and have to be retransmitted at a later time; an algorithm known and applied by all nodes of the system decides which packets should be rejected each time. It is observed that the loss of the rejected packets quite notably reduces network throughput, increases average delay and also requires more buffers per node to store packets that need retransmission.

The Quadro mechanism essentially introduces optical buffering capability to receiving stations. An *optical delay line* (*ODL*) *block* [8] is placed at the receiving part of each node, making it capable of optically queueing some data packets that could not be received due to a receiver collision. Notice that these packets are just lost and have to be retransmitted according to DT-WDMA, as we said before. However, the Quadro enhancement allows *some* of these packets to be received by the destination node at a time slot during which the tunable receiver does not have to receive from another source. The number of lost packets due to receiver collisions is reduced as the size of the optical buffering system increases.

The design of the ODL block allows the implementation of different *reception strategies* by the nodes. The *FIFO* (*First In First Out*) and *LIFO* (*Last In First Out*) reception strategies are studied in [7–8], where the selected data packet for reception is the first or last that was inserted in the ODL block, respectively. The analysis and simulation show that the best throughput performance and lowest average packet delay are achieved when the FIFO reception strategy is implemented. In any case, the Quadro mechanism significantly improves network throughput as compared to the original DT-WDMA scheme. It should be noted that parallel to the reception strategies mentioned before, two different *transmission policies*, namely random and FIFO, are also studied and analyzed in [8]. Transmission policies determine the way a source node selects generated packets from its buffer pool for transmission. Overall, the highest performance is achieved for the combination of the random transmission with the FIFO reception strategy.

Notice that the Quadro mechanism falls into the class of pre-transmission coordination-based protocols with receiver collisions, because it does not actually prevent receiver collisions from occurring. Its main achievement is to reduce their impact on network performance, as compared to the original DT-WDMA scheme, by means of optical buffering at the receivers.

3.3.1.12 Aloha/Delayed Aloha and Slotted Aloha/Delayed Aloha with Multichannel Control Architecture (MCA)

The pre-transmission coordination protocols studied so far are based on the use of a single control channel for coordination between the nodes of the network. In this part, an example of a network architecture and protocol, which assumes multiple control channels and appropriate structure for the nodes, is described in brief.

The main motivation behind such schemes is to relieve part of the electronic processing burden for control information at the nodes of a broadcast-and-select WDM network. In pretransmission protocols a control packet is typically transmitted for each data packet transmission that will follow. Each station is required to process all control packets that are transmitted over the single control channel and correspond to data packet transmissions on any of the available data wavelengths. The maximum processing rate is limited to the speed of the electronic interface of a station, which causes the so-called *electronic processing bottleneck* [18, 33]. The latter is closely related to the number of control channels and the transceiver equipment of nodes. The relatively low cost implementations studied before, assuming e.g. a single fixed-tuned receiver per node for the (only one) control channel used, inherently restrain network performance to lower levels than could possibly be reached otherwise.

Thus, [33], for example, proposes the so-called *Multichannel Control Architecture (MCA)* assuming several control channels, appropriate transceiver equipment per node and simple media access protocols used on control and data channels. The network has N nodes and $W + m$ wavelengths, where wavelengths $\lambda_1, \dots, \lambda_W$ (W in total) are used as data channels and the remaining wavelengths, $\lambda_{c1}, \dots, \lambda_{cm}$ (m in total), are used as control channels. Each node is equipped with a tunable transmitter capable of tuning over all available wavelengths in the system, m receivers fixed-tuned to the control channels (one receiver per channel) and a tunable receiver capable of tuning over the entire range of available data channels (for data packet reception). Hence the system can be characterized as CC^m-TT-FR^mTR.

The optical signal coming from the star coupler to a node comprises the input of a $1 \times (m + 1)$ WDM demultiplexer. The m outputs of the demultiplexer contain the control channels and attach to the m fixed-tuned receivers of the node, while the remaining output is connected to the tunable receiver which appropriately selects the desired data channel each time. Control packet transmission is assumed to last one time unit and data packet transmission L time units.

The operation of two simple media access protocols (examined before for single-control channel networks) is studied in the framework of the MCA architecture, namely Aloha/delayed Aloha and slotted Aloha/delayed Aloha. When a node has generated a packet for transmission or wishes to retransmit a previously collided packet, it first chooses one of the control channels at random and transmits the control packet immediately (unslotted Aloha) or at the next time slot (slotted Aloha). Note that in the case of slotted Aloha, control channels are synchronized and slotted so that one control packet can fit in exactly one time slot. The data packet is transmitted immediately in the following time units only if the corresponding control packet was successful, according to the delayed Aloha scheme. It should be noted that in the analysis of [33], propagation delays and tuning times are taken to be negligible. Observe that all types of collisions may occur in both cases, i.e. control channel, data channel and receiver collisions. A collision implies that the corresponding source node has to wait for a random time and retransmit; in case of a receiver collision this must be done only by the rejected nodes, since it is assumed that one data packet is always received. There are various policies for choosing one data packet for reception among two or more packets involved in a receiver collision and [33] notes the example of accepting the packet transmitted on the data channel with the lowest number (e.g. packet on λ_2 is preferred to one on λ_5).

The performance of these protocols in the framework of the MCA architecture is improved as compared to the single control channel case presented before and in [28]. However, this comes at a considerably higher cost of more transceivers per node and more channels dedicated for control purposes, as a trade-off. More details on the architecture and the related performance analysis may be found in [33].

3.3.1.13 N-DT-WDMA

Another example of a protocol assuming multiple control channels in the network is considered in this section. It can be thought of as an extension of the DT-WDMA scheme presented before, where N control channels are used for coordination purposes. Hence the authors in [18] suggested the name *N-DT-WDMA*.

The main objectives of this quite innovative scheme are the following:

1. Support of *different traffic classes*, including connection-oriented traffic besides connectionless (datagram) traffic that was considered by most of the previous schemes. Established connections may be simplex or full duplex.
2. *Low electronic processing overhead*, in order to relieve nodes from the bottleneck caused by control packet processing in single control channel protocols (this was discussed in the previous protocol).
3. Support of some *transport-layer functions* in an effort to integrate the MAC layer with the transport layer, since the intermediate network layer is essentially absent in WDM broadcast-and-select LANs. This is in accordance with the observation that fewer layers should be included in the layered architecture of high-speed optical networks. The supported transport-layer functions include connection set-up, disconnection, multiple traffic classes and transfer of data [18].

There are two connection-oriented traffic classes supported by N-DT-WDMA, one with *guaranteed bandwidth* (class 1) and the second *without guaranteed bandwidth* (class 2); these may be proper for video and file transfer applications, respectively, for example. The common *connectionless traffic* is still supported (class 3).

As far as the second objective is concerned, this scheme follows the same approach as the previous one, i.e. determines that multiple control channels should be used. In fact it determines that each node has a dedicated control channel, while the data channels are the same as in DT-WDMA, hence $2N$ wavelengths are needed, where N is the number of nodes in the system. This implies that each node maintains and processes control information relating only to its own transmissions and receptions. However, keeping processing overhead at much lower levels as compared to other schemes is costly; thus, the requirement of $2N$ wavelengths for a N-node network is an important disadvantage, considering also that N wavelengths are actually used only for coordination.

The N-DT-WDMA protocol determines that each node is equipped with a fixed-tuned and a tunable transceiver. The fixed-tuned transmitter (FT) and the tunable receiver (TR) are used for data, while the fixed-tuned receiver (FR) and the tunable transmitter (TT) are used for control information. Hence, the system can be characterized as CC^N-TTFT-TRFR. On each wavelength time is divided into cycles of equal length. However, each cycle is divided into m slots on control channels and $n + 1$ slots on data channels, where $n \leq m$. Control slots are assigned to connections (up to m), while the free ones are also used for connectionless traffic. The n data slots of a cycle are used for data transmission, while the remaining one (called *status slot* in the scheme) is used by nodes to transmit their assignments of control slots (for traffic classes 1 and 2) and data slots (for traffic class 1) to connections.

The procedure of connection set-up is based on the status slot and the fact that each node has a dedicated wavelength for transmitting data by means of its FT. It is only outlined in brief here. The basic idea is that a station, say X, wishing to establish a connection with another station Y, first examines the contents of the status slot transmitted by Y, by tuning its TR to Y's data transmission wavelength. Node X is informed in this way about current control slots (for class 1 and 2) and data slots (only for class 1) that are not used by Y for its other connections. From these slots, X suggests control slots, and for class 1 connection set-up also data slots, that are convenient, i.e. not already used by it. One of the suggested control slots, and additionally one of the suggested data slots for class 1 connections, may possibly be accepted by the other side (Y); X will know about that by checking again the contents of status slot transmitted by Y. In case it actually sees assignments for it, connection is considered established and the contents of the status slots are updated appropriately. Note that for full duplex connections, the complement of the described procedure must also be followed by node Y in order to set up a connection with X.

For class 1 traffic, the actual data packet transmission is done on the pre-assigned data slot. In the case of class 2 traffic (no guaranteed bandwidth), a control packet is transmitted from source to destination on the dedicated control slot assigned to the connection. This informs the destination about the data slot that is about to be used by the source. The actual data packet transmission is done on the announced data slot of the following cycle (in a tell-and-go fashion). Receiver collisions may occur when two or more nodes choose to transmit to the same destination in the same data slot; they transmit by their FRs on their dedicated wavelengths, therefore the destination can receive only one packet. The receiver selection policy followed is accepted and known by all nodes, as in DT-WDMA.[21] For class 3 traffic (connectionless) the same procedure as in case of class 2 is followed, with the difference that control channel collisions may also happen, since there is no dedicated control slot. Traffic class 3 data packets may participate in receiver collisions with class 1 or 2 data packets. Note that control channel collisions may also occur during connection set-up (for classes 1 and 2) but this is rare in comparison to the uncollided control packet transmissions after connection establishment.

The suggested protocol achieves high throughput performance and, as discussed, has low processing requirements. However, it is expensive to implement both in terms of number of wavelengths and transceiver equipment per node. Thus, the authors in [18] attempt to generalize the protocol by suggesting ways of reducing the number of control channels to the desired degree (down to one channel, in which case the protocol degenerates to DT-WDMA) and/or the number of transmitters and receivers per node. The former is done by sharing control channels among stations and increasing processing overhead as a trade-off. The latter is achieved by increasing the length of each cycle to $n + m + 1$, thus allowing TT-TR transceiver equipment per station; however, time has to be wasted in each cycle only for control information exchange as a trade-off. Note that the two generalization approaches for allowing less expensive implementations could be used in combination. More details about the protocol and its performance analysis may be found in [18].

3.3.1.14 Random Scheduling Schemes for Multicast Traffic

The schemes presented up to now have considered unicast transmissions between nodes, i.e. transmissions of data packets from one source node to a single destination. *Multicast* transmissions (from one node to multiple destinations) have been receiving much attention lately from higher layer protocols like IP; hence they need to be studied in a WDM broadcast-and-select LAN environment too. Remember that Figure 1.13 illustrates examples of unicast and multicast transmissions in such a network. As far as multicasting is concerned, all recipients are required to be tuned to the wavelength the source node transmitted on, in order to be able to receive the multicast message.

In this section, two simple scheduling schemes for multicast traffic are briefly discussed as an example [29]. The presented schemes fall into the class of *centralized scheduling*, since they are applied by a relatively simple *master/slave scheduler*,[22] which is located at the star. Centralized scheduling generally has the advantage of allowing simplicity at the network nodes, but the hub of the network becomes quite more complex and costly to implement; however, note that the added cost is shared among the various nodes of the network. Another disadvantage of centralized scheduling is the additional delays introduced on account of the communication (including transmission requests in one direction and transmission/reception decisions in the opposite) between the scheduler and the network nodes. There are much more sophisticated scheduling schemes for multicast traffic than the ones studied here. Nonetheless, for those pursuing optimality, it should be noted that this may not be a feasible goal in practice, if we consider e.g. that millions of transmission schedules may possibly have to be produced per second.[23]

The network model considered by the schemes consists of a WDM broadcast-and-select local area network with N nodes and $W + 2$ wavelengths. The nodes are equipped with a tunable transmitter and a tunable receiver capable of tuning (almost) rapidly over all available wavelengths; consequently, the system is CC^2-TT-TR. W wavelengths are used as data channels and the remaining two wavelengths, denoted as λ_c and $\lambda_{c'}$, are the control channels used for communication between the nodes and the centralized scheduler. Specifically, control channel $\lambda_{c'}$ is used by source nodes to send their transmission requests to the scheduler, while control channel λ_c is used by the scheduler to inform sources (destinations) about when and on which channel to transmit (receive). Nodes need not use any synchronization, but just have to execute the assignments decided by the scheduler immediately after receiving them; absence of synchronization at the nodes is another advantage adding to the overall simplicity at the network edges. The examination of both scheduling schemes is based on the assumption that each multicast message is destined to a fixed number of receivers (k), which are *randomly* selected from the N nodes. At each slot (time is slotted for the scheduler with each slot being of length equal to a multicast message transmission) at most W nodes are scheduled to transmit in total.

According to the first scheduling protocol, namely *Random Selection with Persistent Retransmission* [29], whenever a data channel becomes idle, the scheduler selects *at random* one multicast message from the sources that are both ready to transmit and not busy with some other transmission. The protocol determines that this message must be *persistently* retransmitted until it is received by all of its intended recipients. Assuming the system is constantly backlogged (nodes are always ready to transmit), if there is not any multicast message having been received by all of its recipients yet (in a time slot), no new transmission is obviously scheduled for the next time slot.

The problem with the approach above is a kind of *HOL* (*Head-Of-Line*) blocking, that is to say a situation where a multicast message waits for some occupied receiver(s) and thus prevents other messages (possibly with free recipients) from being transmitted on its channel.

It has been observed that when a receiver is occupied in a time slot, the possibility of being occupied again during the next time slot is high. Hence the second scheduling protocol, namely *Random Selection with Backoff Retransmission*, tries to cope with this problem by simply introducing a random delay between retransmissions of a multicast message that has not been received by all intended recipients. This way some new message with available recipients may be transmitted in the mean time. Of course, the old message will have to be retransmitted again later until it is fully received.[24]

Both scheduling schemes may result in receiver collisions which are the ones that actually make retransmissions necessary. The study in [29] additionally proposes three *receiver algorithms*, i.e. algorithms which are applied by receivers in order to select one message among multiple ones destined to them at the same time, and studies the performance of the proposed scheduling schemes in conjunction with the receiver algorithms. The proposed receiver algorithms are the following:

1. *Priority selection* algorithm, which determines that the message with the smallest number of remaining intended recipients is selected.
2. *Random selection* algorithm, which apparently suggests that one message is chosen at random by the receiver.
3. *First-Come-First-Serve (FCFS) selection* algorithm, which always decides that the message originally transmitted in the earliest slot has to be chosen for reception.

According to the performance analysis of [29], the Random Selection with Backoff Retransmission scheduling scheme in conjunction with the first receiver algorithm (priority selection) gives the best performance among other combinations.

As far as access to the control channel is concerned, the protocol used is essentially a variation of the unslotted Aloha scheme and works as follows. Nodes send reservation requests (control packets) according to the unslotted Aloha scheme. These requests have to be repeatedly retransmitted after a random delay until they are answered by the scheduler. Since unslotted Aloha is applied, control packet collisions may occur. When a node's request is answered with the corresponding assignments, the next request of the same node regards some new transmissions. Sequence numbers are used for requests and assignments in order to keep synchronization between the nodes and the scheduler. A more detailed description of the control-channel access protocol may be found in [30].

The performance of the scheduling schemes in this framework is analyzed extensively and is found to be satisfactory, keeping in mind that one of the main objectives is simplicity. Finally, it must be pointed out that the presented schemes are not appropriate for transceivers incapable of tuning (almost) rapidly, nor for circuit-switched traffic. For more details the reader may refer to [29].

3.3.2 PROTOCOLS WITHOUT RECEIVER COLLISIONS

Some of the schemes of the previous subsection determined that in case of a receiver collision, one data packet is selected by the destination node according to a receiver selection policy, while the remaining ones have to be retransmitted (e.g. DT-WDMA). Transmission of the 'rejected' data packets obviously wastes bandwidth and performance is considerably degraded, even though a policy of selecting one packet is kind of a solution to the receiver collision problem as opposed to ignoring it completely (as other schemes do).

In this subsection we outline several pre-transmission coordination-based MAC protocols with the virtue of preventing data packets that may participate in receiver collisions from being transmitted; in this way, performance is improved. Some of the presented schemes cannot avoid the wasteful transmission of control packets, whose corresponding data packets are predicted to cause receiver collisions; these control packets have to be transmitted again later. Note that we consider such schemes fall into this class of protocols, i.e. without receiver collisions, because *data packets are not actually transmitted* to the same destination on different channels at the same time. However, there are several other protocols outlined in this subsection that incorporate more straightforward ways of avoiding receiver collisions, for example, a *scheduling algorithm* based on global status information or a *network architecture that intrinsically precludes any instance of receiver collision.*

Scheduling algorithms were briefly discussed in the introduction to this chapter. As mentioned, if the scheduling algorithm is *distributed*, all nodes typically have to maintain some global status information and update it every now and again, according to info obtained by the shared control channel(s). A characteristic form of global information is a $N \times N$ *traffic demand* (or *backlog*) *matrix* $D = [d_{ij}]$, where d_{ij} represents the number of data packets at node i that are destined to node j and N denotes the number of nodes in the system. The *collapsed* $N \times W$ *traffic matrix* $A = [a_{ic}]$ can easily be derived from D, where each element a_{ic} is the number of data packets at source node i that are to be transmitted on channel c (destined to stations 'listening' to this channel).

Scheduling algorithms can generally be classified as *offline* or *online*. Offline algorithms do not compute the schedule until the entire matrix A is available, while online scheduling schemes start building the schedule on the basis of partial control information available at the time. Online schemes usually have the advantage of reduced delays between reservation and transmission, while, on the other hand, their schedules may not be as efficient as those produced by offline algorithms. Another classification of scheduling algorithms is derived if we consider how the a_{ic} data slots required by source node i on channel c are allocated. Specifically, if the required slots of each source node on a certain channel are allocated contiguously, the scheduling algorithm requires that each node tunes to each channel only once per schedule and is said to be *non-preemptive*; otherwise, the algorithm allows more freedom in the allocation of slots (more tunings may be involved as a trade-off) and is called *preemptive*. The huge number of pre-transmission coordination-based protocols without receiver collisions allows us to outline only a few sample schemes incorporating scheduling algorithms, towards the end of this chapter.

3.3.2.1 Modified Slotted Aloha/Delayed Slotted Aloha with Contention Processing Schemes LCB and LCB-CR

A modified version of the slotted Aloha/delayed slotted Aloha scheme has been proposed in conjunction with two novel *contention processing* schemes, namely *Local Collision Blocking* (*LCB*) and *Local Collision Blocking with Contention Resolution* (*LCB-CR*) [25]. These schemes are applied by means of appropriate hardware at the hub of the network, which is no longer passive as a consequence.[25] The main objective of contention processing schemes is to relieve some electronic overhead from nodes by resolving contentions of control packets and, as opposed to collisions which destroy all participating packets, possibly allowing one control packet pass to the star coupler successfully.

The network considered is a star-coupled WDM broadcast-and-select optical network with Nnodes and $W + 1$ wavelengths, where Wwavelengths $\lambda_1, \ldots, \lambda_W$ are used as data channels and the remaining wavelength λ_0 is used as a control channel for coordination. The system can be characterized as CC-FTTT-FRTR, since each node is equipped with a fixed-tuned transceiver for the control channel and a tunable transceiver for data channels. The time slotting on the control and data channels has some similarities with the one used for the SA/SA$^{(6)}$ protocol [42] presented before, an example of which was shown in Figure 3.9. The differences are that cycle length is equal to L time units and the control channel cycle is further divided into exactly W minislots, while the cycle for each corresponding data channel begins after the maximum round-trip propagation delay R. Note that data packet transmission lasts exactly L time units; hence one data slot has length equal to the length of a cycle. Time slotting of the control channel and data channel λ_2 for the proposed protocol is shown in Figure 3.11 as an example.

Control minislots are pre-assigned to data channels. A ready source node transmits a control packet in one control minislot (chosen at random) of the following cycle. Assume e.g. a source node transmits a control packet in minislot number two of cycle i, which is shown in Figure 3.11. It should be noted that control packets do not contain desired data channel as in Figure 3.2, since this is implied by the chosen minislot. The corresponding data packet is transmitted in the respective cycle of the selected data channel, i.e. immediately after the maximum round-trip propagation delay.[26] In our example, data packet transmission is done during cycle i of data channel λ_2 (Figure 3.11), if both the following two conditions are satisfied:

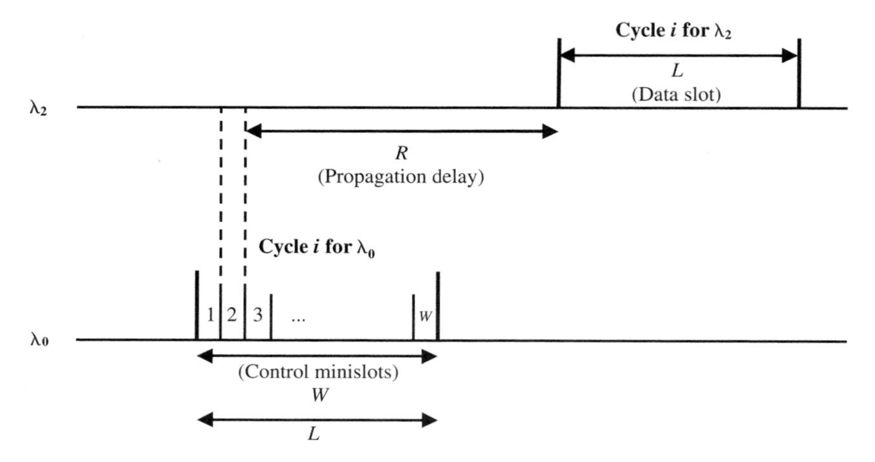

Figure 3.11 Time slotting of the control channel λ_0 and data channel λ_2 for the modified slotted Aloha/delayed slotted Aloha protocol with contention processing schemes LCB and LCB-CR. Cycle i for each channel is shown. Control minislots are pre-assigned to data channels.

1. The control packet returned successfully, because either it was the only one transmitted during this minislot or it was selected by the contention resolution scheme among contenting packets.[27]
2. There was no successful transmission of another control packet designating the same destination in an earlier minislot of the same cycle.

If both conditions are satisfied, the data packet will be transmitted successfully, owing to the fixed pre-assignment of control minislots to data channels; hence, the protocol is data channel collision-free. If one of the two conditions is not satisfied, the procedure is repeated according to the retransmission strategy, e.g. the control packet is transmitted again in the following cycle with a predetermined retransmission probability. Observe that data packets are not transmitted if it is predicted that they will cause a receiver collision; hence, we classify this scheme as pre-transmission coordination *without* receiver collisions, even though the corresponding control packets are actually transmitted and thus some bandwidth is wasted.

Turning to the contention processing schemes, the first one is the so-called Local Collision Blocking (LCB). It is assumed that stations which are geographically close are grouped into G separate groups; signals from nodes within each group are multiplexed and then transferred to the hub by means of an optical fibre. The control channel λ_0 is extracted at the hub and examined separately for each group by means of the appropriate hardware [25]. Control packets that are transmitted in the same minislot by nodes of a certain group are said to be locally collided. On the other hand, a control packet is locally successful if it is the only one transmitted in a control minislot. The locally collided packets are blocked by the hardware mechanism at the hub, while the locally successful ones are allowed to pass to the star coupler. A control channel collision occurs when two or more locally successful packets are transmitted in the same minislot (obviously by nodes belonging to different groups). The participating packets are destroyed and have to be retransmitted.

The latter does not happen in the second contention processing scheme proposed. The so-called Local Collision Blocking—Contention Resolution scheme (LCB-CR) employs an additional mechanism which ensures that one locally successful control packet always passes to the star coupler, while other locally successful packets possibly transmitted in the same minislot have to be retransmitted. This is done by a CRU (Contention Resolution Unit) which has G input ports.

The two schemes are compared to a pure contention scheme (applying no contention resolution) and a CRU-only scheme, which assumes a CRU with N input ports at the hub. As a consequence, in CRU-only, one control packet for each minislot always passes to the star coupler, making this scheme an upper performance bound. However, note that the significantly higher number of input ports calls for much higher electronic processing requirements at the hub and additional implementation cost. The two proposed schemes are more practical solutions that improve network performance in comparison to pure contention. Additional details on the schemes, the required hardware and the performance analysis may be found in [25].

3.3.2.2 TDMA-W

TDMA-W protocol[28] is an interesting pre-transmission coordination-based protocol that avoids all types of collisions [3]. It considers a WDM broadcast-and-select optical LAN of the common star physical topology with N nodes and $W + 1$ wavelengths, where wavelengths $\lambda_1, \ldots, \lambda_W$ are used as data channels and the remaining wavelength λ_0 is used as a control channel. Each node is equipped with one tunable transmitter for both control and data packets and a fixed-tuned transceiver, where the fixed-tuned receiver is always tuned to the control channel. The tunable devices are capable of tuning over all available wavelengths (in fact it is not necessary to tune the tunable receiver to λ_0). Consequently, TDMA-W considers a CC-TT-FTTR system.

Synchronization across all channels is provided on the basis of the control packet transmission time (taken to be one time unit). The protocol has the virtue of supporting variable data packet lengths. One data packet is generally assumed to last L control slots (time units), where L is a positive integer with variable value; hence, data slots of fixed-length are not assumed, unlike many of the schemes examined before. Time on the control channel is further assumed to be divided in cycles, each one with N control slots. One fixed control slot per cycle is pre-assigned to each of the N nodes in a TDM fashion, which implies elimination of control packet collisions.

According to the protocol, a ready station has to transmit a control packet in its slot of the control channel cycle and the corresponding data packet after α slots, where α is the switching latency; $\alpha = \max\{t_s, t_r\}$, where t_s is the tuning time for the TT of the source node and t_r is the time required by the destination to receive and examine the control packet and also tune its TR to the proposed data channel. Note that besides the three elements of Figure 3.2 (source, destination, data channel) control packets additionally include a value L corresponding to the size of the respective data packet (in time slots).

The key feature of the protocol is that each node maintains two *status tables*, which respectively help in avoiding data channel and receiver collisions. The first table has W entries, one for each data channel, while the second has N entries, one for each destination node. These are similarly updated by each node at the end of each control slot after examining the received control packet. Specifically, $L + \alpha$ is added to both entry i of the first table and entry j of the second, where λ_i is the data channel, j is the destination and L is the packet length that were declared in the control packet; obviously, $i \in \{1, \ldots, W)$ and $j \in \{1, \ldots, N\}$. All other positive entries in the tables are decreased by one at the end of each slot.

If we suppose a node is ready to transmit, at the beginning of its control slot it checks in the second table if the entry of the desired destination is less than or equal to zero (idle destination, no receiver collision) and then it scans the first table for a data channel entry with value less than or equal to α (idle channel, no data channel collision). Note that if a data channel is busy only for the next α slots, it can be used since actual transmission of the new data packet will start after the switching latency α, as described before; this overlapping reduces the performance penalty of large switching latencies. If either the destination is not idle or there is not any idle data channel, the node retries in the next cycle.

The performance of the TDMA-W protocol is analyzed and it is shown that it generally performs better than DT-WDMA, even though it requires fewer wavelengths ($W \leq N$ as opposed to $W = N$ for DT-WDMA). Yet, an array of VLSI chips per node for keeping the two status tables is required, as proposed in [3]. In the same paper, the interested reader may find more details on the protocol's architecture and performance analysis.

3.3.2.3 RCA

The first scheme employing relatively simple node architecture and avoiding the wasteful transmission of data packets predicted to participate in receiver collisions is probably the *Receiver Collision Avoidance* (*RCA*) protocol [21]. Unlike most pre-transmission coordination-based schemes without receiver collisions, this one only requires each node be equipped with one tunable transmitter (TT) and one tunable receiver (TR), each used both for control information and data; hence, it is considerably less expensive to implement than other protocols of this category. One wavelength is used as control channel for coordination between the N nodes of the system, while W additional wavelengths are used as data channels; thus, the system assumed by RCA is characterized as CC-TT-TR. Note that the absence of an additional receiver fixed-tuned to the control channel in such systems, introduces one more eventuality for receiver collisions. Specifically, receiver collisions may also occur when a control packet cannot be received by a node, because the tunable receiver of the node is receiving data at that time.[29]

The RCA protocol assumes time is slotted across all channels on the basis of control packet transmission time, i.e. one slot is equal to the control packet transmission time. Data packets are taken to be of fixed length (unlike TDMA-W) exactly equal to W slots. Moreover, time on the control channel is divided into cycles of length W slots too, each pre-assigned to a data channel in a fixed manner. Note also that the protocol has the advantage of making realistic assumptions about round-trip propagation delay and tuning times of transceivers, both considered non-negligible and respectively equal to R and T slots.

Access to the control channel is provided via a modification of the slotted Aloha scheme. The difference with the original slotted Aloha is that a node wishing to transmit first checks some appropriate values in two data structures maintained by each node. The first data structure is a so-called *Node Activity List* (*NAL*), which keeps entries for the most recent $2T + W$ slots; for slots that contained an uncollided control packet, it has the corresponding destination (found in the control packet). Observe that this list may not be properly updated at times the TR is receiving data, hence not tuned to the control channel. The second data structure maintained by each node is a so-called *Reception Scheduling Queue* (*RSQ*); this has entries for each scheduled reception for the node. Namely, for each scheduled reception, it has the data channel and the time the node must start tuning its TR to it.

A control packet is transmitted according to slotted Aloha scheme, if NAL does not contain any entry for successful control packet with the same destination for the past $2T + W$ slots and the RSQ at the source has been empty for the same time interval (no scheduled receptions for source, TR has been on control channel implying NAL values are updated). Notice, for example, that if there was at least one entry with the same destination in NAL, the TR of the destination node would not have enough time to tune back to the control channel and receive the transmitted control packet (data packet transmission lasts $T + W$ slots and T more slots are needed to tune back to control channel). Control packet is transmitted if the above are true; otherwise, the structures are checked again at the beginning of the next slot.

The RCA protocol's access policy for the data channels is similar to delayed slotted Aloha in that the data packet is not actually transmitted until the control packet arrives back successfully. In addition to original delayed slotted Aloha, RCA determines that receiver collisions are detected by the source (by means of its TR) in the period in-between control packet transmission and reception back from the star after a round-trip propagation delay. The TR of the source has to be constantly tuned to the control channel in the above mentioned period for this purpose.

For example, if the original control packet was transmitted by node i in time slot $[0, 1)$ it will arrive at all nodes (destination too, say node j) in slot $[R, R + 1)$. Let us estimate by what time the TR of j must be tuned to the selected channel, in order to receive the corresponding data packet without a problem. Tuning at source node i is done in time interval $[R + 1, R + T + 1)$ and data packet is transmitted during interval $[R + T + 1, R + T + W + 1)$, which implies it arrives after a round-trip delay to node j in $[2R + T + 1, 2R + T + W + 1)$. Hence, the TR of j must already be tuned to the selected data channel by time $2R + T + 1$.

It can be observed that if another control packet is successfully received in a slot of time interval $[R - T - W + 1, R)$, a receiver collision will occur. Let us take the extreme case of a successful control packet, which designates the same destination j and is transmitted by another source node k, is received by j (and all nodes) in time slot $[R - T - W + 1, R - T - W + 2)$, i.e. in the leftmost slot of $[R - T - W + 1, R)$. Node k will have to tune to the right channel for data packet transmission in time interval $[R - T - W + 2, R - W + 2)$. Data transmission by k is done during time interval $[R - W + 2, R + 2)$ and arrives after a round-trip from the star to destination j at time interval $[2R - W + 2, 2R + 2)$. In order to receive the data packet from i too, the destination node j has to additionally tune to the right channel in interval $[2R + 2, 2R + T + 2)$. But we saw before that it should have been tuned to the data channel by time $2R + T + 1$, hence a receiver collision is predicted to occur.

Generally, for a control packet transmitted by node i at time slot $[t, t + 1)$, receiver collision is detected if a successful control packet with the same destination is received in a slot of time interval $[t + R - T - W + 1, t + R)$; this may be seen by node i in the (constantly updated) NAL, until its transmitted control packet gets back, in time slot $[t + R, t + R + 1)$. In this case the actual data packet is not transmitted by node i, which will have to retry accessing the control channel in the following slot according to the modification of the slotted Aloha (described before). Assuming no receiver collision is detected, source i learns in slot $[t + R, t + R + 1)$ whether its control packet experienced a control channel collision, If so transmission of data packet is aborted; otherwise, it transmits the data packet during interval $[t + R + T + 1, t + W + R + T + 1)$, i.e. after a tuning period of T slots. This will arrive at the destination after R slots.

In an analogous way, a node has to detect whether uncollided (successfully received) control packets declaring it as a destination, imply receiver collisions. Such control packets are simply ignored so as to avoid unnecessary tuning. The last entry of the RSQ list is helpful for this. Specifically, it is examined for a certain control packet, if the corresponding data packet can be received after a round-trip propagation delay without a receiver collision with the last scheduled reception in RSQ. If receiver collision is ascertained, the control packet is ignored; otherwise, a new entry is placed in RSQ, since we have a new scheduled reception for the node.

The protocol's performance is not very high; anyway it is satisfactory, especially if we consider the limitations due to the simple node architecture and the realistic assumptions about tuning and propagation delays. For additional details about RCA and its performance, please consult [21].

3.3.2.4 WDM/SCM (NG-TDMA)

The *WDM/SCM* [45] or *Node Grouping-Time Division Multiple Access* (*NG-TDMA*) protocol [46] (old version of WDM/SCM) is a pre-transmission coordination-based scheme without receiver collisions. It assumes a slightly differentiated physical topology, namely a *double star* topology, which is different from the classic star topology of Figure 1.11 in that, instead of nodes, *passive-star subnetworks* are interconnected through the central star coupler. Let the number of subnetworks (groups of nodes) be G. Each subnetwork is assumed to comprise M nodes, physically located close to each other; note that the scheme takes advantage of the latter, as we will see.

The number of wavelengths is G, equal to the number of groups (subnetworks), with each wavelength pre-assigned to a certain group and comprising the group's dedicated home channel for data reception, as in some pre-allocation schemes, e.g. wavelength λ_1 is the home channel of group 1, wavelength λ_2 is the home channel of group 2, and so on. Even though there is no separate wavelength used as a control channel for coordination, the protocol is pretransmission coordination based. This is achieved by means of the *subcarrier multiplexing* (*SCM*)[30] technique, which allows the placement of control and data SCM channels within the same wavelength. Specifically, the home wavelength λ_i of group i, is subdivided into $M + 1$ SCM channels, f_0, $f_1, \ldots, f_M$, where channel f_0 is used as a control channel for the group and the rest are pre-assigned to the M nodes of the group for reception. It is sufficient for each node to have one tunable transmitter capable of tuning to the control WDM/SCM channel of the group and all WDM/SCM data channels in the system, and two fixed-tuned receivers; one is assumed fixed-tuned to the control WDM/SCM channel and the other fixed-tuned to the node's WDM/SCM data channel (home channel for reception), both contained within the home wavelength of the group. It is worth noting that since each node has fixed-tuned receivers only and one dedicated home WDM/SCM channel for reception, receiver collisions are intrinsically avoided.

All wavelengths are divided into TDM-frames (cycles), where each frame consists of exactly G slots, one for each group; hence, the name NG-TDMA. This results in the elimination of collisions between data packets transmitted by nodes belonging to different subnetworks. The elimination of collisions between data packets transmitted by nodes of the same group is explained later. Observe that access delays are significantly shorter in this protocol than in I-TDMA, where each cycle consists of more slots ($N = G \cdot M$), one for each node. Each slot is further divided in a control sub-slot and a data sub-slot for control and data packets, respectively, transmitted by nodes within the corresponding group. Each control sub-slot is additionally divided into M minislots, one for each node within the group. Consequently, control packet collisions are also eliminated.

The WDM/SCM protocol works as follows. Each ready node attempts to reserve the destination group's home wavelength for transmission (either its own group's home channel or not), by means of a control packet transmitted on the control WDM/SCM channel f_0 (contained within its own group's home wavelength). This is done in order to avoid collision with data packets transmitted by nodes of the same group to the same destination group, that is to say on the same home wavelength. The control packet is received by all nodes of the same group after a round-trip propagation delay from the group's local star coupler. Therefore, notice that the protocol generally determines that control packets are only 'travelling' within groups and not within the entire system, which implies significantly less impact of control packet propagation delays on network performance than other schemes (because nodes within a group are geographically located close to each other). The group's round-trip propagation delay is known and nodes transmit their control packets (in their pre-assigned minislots) in a previous slot, so that there is enough time for all control packets to be received back just before the group's own slot in the TDM cycle.

After receiving control packets back, the protocol decides which node will transmit in a way that eliminates collisions. If only one node of the group wishes to transmit on a specific wavelength, it is given permission to transmit. If two or more nodes of the group want to transmit on the same wavelength (i.e. to nodes belonging to the same group), the protocol determines that only the source node with the earliest packet-generation time should transmit. The same decision is made by all nodes; therefore nodes that were not allowed to transmit have to try again in the group's slot of the next TDM cycle. Note that the data packet's generation time is assumed to be included in control packets (besides source node and destination group).

The protocol's performance is analyzed and it is shown to be improved, in comparison to the previously examined I-TDMA and TDMA-W protocols. More details may be found in [45–46].

3.3.2.5 MAC Protocol with Centralized Scheduling Based on a Look-Ahead Capability

The protocol presented here comprises a centralized scheduling scheme based on a look-ahead capability [30]. It assumes exactly the same network model, number of nodes and wavelengths as the centralized scheduling algorithms for multicast traffic of [29], which were presented in 3.3.1.14 before; hence, these characteristics are not repeated here. The node structure can be TT-TR. However, in order to achieve full utilization of available bandwidth, it will be assumed that the system is CC^2-TTFR-TRFR, i.e. each node has also one FT fixed-tuned to control channel λ_c for transmitting requests and one FR fixed-tuned to control channel $\lambda_{c'}$ for receiving assignments from the centralized master/slave scheduler.

The nodes of the network are assumed to tune and transmit/receive data immediately after receiving the corresponding assignment from the scheduler. Therefore, they do not need to maintain any synchronization and operate asynchronously, which makes them simpler than nodes of most other schemes. In a way that resembles the one described in the RCA protocol but implemented differently (in a centralized manner), the scheduler takes into account the round-trip propagation, tuning and processing delays of nodes and issues the assignments just in time so that nodes have enough time to tune and transmit/receive. In this way, the impact of these delays on performance is reduced. The overall delays may vary from node to node (since non-equidistant stations from the hub are assumed) and are estimated by the scheduler by measuring the amount of time it takes for nodes to respond to its assignments. Nodes access the control channels via a modification of the unslotted Aloha protocol, which is identical to the one described before in 3.3.1.14 for the case of centralized scheduling algorithms for multicast traffic of [29].

The scheduling algorithm adopted by the protocol is based on N input queues (one for each node) that keep the incoming requests of nodes. Even though nodes operate asynchronously, the centralized scheduler assumes time is slotted and produces one schedule per slot. The scheduler selects a request queue at random and examines whether the request at the top of the queue can be scheduled, i.e. checks if it regards a transmission to a receiver that has not been assigned yet. According to the *plain First-Come-First-Serve (FCFS)* algorithm, only the head of the queue is examined; the so-called Head-Of-Line (HOL) effect limits performance, as a consequence, for FCFS. The scheduling algorithm of the proposed protocol, however, is assumed to have a *look-ahead capability*, i.e. it may look ahead into each queue until a certain depth k. This means that it is capable of searching each queue for a request that can be scheduled until depth k. The algorithm continues examining input queues, until either all available wavelengths have been assigned or there are no queues left to examine. Of course, at most one transmission is scheduled in each slot. Furthermore, the algorithm ensures that each destination is assigned to receive from at most one source node per slot. Hence, the protocol is both data channel and receiver collision-free. It should be noted that the implemented scheduling algorithm brings about a considerable performance improvement while it is only a bit more complex than the basic FCFS.[31] Of course, there are much more sophisticated scheduling algorithms that achieve full utilization of bandwidth, but most of them come with rather impracticable computational requirements, if we consider that the scheduler may have to produce millions of schedules per second.

The performance analysis, simulation and the corresponding results of the described protocol, as well as additional details about its operation, may be found in [30]. Finally, a notable advantage of the scheme that is worth mentioning is that it may be applied in WDM passive optical networks (PONs),[32] owing to the assumed simple implementation of nodes.

3.3.2.6 Contention-Based Reservation Protocol with Load Balancing

The protocol outlined in this section[33] is presented as a sample pre-transmission coordination-based scheme without receiver collisions, which makes some realistic assumptions that are usually absent from several other schemes [20]. Namely, the scheme assumes the following:

- Nodes have arbitrary distances from the star coupler, implying different propagation delays, which are non-negligible.
- Traffic offered by nodes is generally non-uniform (asymmetric) and comprises variable-length messages. The protocol adopts a *load balancing technique* in order to deal with this problem, as we will see.
- The tuning times of lasers and filters used by the stations are non-negligible.

The number of nodes and wavelengths in the WDM broadcast-and-select network is taken to be N and $W + 1$, respectively, where, like in many previous schemes, one wavelength is used as the control channel and the remaining W as data channels. Time is slotted across all channels and one time slot is assumed equal to one data packet transmission time; moreover, each time slot in divided into L minislots on the control channel. In order to deal with the problem of varying distances of stations from the star coupler, the protocol provides synchronization of nodes by generally assuming that, for stations located closer to the hub, the same slot begins later than for stations located further from the hub. That is to say, the stations wait for an appropriate time interval so that they are synchronized with the station located furthest than all from the star (i.e. the corresponding slots are delayed appropriately). This is done, because all stations must use the same control information at the same time in order to produce a schedule for using data slots. Each station is assumed to have a buffer for storing control packets, which, as expected, arrive back from the star earlier than all stations located further than it from the star. A certain control packet is retrieved from the buffer and processed by a node, only after the appropriate *synchronization waiting time*, which is twice the difference of its distance from the furthest distance from the hub of the network. The furthest distance is $R/2$, if R denotes the round-trip propagation from the star coupler to the furthest station in slots. Consequently, control packets transmitted in the L minislots of slot x are only retrieved from buffers and processed just before slot $x+ (R + 1)$; control packets transmitted during slot $x+ (R + 1)$ are processed by all nodes just before slot $x+ 2(R + 1)$ and so on. This can be seen as a division of the control channel into $R + 1$ subchannels, $S_0, S_1, \ldots, S_R$ [20].

The system considered by the protocol is CC-FTTT-FRTR, where the fixed-tuned transceiver is constantly tuned to the control channel. The protocol has the following two operation modes:

1. *Normal operation mode*: during this, the system behaves as CC-FTTT-FR^2, since the TR for data of each node is assumed to be constantly tuned to a certain data channel decided in the second mode. The period in which the protocol operates in normal mode, is fixed and known in advance by all nodes. After this, the system enters the second operation mode for a while, before

returning again to normal operation mode and so on (it alternates between these modes).

2. *Wavelength adjust mode*: this starts after a normal operation mode. During this mode, the nodes adjust their tunable receivers, i.e. decide where to tune them for the next normal operation mode. The decisions are based on history (load destined to each station in the previous normal operation mode), which is maintained by means of appropriate variables (N load counters). Note that during this mode no control packets are transmitted, but unfinished data packet transmissions from the previous mode are completed; when this happens (the tuning decisions are made in the mean time) the system returns to normal mode. Apparently, the objective of introducing the wavelength adjust operation mode is to balance the offered load, which was originally assumed to be non-uniform. Details on the load balancing algorithm are found in [20].

Next, we briefly describe how the protocol operates in the normal mode. Each ready station is assumed to choose one of the $R + 1$ control subchannels at random and transmit a control packet according to the slotted Aloha protocol with some transmission probability[34] in the next minislot (after packet generation) of a slot belonging to the selected subchannel. The control packet contains *source address* (*SA*), *destination address* (*DA*) which also implies the data channel (remember that in normal mode TRs are fixed-tuned to data channels known by all nodes), and *number of packets* (*NPK*) since variable-length messages are assumed.

If the control packet is successfully received, the protocol assures a reservation is made for the corresponding data message in the following way. Each node maintains W separate counters, each for the length of the corresponding data channel's 'virtual queue' (the queue length for a channel is actually the number of slots after which the channel will be available). Obviously, these lengths are reduced by one after each slot by all nodes. Furthermore, each source node maintains three arrays CH, CD and RC for its own successfully transmitted control packets (i.e. its own reservations). For its ith successful reservation, CH(i), CD(i) and RC(i) respectively represent the reserved data channel, the number of slots after which transmission should start (also reduced by one after each slot and implying transmission when zero) and the length in slots of the corresponding message. According to the protocol, a control packet that was received successfully by all nodes is retrieved from the nodes' buffers and examined (after the synchronization waiting time, which varies from node to node) at the same time. If this is the kth successful reservation for the source node, it performs the following operations: CH(k) $= \lambda_j$ (if λ_j corresponds to the DA of the control packet), CD(k) $=$ length of virtual queue for λ_j and RC(k) $=$ NPK of the control packet. In addition, the length of the virtual queue for channel λ_j is incremented by NPK slots by all nodes, including the source node, which does this after setting the value for CD(k) as described above.

Notice that since the TRs are essentially fixed-tuned to a certain wavelength during normal operation mode, the protocol *intrinsically avoids receiver collisions*. However, a certain station may be scheduled to transmit to different destinations during overlapping time intervals.[35] In this case, it can only choose and transmit one message through its TT and reattempt to make reservations for the non-selected messages in subsequent slots. This introduces additional delays for the transmissions that were not selected. Moreover, when non-negligible tuning time is assumed for the TT, a source node may possibly not use slots reserved for a certain transmission, because it tunes its laser at that time. Reservations have to be made again for the corresponding transmissions and, consequently, additional delays are introduced. The protocol's performance is analyzed and found to be good; however, note that this comes at the cost of considerable processing requirements and a quite expensive node structure.

3.3.2.7 MAC Protocol with Guaranteed Bandwidth Provision Capability

The pre-transmission coordination-based protocol outlined here has the additional capability of providing bandwidth guarantees to certain nodes of the network [11].[36] Like many other protocols, this scheme assumes one separate control channel shared by all nodes (λ_0), W data channels ($\lambda_1, \ldots, \lambda_W$), N nodes and a FTTT-FRTR node structure where the fixed-tuned transceiver is always tuned to the control channel. However, it innovates by requiring that there is an additional *control node*, whose task is described later on.

Time is slotted across all channels in the same way and one time slot is equal to a data packet transmission; this may include tuning also, if non-negligible tuning times are considered. One slot is further divided on the control channel in X bits followed by m minislots. The X-bit pattern comprises the *reservation information.* Generally, it is assumed that a variable part of the available bandwidth is dynamically allocated to some nodes for guaranteed bandwidth services (called 'guaranteed nodes' from now on) through *reservation* and the remaining part is left for all other nodes to make use of it on a *contention-basis* for their on-demand service requirements. Each bit is assigned by the control node to a guaranteed node; thus, the maximum number of such nodes in the system is fixed and equal to X. The status of each bit is dynamically set by the control node[37] and determines whether a control minislot is reserved for the corresponding guaranteed node or not (let us assume that a value of 1 implies reservation of a minislot). The first control minislots (of the m, in total, minislots of a slot) are called *reservation minislots* and are allocated to guaranteed nodes. Their number is not fixed, but changes according to the number of reservation bits that were set to 1 by the control node for the specific slot. Minislots are allocated to guaranteed nodes in the order that 1s appear in the X-bit reservation pattern. The remaining minislots of a slot are shared by all remaining nodes (including guaranteed nodes whose assigned bit was 0 in the X-bit reservation information) in a contention-basis (hence called *contention minislots*). Thus, priority is given to guaranteed nodes, whose control packet transmissions are always uncollided. On the other hand, other nodes have to choose randomly one of the remaining minislots and transmit without being sure about success or collision of their transmission.

Transmitted control packets (which always include source and destination address) within a certain slot are received by all nodes after a round-trip propagation delay. A data channel is allocated for the data packet transmission corresponding to an examined control packet, if the latter satisfies the following two conditions:

1. The control packet returned successfully, i.e. without a collision. Note that this is always the case for control packets transmitted in reservation minislots, as mentioned before.
2. There was no successful transmission of another control packet designating the same destination in an earlier minislot of the same slot. In this way receiver collisions are eliminated.

The above access mechanism is actually a modification of delayed slotted Aloha for the data channels, very similar to the one adopted by the protocol in Section 3.3.2.1. According to the protocol, data channels are allocated to transmissions in the order that the corresponding control packets are found to satisfy the above two conditions. Hence, λ_1 is allocated to the data packet transmission corresponding to the first acceptable control packet, λ_2 is allocated to the data packet transmission corresponding to the second acceptable control packet and so on. This allocation eliminates data channel collisions as well. The actual data packet transmissions begin in the following slot.

It remains to comment on the different ways the control node may manage reservations, i.e. determine how the bits should be set in the reservation information of each slot. The simplest policy would be some fixed predetermined setting of reservation bits, which is constantly repeated. One step ahead is a measurement-based policy, where the control node sets the bits according to the usage of reservation minislots in prior slots. The most efficient policy would be to have the reservation bits directly managed by the guaranteed nodes themselves according to their needs;

however, this is currently infeasible, as it requires bit-by-bit synchronization at the optical layer [11].

One advantage of the protocol is that, even at extremely high load conditions, it sustains a satisfactory throughput level, because it ensures that guaranteed nodes always have a certain amount of bandwidth for their needs. The remaining bandwidth is fairly shared by all other nodes in a contention-basis, as we said before. Moreover, the protocol assures that when the reserved bandwidth is not used by certain guaranteed nodes, instead of just being wasted, it is also added to the amount of bandwidth that is shared by all other nodes. This is achieved because control minislots, not data channels, are actually reserved for guaranteed nodes. Additional details, performance analysis and the related simulation results for the protocol may be found in [11].

3.3.2.8 AP-WDMA

Another interesting protocol that avoids receiver collisions is the *Accelerative Pre-allocation—Wavelength Division Multi-Access* (*AP-WDMA*) protocol [43]. This has several common features with pre-allocation schemes, such as the pre-assignment of home channels to transmitters and the fixed assignment of slots to home channel-destination pairs. However, it differs in making use of a separate control channel for *accelerating* (when possible), i.e. rescheduling earlier, certain transmissions that would take place later according to the fixed assignment of slots to channel-destination pairs in a typical pre-allocation based scheme. This improves performance of AP-WDMA in comparison to the corresponding pre-allocation based protocols. At the same time, the scheme retains the main advantages of pre-allocation schemes and avoids the increased complexity noticed in several other pretransmission coordination-based protocols.

The protocol assumes a WDM broadcast-and-select star-coupled network with N nodes, W data channels ($\lambda_1, \ldots, \lambda_W$) and one control channel (λ_0), where generally $N \geq W$. Each node is equipped with a transceiver for control information fixed-tuned to λ_0; note that control information comprises control packets, through which sources request permission from destinations to transmit, and acknowledgements sent back from destinations to sources and implying that they should proceed to the actual data transmission. Besides the fixed-tuned transceiver, each node has a fixed-tuned transmitter and a tunable receiver for data; hence, the system becomes CC-FT^2-FRTR. AP-WDMA adopts the fixed allocation of home channels to transmitters (FTs) encountered in FT-TR pre-allocation-based schemes. Pre-allocation of home channels follows some policy (known in advance by all nodes, e.g. interleaved allocation), which determines how the available home channels are shared among source nodes, when $N > W$.

Synchronization is provided across all channels in the system. One time slot can host a packet transmission on the data channels (besides some additional time for tuning purposes). For the control channel the same time slot hosts a *control frame* and an *acknowledgement frame*, during which control packets and acknowledgements are transmitted, respectively. Each control and acknowledgement frame are further divided into (control and acknowledgement) minislots in a time-interleaved fashion, such that e.g. the ith control minislot may be used only by station i to transmit its control packet. Note that both the control and acknowledgement frames along with the necessary round-trip propagation delays are included in a single time slot. In the general case of non-negligible (and relatively long) propagation delays, the information in a time slot of the control channel corresponds to data transmission during the next time slot on data channels.

In the general case $N > W$, channels are pre-assigned to destinations for reception in an analogous way channels are pre-assigned to sources for transmission in I-TDMA (as shown in the example of Table 3.2(a)). Notice that the disagreement in the assignments is because I-TDMA assumes TT-FR node structure, while FT-TR is the case for AP-WDMA (concerning data only, of course). Thus, each node of the network is pre-assigned exactly one slot per cycle to receive from every channel (every source, for the $N = W$ case of self-routing).[38] As mentioned in the previous paragraph, however, AP-WDMA requires that, for transmission in any time slot, the relevant control and acknowledgement information should be exchanged in the previous slot among source and destination. The key idea of the protocol is that if it is observed by the absence of control packets (requests) that the following time slot will not be used by a destination for reception on its pre-assigned channel, another transmission (originally scheduled for later) may take place earlier to avoid waste of bandwidth.

This becomes clearer if we consider the protocol's channels/destinations allocation map of Table 3.10 for the same example network examined for I-TDMA with $N = 7$ and $W = 5$. Nodes and channels are numbered from zero to $N - 1$ (6) and $W - 1$ (4), respectively. For example, destination node 3 may receive on channel 2 (from source nodes that have it as home channel) during the second time slot of each cycle. In the table the shaded cells relate to the control information of time slot 1 of the 1st cycle. Slots that were found to be idle are shown in light shade (five in total). For example, there was not any control packet for destination node 1 transmitted during time slot 1 by a source with channel 0 as home channel. According to the protocol an early transmission is possible. Each source node transmitting on this home channel (0) can send in its control minislot

Table 3.10 Channels/destinations allocation map of AP-WDMA for an example network with $N = 7$, $W = 5$. The shadowed cells relate to the control information of *time slot 1* of the 1st cycle. Idle slots are shown in light shading and the pre-allocated slots that can be moved earlier (in time slot 2) are shown in dark shading.

Channels	Destinations									
0	0	1	2	3	4	5	6	0	1	...
1	1	2	3	4	5	6	0	1	2	...
2	2	3	4	5	6	0	1	2	3	...
3	3	4	5	6	0	1	2	3	4	...
4	4	5	6	0	1	2	3	4	5	...
Time slots:	1	2	3	4	5	6	7	1	2	...
Cycles:	← 1st cycle →							← 2nd cycle →		

(of time slot 1) one control packet indicating the earliest next possible pre-allocated destination.

Suppose source nodes 0 and 5 share channel 0 for transmission. Suppose also that during time slot 1, both have no packet for destination 1 and one packet for destination 2, which is the earliest next pre-allocated destination and thus, transmit a control packet indicating 2 as destination in their control minislots of time slot 1. Now both have to look at acknowledgement minislot number 2 (used by node 2) included in the acknowledgment frame of time slot 1 too, as explained before. According to AP-WDMA, in this case destination node 2 has to choose one source at random and acknowledge its request. The result will be that destination 2 is in fact rescheduled in an earlier time slot (2 instead of 3). Such rescheduled destinations are shown in dark shading in our example. Notice that on channel 1, destination 4 cannot be moved from time slot 4 back to time slot 2 (therefore not in dark shading), because it has a pre-allocated reception on channel 3 in slot 2. Thus, receiver collisions are avoided. Note that advanced destinations 2 and 0 (dark shading) can be replaced by the corresponding stations of the next cycle.

11111111111111

The key mechanism that determines the actual rescheduling of transmissions in earlier slots (acceleration) is request acknowledging, which is generally outlined next. Each destination examines all requests and acknowledges only one corresponding to the earliest next pre-allocated time slot. If two or more requests imply the earliest next pre-allocated slot, one request is acknowledged at random (this is essentially a selection of the corresponding source node for transmission). Thus, in the example of Table 3.10, destination 4 chooses to acknowledge in time slot 1 a request by a node transmitting on channel 3, because this request corresponds to the earliest next pre-allocated slot (i.e. slot 2). It should be noted that to implement this mechanism, each node needs N memory spaces to store the ids of source nodes requesting transmission to it. Each destination node simply sends at most one acknowledgement per slot to a source node selected appropriately, in the way described before.

All in all, AP-WDMA indeed accelerates transmissions to a certain extent by avoiding what a purely pre-allocation based FT-TR scheme would do, i.e. waste pre-allocated slots. The performance of AP-WDMA is analyzed for various parameters and traffic types and is compared to the one of I-TDMA* in [43]. In the same article the reader may find a more detailed presentation of the protocol and a discussion on the effect of the policy used for allocating channels to destinations (three allocation schemes are considered) on the performance of the protocol and the degree of load balancing among available channels.

3.3.2.9 DAS and HTDM

In this section we consider two sample pre-transmission coordination-based schemes that incorporate an offline scheduling algorithm, that is to say an algorithm which computes a schedule when the entire traffic matrix is available (in contrast to online scheduling algorithms, according to the relevant discussion of the introduction to this subsection in 3.3.2).

Both schemes consider the same network model as the DT-WDMA protocol and could be viewed as an extension of it. Hence, a WDM broadcast-and-select optical LAN with N nodes, $W = N$ wavelengths used as data channels and one wavelength used as control channel for coordination are assumed. The node structure is also the same as in DT-WDMA, i.e. a fixed-tuned transceiver per station for transmission/reception of control information, a tunable receiver capable of tuning over all data channels for data packet reception and a fixed-tuned transmitter for data transmission over a home channel available to each station for exclusive use (since $W = N$). Consequently, the system is classified as CC-FT^2-FRTR. Synchronization is assumed over all channels (including the control channel) with one time slot being equal to the data packet transmission time (for simplicity taken to be fixed). On the control channel each time slot is further divided into N minislots, each pre-assigned to a distinct node for control packet transmission. Thus, access on the control channel is TDM-based and therefore collisionless. Round-trip propagation delay is generally assumed to be non-negligible and equal to R time units (anyway note that the protocols' performance is also examined for zero propagation delay in [6]). Another common assumption made by both schemes is that each source station maintains N queues, one for each possible destination including itself for simplicity.

The first protocol is the so-called *Dynamic Allocation Scheme* (*DAS*) [6], which dynamically allocates time slots to source-destination pairs according to the schedule computed by a relevant (scheduling) algorithm. The core of DAS[39] is the algorithm itself, namely *Random Scheduling Algorithm* (*RSA*), which is executed at the beginning of each time slot. This is done by all nodes in the system,[40] which therefore compute an identical schedule determining for each node, which packet (if any) should be selected (among the available ones in its N destination queues) for transmission in the upcoming slot. More specifically, during each step RSA chooses *at random* a source node and one of its non-empty queues corresponding to a destination that was not selected in any previous step (in order to eliminate receiver collisions). Random choices of source node and destination queue during each step are the same at all nodes, assuming all use a common random number generator and the same seed.[41] If all destination queues are empty or correspond to destinations that were previously selected by the algorithm, no packet is scheduled for transmission for this node; otherwise, the packet at the head of the queue is scheduled for transmission in the upcoming slot. The algorithm continues until a decision is made for all nodes.

Obviously, each station needs global queue status information in order to be able to execute the RSA algorithm. That is to say, in the beginning of each slot, every node should somehow know the status of all (N) destination queues maintained by every other node in the system (a total of N^2 queues including its own). This can be seen as a $N \times N$ traffic demand matrix $D = [d_{ij}]$, where d_{ij} is the number of packets that source node i has in its destination queue for node $j(i, j = 1, \ldots, N)$. The necessary information to construct and update the traffic demand matrix is transferred over the control channel. This control information essentially consists of N^2 bits per time slot, where N bits are transmitted by each node in its pre-assigned control minislot. If the ith bit transmitted by a certain node is one, this means that a new packet was generated in its queue for destination i, while a zero would imply no new packet arrival.[42] Every node processes all N control packets (N^2 bits) during each slot and updates (increments some values of) the traffic demand matrix appropriately. The correct values of the traffic matrix are decreased after scheduling of the corresponding transmissions.

The second protocol, namely *Hybrid TDM* (*HTDM*) [6], tries to reduce the processing requirements, which are obviously very high for DAS. It considers that time slots are pre-assigned to source-destination pairs, such as in I-TDMA for

Table 3.11 Sources/destinations allocation map of HTDM protocol for an example network with four nodes and data channels ($N = W = 4$). Parameter M is taken to be equal to two making each cycle six slots long.

Sources	Destinations									
1	1	2		3	4		1	2		...
2	4	1		2	3		4	1		...
3	3	4		1	2		3	4		...
4	2	3		4	1		2	3		...
Time slots:	1	2	3	4	5	6	1	2	3	...
Cycles:	← 1st cycle →						← 2nd cycle →			

$N = W$ (with the slight difference that here each node is assumed to transmit to itself also). However, HTDM considers a longer cycle length equal to $N + M$ slots, where M is an integer. It is assumed that after $\lfloor N/M \rfloor$ slots (pre-assigned to source-destination pairs) within a cycle, one slot is left open i.e. available for transmission to any destination. An example source/destination allocation map for HTDM is illustrated in Table 3.11 for $N = W = 4$ and $M = 2$. We observe that cycle length is six slots ($N + M$) and an open slot follows after two ($\lfloor N/M \rfloor$) pre-assigned slots; overall there are two open slots per cycle, namely slots 3 and 6. The RSA algorithm is invoked by each node in the beginning of each open slot (only) to determine an appropriate schedule for transmissions in this slot.

According to HTDM, control information is transmitted only at the end of each open slot by the nodes. This information regards the status of destination queues at the time, considering also the transmissions that are pre-assigned for the next R time units (time for the control information to reach all nodes in the system). This information may not be perfectly updated when RSA is executed, but it is observed that the errors caused are generally not significant. The good thing for HTDM is that control signalling overhead and processing requirements at the nodes are noticeably reduced in comparison to DAS. Also, for a certain packet size, the HTDM scheme generally allows more nodes in the system.

Even though HTDM performs worse than DAS, it is proposed as a more practical solution, since it has more reasonable processing requirements. Note that the performance of HTDM is still found to be higher than I-TDMA, especially for non-uniform traffic (when uniform traffic is considered, it is higher only under low loads). For more details on the schemes and their performance analysis the reader may refer to [6].

3.3.2.10 EATS, RO-EATS and MSL

The pre-transmission coordination-based protocols presented here assume the same network model and data structures maintained by each node. Based on these data structures, each protocol follows a different (scheduling) policy in order to determine how transmissions of variable-sized messages should be made.

Specifically, it is assumed that the protocols are applied to a WDM broadcast-and-select local area network with N nodes, W data channels ($\lambda_1, \ldots, \lambda_W$) and one control channel (λ_0) for coordination. Each node is equipped with a transceiver fixed-tuned on λ_0 and a tunable transceiver for data; hence the system is CC-FTTT-FRTR for the three schemes. Each data message is assumed to consist of one or more fixed-size data packets, where each data packet transmission time is equal to one data slot (time is slotted across all data channels). The control channel is divided into frames with length generally independent from the length of a data slot. Each control frame is further subdivided into N control minislots, which implies that the control channel is accessed in a TDM-fashion (ith minislot is pre-assigned to node i). Furthermore, each node is assumed to have one queue for storing all of its generated messages before transmission to the appropriate destinations. All schemes consider non-negligible round-trip propagation delay and tuning time for the tunable transceivers, taken to be R and T time slots respectively.

Before outlining the three schemes, let us describe the data structures (global status information) maintained and updated by each node of the network, which are common for the three protocols. The first is the *Receiver Available Time* (*RAT*) table, which is an array of N elements: $\text{RAT}[j] = t$ implies destination node j will be available (idle) after t time slots, where $j \in \{1, \ldots, N\}$. The second data structure is the *Channel Available Time* (*CAT*) table, an array of W elements: $\text{CAT}[c] = t$ implies that channel c will be available (idle) after t time slots, where $c \in \{1, \ldots, W\}$. Obviously, after each time slot the values of all elements in both tables are reduced by one.

The *Earliest Available Time Scheduling* (*EATS*) [22] protocol determines that each node examines the control packets one-by-one by means of its FR and schedules the corresponding transmissions to the earliest available channel (derived from the CAT table) just at the time the corresponding receivers will be available (this is derived from the RAT table). The same scheduling decisions are made by each node for every control packet received (including those transmitted by other nodes). Hence, every station in the system knows exactly when and on which channel to transmit its own messages and the protocol becomes free of any type of collisions.

More specifically, the scheduling decision for a transmission corresponding to a received control packet, i.e. the choice of a data channel c and the earliest time

Table 3.12 Example control packet transmissions for two control frames in a network with $N = 4$, $W = 3$ applying one of the three schemes (EATS, RO-EATS, MSL). Control packet (l,j) means the corresponding data message is l slots long and is destined for node j. Source nodes are implied by the control minislots (node i transmits in the ith minislot).

1st Control Frame	(8,3)	(2,1)	(4,3)	(7,2)
2nd Control Frame	(2,2)	(5,3)	(10,4)	(8,1)
Minislots:	1	2	3	4

slot $t_s(c)$ during which the source node can start transmitting on c, is done in the following way. Assume the control packet indicates message length equal to l slots and node j as destination. According to EATS, the data channel with the minimum value in the CAT table is selected to host the transmission; let this be channel c. The destination node will be able to receive at time $\text{RAT}[j] + T$ (tuning to channel c is necessary). Thus, so far as the receiver is concerned, a transmission could start at time slot $t_1 = \text{RAT}[j] + T - R$, since R slots is the round-trip propagation delay. However, $t_s(c)$ might have to be set equal to CAT[c], if the channel actually becomes available later than t_1. Hence, the transmission of the message is actually scheduled at time slot $t_s(c) = \max\ \{t_1, \text{CAT}[c]\}$. After the scheduling of a message, it remains for each node to update the two tables: $\text{CAT}[c] = t_s(c) + l$ and $\text{RAT}[j] = t_s(c) + l + R$.

Table 3.12 shows example control packet transmissions for two control frames in a network with four nodes ($N = 4$) and three wavelengths ($W = 3$) applying the EATS protocol.[43] Control packet (l, j) means that the corresponding data message is l slots long and is destined for node j. Notice that the source nodes are implied by the control minislots as explained before, e.g. source node 2 always transmits in the second minislot of each control frame.

The final schedule of the corresponding data messages decided by EATS is shown in Figure 3.12(a). Note that this is computed step-by-step (separate scheduling decision for each received control packet). Length and destination of each data message are shown above it in parentheses (in fact this is the corresponding control packet of Table 3.12 for each message). The numbers in grey font indicate wasted slots on each channel. Tuning time of transceivers is assumed to be one time slot ($T = 1$), while the round-trip propagation delay is taken to be two time slots ($R = 2$). Scheduling is computed as described before, considering that the initial value of all elements in tables CAT and RAT is zero. Note also that the initial wasted slots for tuning on each channel are not shown for simplicity.

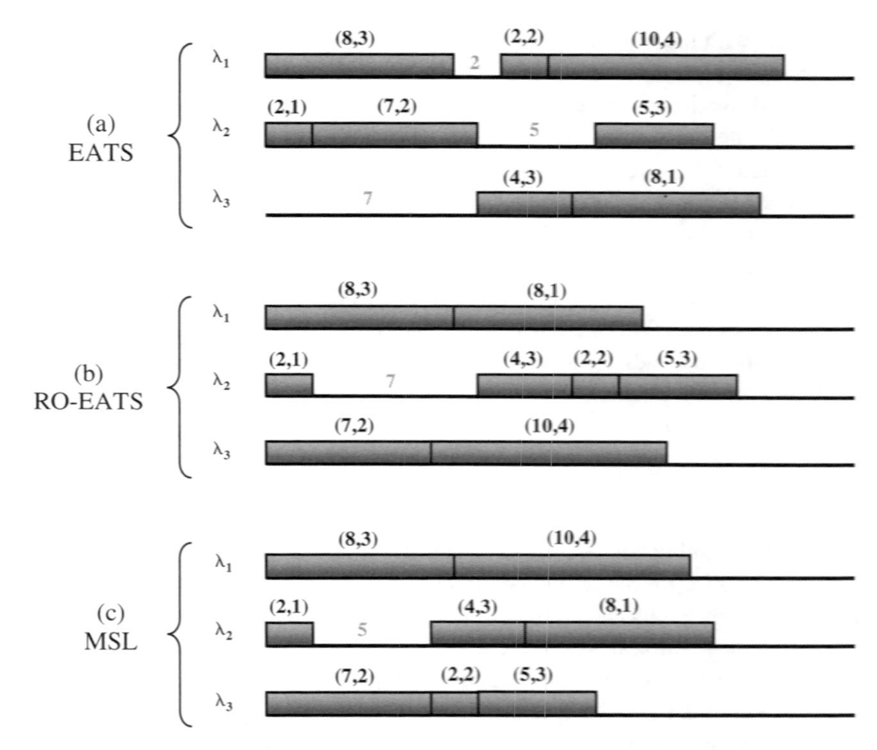

Figure 3.12 Scheduling of data messages corresponding to the control packets of Table 3.12 according to: (a) EATS, (b) RO-EATS, and (c) MSL. It is assumed that $T = 1$ slot and $R = 2$ slots. The numbers in parentheses above each message represent its length and destination, while numbers in grey font indicate wasted slots for a channel. Initial wasted slots because of tuning are not depicted.

The *Receiver Oriented—Earliest Available Time Scheduling* (*RO-EATS*) scheme [26] comes as an extension of EATS considering not only the channel that should be selected for transmission, but also the order in which the corresponding messages declared in a control frame should be scheduled. It is assumed that the RO-EATS scheduling algorithm is executed by all nodes, when they receive the entire control frame (not a single control packet like EATS).

The main idea of the protocol is to schedule first messages that are destined for the least used destination nodes (destinations with lowest value in RAT table). More specifically, the RAT table is sorted in ascending order (lower values first). It is examined whether there is any control packet in the control frame indicating the first destination in the sorted RAT table (the one which will become available sooner, i.e. with the lowest value). If so, the corresponding message is scheduled first according to EATS (RAT and CAT tables are updated accordingly) or else the next destination in the sorted RAT is examined and so on until a destination that is indicated in some control packet is found; the corresponding message for this destination is scheduled first according to EATS (RAT and CAT tables are updated accordingly). This process continues until all messages referred in the current control frame are scheduled.

As an example, the scheduling of data messages corresponding to the same control packets of Table 3.12 according to RO-EATS is shown in Figure 3.12(b). As we said, RO-EATS is executed when the entire control frame is received and probably schedules messages in an order different from the arrival order of the corresponding control packets in the frame. In our example, messages are scheduled in the following order:

- (8,3), (2,1), (7,2) and (4,3) for the first control frame, and
- (10,4), (8,1), (2,2) and (5,3) for the second control frame.

Notice that, unlike EATS, the RO-EATS protocol tries to avoid scheduling messages destined for the same node consecutively, because each message in such a row (allowed by EATS) has to wait for the previous one to stop keeping the corresponding destination busy, and, consequently, slots are wasted on the selected channel. For that reason, RO-EATS achieves higher throughput and lower average message delay than EATS (which in turn outperforms many other schemes, such as TDMA-W presented before). Note that this is also the result of the performance analysis in [26] and it is also obvious from the example of Figure 3.12, where only seven slots are wasted (as compared to 14 for EATS) and the schedule is shorter. However, complexity of RO-EATS is a bit higher, as it requires sorting the RAT table after each message is scheduled. It should be noted that the authors in [26] extend the protocol by allowing each source node to make reservations for l messages in its queue by means of a single control packet; l obviously determines the control packet length and is a system design parameter (in our example it was taken to be one).

Last of all, we shall consider the *Minimum Scheduling Latency* (*MSL*) scheme [12]. Like EATS, the corresponding MSL algorithm is executed by all nodes after the reception of each control packet. MSL tries to cope with the previously described drawback of EATS not by appropriately modifying the order in which messages are scheduled (as RO-EATS does), but by modifying the actual choice of data channel. Remember that EATS first selects a channel c (the one with the lowest value in CAT table) and then computes the earliest time slot during which the source node can start transmitting on this channel $t_s(c)$. MSL determines that

$t_s(c)$ should be computed for all channels first ($c = 1, \ldots, W$), in the way described for EATS before. The channel with the minimum earliest time the source can start transmitting its message (t_s) is selected. If more than one channel has the minimum t_s, the one with the maximum value in the CAT table is selected. Let the selected channel be k. Transmission of the message is scheduled on this channel at time slot $t_s(k)$ and after that the values of RAT and CAT tables are updated. This protocol essentially selects the channel with the minimum *scheduling latency*, i.e. the one for which the shorter time interval will pass from the time it becomes available till the time $t_s(c)$ it can actually start transferring the message.

The scheme has higher processing requirements but improves the performance of EATS quite considerably. Its operation and the performance improvement become clearer if we consider the scheduling of messages corresponding to the same control packets of Table 3.12. The messages are scheduled in the order of control packet arrival like EATS, but the schedule is much better (Figure 3.12(c)). The performance of the scheme is analyzed and compared to EATS in [12], but not RO-EATS. Hence, even though MSL performs better than RO-EATS in our example, this may not be considered a formal conclusion for all possible traffic loads and network parameters. More details on the three schemes may be found in [22], [26] and [12] respectively.

Next we outline two protocols proposed in [12] as well, which are intended to handle real-time traffic and are respectively based on EATS and MSL for channel selection.

3.3.2.11 PDS-EAC and PDS-MSL

The two schemes presented here consider networks that provide service to both nonreal-time and real-time traffic. Their main objective is to handle real-time traffic in a way that helps in meeting the related quality of service (QoS) requirements.[44] The proposed protocols try to cope with the head-of-line blocking of new (unscheduled) real-time packets caused by already scheduled nonreal-time packets; e.g. this blocking occurs when a real-time packet finds the desired destination is already scheduled to receive a nonreal-time packet. This would clearly block the real-time packet until completion of the scheduled packet transmission, which is generally undesirable.

The proposed solution in favour of real-time traffic is to allow real-time packets to preempt nonreal-time ones and be transferred earlier. After a packet is preempted, the real-time packet is scheduled appropriately and the preempted packet is rescheduled for transmission at a later time. The distinction between packets is implemented by assigning priorities: high to real-time packets and low to others.

Priority is an additional element included in each control packet and, along with the values in tables RAT and CAT (described in the previous schemes and used here as well), comprises the basis of scheduling decisions.

The proposed protocols are called *Priority Differentiated Scheduling—Earliest Available Channel* (*PDS-EAC*) and *Priority Differentiated Scheduling—Minimum Scheduling Latency* (*PDS-MSL*) [12]. They are respectively based on the EATS[45] and MSL schemes for channel assignment, but additionally incorporate priority differentiation since they are intended to handle real-time traffic. The network model, node structure (FTTT-FRTR), message length (variable), time-slotting of channels, control channel access protocol (TDM) and data structures (tables CAT and RAT) described for the previous three protocols are assumed to be identical by these schemes. Therefore we only describe an additional data structure, namely *Scheduled Packet Job List* (*SPJL*), which is assumed to be maintained and updated by each node in the context of PDS-EAC and PDS-MSL. The SPJL list contains the details of each scheduled transmission, that is to say source and destination ids, message length l, selected data channel c, transmission starting time $t_s(c)$, transmission ending time and priority.

The operation of PDS-EAC and PDS-MSL can be outlined as follows. Each node executes the PDS-EAC (PDS-MSL) algorithm upon reception of every single control packet. If it is a low priority packet, it is scheduled exactly in the same way as EATS (MSL) protocol determines and, moreover, one new entry is added in the SPJL list. Otherwise, the following procedure is followed for a high priority packet. First, all values in tables CAT and RAT are reset (i.e. set to zero) since packets have to be rescheduled and, additionally, the SPJL list is sorted in ascending order of transmission starting time $t_s(c)$. The values of tables RAT and CAT are calculated again according to entries of the sorted SPJL only for packets that should not be preempted by the high-priority packet; these include packets with high priority, or low-priority packets for which at least tuning of the TT has already started (i.e. packets with $t_s(c) < T$). Subsequently, the high-priority packet is scheduled according to EATS (MSL) taking into consideration the new values of tables RAT and CAT, and a new entry in the SPJL list is added for it. It remains to reschedule all the packets whose entry in the SPJL had transmission starting time $t_s(c) \geq T$, i.e. the preempted low-priority packets. This is done according to EATS (MSL) and an entry in SPJL is added after each packet is scheduled. It should be noted that after a low-priority packet is preempted for a predetermined number of times, its priority is set high by the protocols, so that additional preemptions are avoided. Obviously, this improves the fairness of the schemes.

The performance of PDS-EAC and PDS-MSL is analyzed and compared to the one of EATS and MSL in [12]. It is found that the PDS protocols handle real-time packets better than the two other schemes (which were not designed to do so in the first place). Observe from the relevant descriptions of the protocols, however, that this comes at the cost of higher processing requirements as a trade-off. If we consider traffic as a whole, i.e. without distinction between high and low priority packets, the highest performance in terms of throughput and mean packet delay is achieved by PDS-MSL followed by MSL. For a more detailed description of PDS-EAC and PDS-MSL along with a performance analysis and simulation under various parameters, the reader may consult [12].

3.3.2.12 FATMAC, HRP/TSA, OIS and POSA

The protocols outlined in this section are examples of schemes incorporating *on-line scheduling* algorithms. As discussed in the introduction to this subsection, on-line scheduling schemes start building the schedule before the entire traffic matrix is available, as opposed to offline scheduling ones that require complete global status information before producing a schedule.

The presented schemes assume a WDM broadcast-and-select local area network with N nodes and W wavelengths. Even though there is *no separate control channel* used for coordination, the protocols are still pre-transmission coordination based.[46] Each node is assumed to be equipped with an array of W fixed-tuned transmitters (or equivalently a tunable transmitter) and a fixed-tuned receiver, used for both control and data packets. Channels are pre-allocated to receivers according to some policy, e.g. interleaved channel allocation as in I-TDMA, and time is assumed slotted across all channels on control packet boundaries. Furthermore, transmission is organized into cycles, where each cycle consists of a *control* (or *reservation*) *phase* and a *data transfer phase*. During the control phase of each cycle taken to be N slots long, each source node is assigned a unique slot for *broadcasting* its control packet to all channels (simultaneously) by means of its transmitter array (hence access is TDM-based). Control packets are received by all nodes on their corresponding home channel by means of their FR and are assumed to make reservations for the data phase. The actual way of scheduling the corresponding data packets is the main difference between the schemes.[47]

The *FATMAC* protocol [38] assumes that each source node has only one queue and is able to reserve access during the control phase only on one channel, i.e. only for one data packet transmission during the subsequent data phase. The arbitration mechanism of FATMAC determines that the nodes which reserved access on the same channel should transmit in the order of the transmitted control packets. Thus, data channels provide First-Come-First-Serve (FCFS) service to reserved data packet transmissions. Observe that a predetermined assignment of control phase slots to nodes would be a bit unfair to nodes that transmitted in the last slots of the control phase. Therefore the assignment of control phase slots to nodes should preferably vary with time. Since all nodes employ the same arbitration mechanism and examine all control packets, each node can easily estimate by the end of the preceding control phase exactly when it should start transmitting within the data phase. The length of the data phase depends on the number of control packets within the previous control phase and the actual destinations (home channels) indicated in the control packets. For example, if N control packets designating the same home channel are transmitted in the control phase, the subsequent data phase will have maximum length.

The *Hybrid Random Access/Time Slot Assignment* (*HRP/TSA*) scheme [40] extends FATMAC by allowing a node to place reservations for access to multiple channels in the same control phase. For this purpose, each node maintains W queues for data packets and one queue for control packets. Similar to FATMAC, by the end of the control phase, all nodes compute in a distributed manner an identical transmission schedule for the data phase, which is free of collisions and entails minimum number of transmitter tuning operations. The latter is implied by the fact that HRP/TSA produces non-preemptive schedules[48] by means of an online scheduling algorithm, which is slightly more sophisticated than the plain FCFS reservation service policy of FATMAC. The basic input to this algorithm is the $N \times W$ (collapsed) traffic matrix[49] $A = [a_{ic}]$, where a_{ic} is the number of slots demanded from source node i on channel c ($1 \leq i \leq N$ and $1 \leq c \leq W$). The algorithm is based on decomposition of traffic matrix A into n transmission matrices S_k, $A = \sum_{k=1}^{n} S_k$, where each one has at most one non-zero entry in each row and each column. Note that rows and columns of transmission matrices correspond to source nodes and channels, respectively, as in the original traffic matrix. Each transmission matrix represents a separate schedule, which is a part of the overall schedule for A. The length of each partial schedule is equal to the largest entry of the corresponding transmission matrix. Hence, the overall schedule is the partial schedule of transmission matrix S_1 followed by the one of S_2 and so on up to S_n.

Since we are considering online scheduling, notice that execution of the algorithm begins when the first reservation arrives (entire traffic matrix not available yet). Thus, the algorithm begins by processing the reservation requests of node 1 sequentially, $a_{11}, \ldots, a_{1W}$ (first row of traffic matrix A), continues with requests of node 2, $a_{21}, \ldots, a_{2W}$ (second row), and so on, up to the last node's requests which arrive last in the TDM access-based control phase.

The key idea of the algorithm is the following. For each request a_{ic} it searches (using one of two methods outlined later) for an already created transmission matrix S_k, which can include a_{ic}, i.e. equivalently, has zero entries for all elements in row i and column c. If one such transmission matrix S_k is found, element a_{ic} is added to it and node i is considered to have reserved a_{ic} slots on channel c during the partial schedule (corresponding to the transmission matrix). Otherwise, a new transmission matrix is created and a_{ic} is added to it. The above mentioned 'searching' for an appropriate already created S_k is performed consistent with one of the following two methods [40]:

- *First fit*. Sequential search (starting from S_1) until the first applicable transmission matrix is found.
- *Best fit*. Search for the 'best' transmission matrix for request a_{ic}, where 'best' is considered to be the transmission matrix with corresponding (partial) schedule of length closest to the value a_{ic}. This method implies less wasted slots for the partial schedule than first fit, at the cost of some additional computational complexity.

As an example, consider the decomposition of the following 3×3 traffic matrix (corresponding to a network with three nodes and three wavelengths) according to the HRP/TSA protocol, when the scheduling algorithm applies the first fit method:

$$A = \begin{vmatrix} 2 & 4 & 3 \\ 2 & 3 & 0 \\ 5 & 1 & 0 \end{vmatrix} = \underbrace{\begin{vmatrix} 2 & 0 & 0 \\ 0 & 3 & 0 \\ 0 & 0 & 0 \end{vmatrix}}_{S_1} + \underbrace{\begin{vmatrix} 0 & 4 & 0 \\ 2 & 0 & 0 \\ 0 & 0 & 0 \end{vmatrix}}_{S_2} + \underbrace{\begin{vmatrix} 0 & 0 & 3 \\ 0 & 0 & 0 \\ 5 & 0 & 0 \end{vmatrix}}_{S_3} + \underbrace{\begin{vmatrix} 0 & 0 & 0 \\ 0 & 0 & 0 \\ 0 & 1 & 0 \end{vmatrix}}_{S_4}.$$

When the scheduling algorithm applies the best fit method, the same traffic matrix is decomposed into transmission matrices in the following way:

$$A = \begin{vmatrix} 2 & 4 & 3 \\ 2 & 3 & 0 \\ 5 & 1 & 0 \end{vmatrix} = \underbrace{\begin{vmatrix} 2 & 0 & 0 \\ 0 & 3 & 0 \\ 0 & 0 & 0 \end{vmatrix}}_{S_1} + \underbrace{\begin{vmatrix} 0 & 4 & 0 \\ 0 & 0 & 0 \\ 5 & 0 & 0 \end{vmatrix}}_{S_2} + \underbrace{\begin{vmatrix} 0 & 0 & 3 \\ 2 & 0 & 0 \\ 0 & 1 & 0 \end{vmatrix}}_{S_3}.$$

The corresponding schedules produced by HRP/TSA with the first fit and the best fit methods are shown in parts (a) and (b) of Figure 3.13 respectively. Observe that they consist of the partial schedules corresponding to the transmission matrices generated by decomposition of the original traffic matrix A. For example, the partial schedules for best fit are three (since three are the generated transmission matrices) and have lengths equal to three, five and three slots respectively (Figure 3.13(b)). Note that in Figure 3.13 the two numbers (s, l) in parentheses above each packet correspondingly denote the *source node* and *message length* in time slots.[50] Observe that best fit generally leads to shorter schedules and less

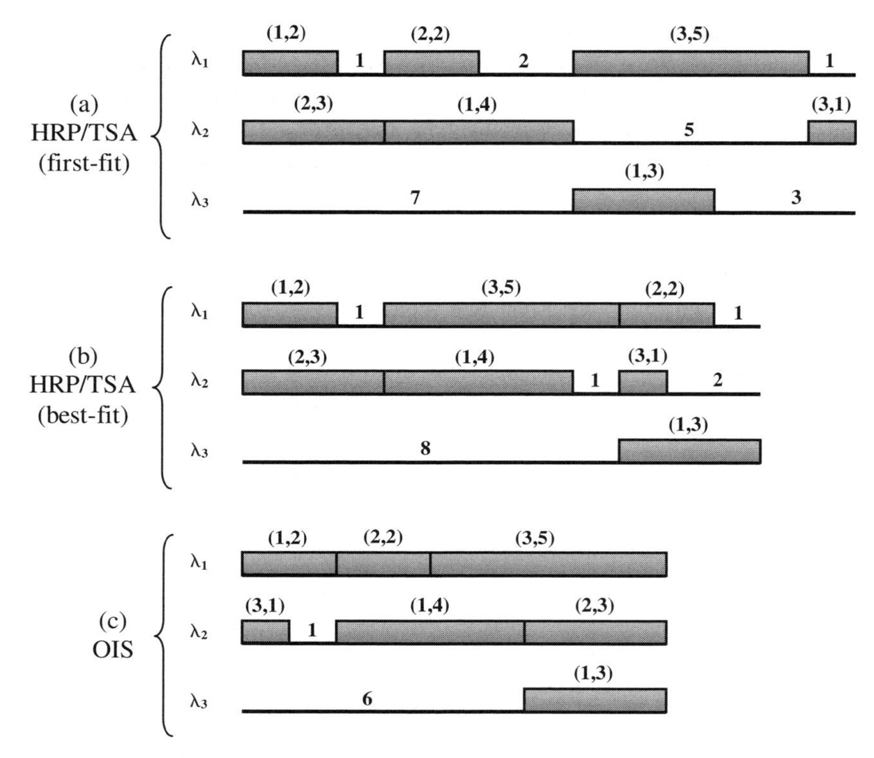

Figure 3.13 Example of scheduling according to the three presented protocols: (a) HRP/TSA with first fit; (b) HRP/TSA with best fit; (c) OIS. The two numbers in parentheses (s, l) above each message correspond to the source node and the length in time slots respectively. Numbers in-between several messages (rectangles) represent wasted time slots.

wasted slots (i.e. lower delay and higher throughput). It should be noted that for different numbers, the produced scheduling would most probably be somewhat improved, but the example was chosen such that the performance difference between HRP/TSA with first fit and HRP/TSA with best fit (and also from OIS presented next) becomes clear for a small number of nodes and channels.

The HRP/TSA protocol's performance is analyzed for uniform and non-uniform traffic and compared to that of FATMAC. HRP/TSA generally performs better but not in all cases; for the corresponding analytical and simulation results, please refer to [40]. As far as computational complexity is concerned, it should be recalled that the principal objective of these online scheduling protocols is a good performance (as high as possible channel throughput and as low as possible packet delay) using practical scheduling algorithms that run in linear[51] time in terms of the number of nodes N. For a description of the data structures that make time complexity of both first fit and best fit algorithms of HRP/TSA drop from quadratic to linear time in terms of N, please consult [40].

The third protocol examined in this part is called *Online Interval-based Scheduling (OIS)* [41] and is based on the same assumptions as the previous protocols (network model, node structure, etc.). Moreover, like HRP/TSA, OIS assumes each can place reservations for access to multiple channels in the same control phase; hence, a collapsed traffic matrix A is input to the scheduling algorithm (like for HRP/TSA). As the name suggests, this scheme incorporates online scheduling on the basis of available time intervals on channels and for each examined node that requests reservation. All requests of each node (row of A) are examined sequentially before proceeding to the next node (row), as in the previous scheme. The OIS protocol innovates in abandoning the practice of traffic matrix decomposing (into transmission matrices) adopted by HRP/TSA and many other scheduling schemes; this approach was originally proposed for *Satellite-Switched/TDMA* systems and is not necessary in WDM broadcast-and-select star networks. We focus on the main idea of OIS, whose operation will become clear if we consider the same example traffic matrix used for HRP/TSA before. Details of the actual data structures necessary for the protocol's implementation by each node may be found in [41].

As before, each node runs the same scheduling algorithm and needs to maintain the same status information (distributed scheduling). Hence, all nodes eventually arrive at identical scheduling decisions and the protocol is free of collisions. Execution of the online scheduling algorithm begins upon reception of the first request (entire traffic matrix not available).

In order to be able to execute the scheduling algorithm of OIS, each node needs to maintain a list of time intervals that are available on every data channel. For example, the list of available intervals for a certain channel c at some specific time might contain intervals $[3, 5]$, $[8, 12]$ and $[15, \infty)$; this implies that intervals $[1, 3]$ $[5, 8]$ and $[12, 15]$ have been assigned to source nodes and no slot included in any of these should be assigned to any other transmission during the oncoming reservation phase. Furthermore, for each node whose request is being processed, nodes maintain one additional list of intervals that have not yet been assigned to the specific node for transmission. For example, let the list of intervals for node i contain $[9, 13]$ and $[15, \infty)$, which means that node i has already been scheduled to transmit at intervals $[1, 9]$ and $[13, 15]$. If we assume request $a_{ic} = 4$ is processed, i.e. node i requests four slots on channel c, and considering the list of intervals given for channel c before, notice that node i cannot be scheduled to transmit during

[8, 12] even if it is available on c, because this overlaps with interval [1, 9] already assigned to i for transmission. Hence, the transmission is eventually scheduled during interval [15, 19] which is free for both channel c and node i. After that, the lists are updated appropriately.

Note that at the beginning of a control phase and before any request is processed, all lists are initialized to contain interval $[1, \infty)$ only. As mentioned before, all requests of each node (row of traffic matrix A) are examined sequentially before proceeding to the next node (row).

The operation of OIS becomes clearer if we consider for example the 3×3 traffic demand matrix A that was also used for the scheduling example of HRP/TSA before (implying a network with three nodes and three wavelengths):

$$A = \begin{vmatrix} 2 & 4 & 3 \\ 2 & 3 & 0 \\ 5 & 1 & 0 \end{vmatrix}$$

Scheduling of the reservation requests (elements of A) according to OIS is illustrated in Figure 3.13 (c). After the first request of the first row (node 1) is processed, both the list of available intervals for channel λ_1 and for node 1 become $[3, \infty)$. The second request is for four slots on channel λ_2. Since the node will be busy for the first two slots, the corresponding transmission is scheduled for interval [3, 7] making the list of intervals for channel λ_2 contain [1, 3] and $[7, \infty)$. The list for the node 1 now contains $[7, \infty)$. After the third request, the list for channel λ_3 contains [1, 7] and $[10, \infty)$ and we are finished with node 1 for this cycle. Subsequently, the second row of A is processed (requests of node 2). After the first request for 2 slots on λ_1, the list of intervals for λ_1 is $[5, \infty)$, while for node 2 the list contains [1, 3] and $[5, \infty)$. The second and final request of node 2 is scheduled for interval [7, 10] which makes the list of intervals for λ_2 contain [1, 3] and $[10, \infty)$. Since no slots are requested on λ_3, node 2 is satisfied and we proceed to the next row corresponding to requests of node 3. The first request for five slots on channel λ_1 is scheduled at [5, 10]; notice that we are finished with this channel and its list need not be updated any more. The list of available intervals for node 3 contains [1, 5] and $[10, \infty)$. The second and final request of node 3 on channel λ_2 is scheduled at [1, 2] which was available on this channel. Thus, the protocol arrives at the final schedule depicted in Figure 3.13 (c).

Observe that the schedule computed by OIS is shorter in length than the two variants of HRP/TSA (9 slots instead of 13 slots for HRP/TSA with first fit and 11 slots for HRP/TSA with best fit) and results in a reduced number of wasted slots (7 wasted slots instead of 18 for HRP/TSA with first fit and 13 for HRP/TSA with best fit).[52] Performance of the scheme is generally higher than protocols which apply decomposition of the traffic matrix into transmission matrices. Performance analysis of OIS and comparison with such schemes may be found in [41]. It should be noted that the scheme runs in linear time in terms of the number of nodes (N), which makes it computationally feasible. For the proof of this along with more details on OIS and a slight modification when the tuning latency is assumed to be non-negligible, the reader may consult [41].

Finally, we consider a scheme called *Predictive Online Scheduling Algorithm* (*POSA*) [23], which is essentially an extension to OIS based on *traffic prediction* according to the history of recent reservation requests. The main objective of this scheme is to eliminate the possible delay introduced by the scheduling computation between the control and data phases of each cycle. For this purpose, POSA assumes the reservation information of each control phase is input to a *predictor* whose operation is mainly based upon the *Hidden Markov chain Model* (*HMM*).[53] After an initial *learning period* of several cycles for the duration of which the protocol operates just like OIS (with the enhancement that the predictor is informed about reservation requests as we said), the network uses one cycle to transit to the *prediction period*. During this period, the schedule computed for the data phase of a certain cycle, instead of taking into account the reservation requests of the same cycle's control phase (as in OIS), is based on the *predicted reservations* (output of the predictor) provided to the scheduling algorithm in the previous cycle. The current control phase is used to train the predictor and affects scheduling decisions for subsequent cycles. As a result of applying POSA, during the prediction period the scheduling algorithm has more time to compute every schedule. The schedule is generally available by the start of the data phase consistent with the algorithm's objective and POSA brings about some performance improvement (presupposing adequate accuracy of predictions which is claimed in [23] based on empirical results), if the average duration of the control and data phases is at least equal to the time needed for predicting reservations and computing the corresponding schedule.

For the related performance analysis, simulation and experimental results please refer to [23]. In the same study, the reader may find an analysis of POSA's time complexity and a complete description of the predictor along with the necessary discussion on implementation issues.

NOTES

1 A more appropriate term to use instead of receiver collision here is 'contention' of the arriving messages.

2 Access delay is defined as the delay between the time a packet arrives for transmission at a node and the time it is actually transmitted.

3 Time slots needed for transmissions from i to j may be considered instead, which is equivalent (since actually one data packet lasts one time slot). The matrix may also be referred to as a *backlog* matrix.
4 This *inband polling protocol* has been used in *Rainbow I* and *Rainbow II* optical WDM network testbeds. Some additional information concerning these systems will be given later in Section 3.1.
5 The same allocation map applies also for I-TDMA* with the same assumption of zero tuning latency. I-TDMA* is an extension of I-TDMA which will be discussed later.
6 Obviously, T was equal to one in the cases examined so far, since packet transmission (of duration equal to one slot) was the only time-consuming process.
7 We assume that $W < N$ which is the most usual and general case.
8 We imply communication in a single hop of course.
9 This is obvious from the algorithm and particularly steps 2 and 3.
10 If it is ready to transmit to more that one node, i.e. its queue has more than one packet buffered, it chooses to transmit the packet at the head of the queue.
11 Refer also to the short introduction to Section 3.2.
12 The notations used in [16] are slightly different, but we will keep on using the same notations for nodes and wavelengths as in previous schemes.
13 That is to say, points $(t_0 - 1)$ and $(t_0 + 1)$ are not included.
14 We assume that cycle i is the current cycle.
15 As in the basic slotted Aloha/slotted Aloha scheme, we assume that a data packet transmission (and slot) is L control slots long.
16 This period is obviously a control slot for slotted Aloha/CSMA.
17 One data packet is assumed L times a control packet as in previous schemes.
18 In [16] it is actually called CSMA/N-Server switch, since the number of data channels is denoted there by N. We renamed the data channel scheme as W-Server switch to keep uniformity with our symbolism of the number of data channels as W.
19 For example, it is appropriate for file transfers and generally long 'messages' that can be broken up into several data packets.
20 This presupposes negligible (zero) tuning times for the tunable receiver of each node. The protocol can also support tuning times that are a part of the cycle duration [5].
21 Unacknowledged transmissions, possibly due to a receiver collision, have to be retransmitted. The way of acknowledging transmissions is discussed in [18].
22 This scheduler is also used in [30–31].

23 This is the case for a network operating at 10 Gb/s per WDM channel [29].
24 Note that the impact of the HOL blocking on network performance could be alleviated as well by applying a more costly (hardware) solution, namely equipping each node with additional receivers.
25 Several adaptive protocols assuming the hub performs active operations are examined in Chapter 4. The random scheduling schemes for multicast traffic outlined before also assume the hub is not passive but includes a master/slave scheduler [29].
26 Note that network processing delays, like the delays introduced by the contention processing schemes implemented at the hub, are included; the actual contention processing schemes applied in the framework of this protocol are described later.
27 Waiting for a successful control packet before proceeding to transmit the data packet resembles the original slotted Aloha/delayed Aloha scheme.
28 The original name is *TDMA-C*, where C is the number of data channels in the system. However, here we use W to denote this, in order to keep uniformity with the notations used in previous schemes.
29 This is called *receiver collision on control packet* in [21], contrary to the usual *receiver collision on data packet*, where data packets are only involved.
30 The SCM multiplexing technique was briefly described in Section 1.2.
31 In an example network analyzed and simulated in [30], it is observed that a look-ahead window of just four packets increases throughput performance to over 80% as compared to the FCFS algorithm.
32 WDM Passive Optical Networks were briefly introduced in Subsection 1.5.4.
33 Unfortunately it was not given a name in [20].
34 If p is the transmission probability, the node postpones control packet transmission with probability $1 - p$; in the latter case it reattempts to transmit in the next slot of the chosen control subchannel. The transmission probability updating scheme is not discussed here. Anyway its main idea is that transmission probabilities have to be decreased when an increased number of control channel collisions are observed and vice versa.
35 This is called a *source conflict* [20], because the source node can only tune and transmit one of the scheduled messages. In fact, it is the complementary to a receiver collision, where a destination node can only tune and receive at most one of the data packets destined to it. However, the difference is that in a receiver collision there is wasteful transmission of data packets, while in a source conflict a delay is only introduced for the packets that could not be transmitted.

36 Remember that the N-DT-WDMA protocol supported traffic with bandwidth guarantees, as well.
37 The general approaches to the reservation bits actual management and setting by the control node are discussed later.
38 The discussion on optical self-routing and partial self-routing (for N = W and N > W respectively) of the chapter's introduction and Subsection 3.1.1 apply in this case as well.
39 This also applies for the next protocol (HTDM), as we will see.
40 Hence we have distributed scheduling.
41 Notice that a common random generator with the same seed is also used by the R-TDMA protocol (presented in Subsection 3.1.3) and also by LABP (presented in 4.1.2), CPF (4.4.2) and CWC (4.4.3); moreover, in reference [15].
42 One bit is enough to indicate an arrival for a destination queue, if we assume that at most one new packet per time slot may be generated by each source node for a certain destination.
43 The same values will be used as an example for RO-EATS and MSL too.
44 We have already considered a couple of schemes that ensure higher priority is given to certain traffic types than others, such as N-DT-WDMA in 3.3.1.13 [18] and the protocol in 3.3.2.7 [11].
45 The first scheme could be called PDS-EATS, but the original name in [12] is more preferable in order to avoid using the word 'scheduling' twice.
46 The schemes are applicable to environments where the number of wavelengths is relatively small and it is crucial to use all wavelengths for data transmission. When control information is transferred via the same channels used for data, we say that we have *in-band* control signalling (as opposed to *out-of-band* in case separate control channels are used).
47 Note that there are other differences too, for example as far as the assumed number of queues per node is concerned.
48 Preemptive and non-preemptive scheduling was discussed in the introduction to this Subsection in 3.3.2.
49 The simple data structures needed to be maintained by each node to represent the collapsed traffic matrix are described in [40]; however, we focus only on the scheduling algorithm itself.
50 This is different from Figure 3.12 where numbers in parentheses above each message represent its length and destination respectively. That is because the protocols presented in this part have different node structure with a FR for data, which means that the destination is implied by the transmission channel. Hence, the important information is the source node, the length of the message and the data channel.

51 For a survey on several earlier scheduling schemes which produce better schedules but at the cost of much higher computational complexity, please refer to [40]. Such solutions are obviously impractical for the networks considered in this book, at least with current technology.
52 Note that the improvement may not always be that dramatic as in this numerical example; anyway, performance is quite considerably improved in most cases.
53 For information and references on this model, which are obviously beyond the scope of this book, please consult [23].

REFERENCES

[1] Abramson N, "The aloha system—another alternative for computer communications", in *Proceedings of Fall Joint Computer Conference (FJCC), AFIPS Press*, Montale NJ, vol. 37, pp. 281–285, 1970.
[2] Banerjee S and Mukherjee B, "Fairnet: A WDM-based multiple channel lightwave network with adaptive and fair scheduling policy", *IEEE Journal of Lightwave Technology*, vol. 11, pp. 1104–1112, May/June 1993.
[3] Bogineni K and Dowd PW, "A collisionless multiple access protocol for a wavelength division multiplexed star-coupled configuration: Architecture and performance analysis", *IEEE Journal of Lightwave Technology*, vol. 10, no. 11, pp. 1688–1699, November 1992.
[4] Bogineni K, Sivalingam KM and Dowd PW, "Low complexity multiple access protocols for wavelength division multiplexed photonic networks", *IEEE Journal on Selected Areas in Communications*, vol. 11, pp. 590–604, May 1993.
[5] Chen MS, Dono NR and Ramaswami R, "A media-access protocol for packet-switched wavelength-division metropolitan area networks", *IEEE Journal on Selected Areas in Communications*, vol. 8, no. 6, pp. 1048–1057, August 1990.
[6] Chipalkati R, Zhang Z and Acampora AS, "High speed communication protocols for optical star coupler using WDM", in *Proceedings IEEE INFOCOM'92*, Florence, Italy, vol. 3, pp. 2124–2133, May 1992.
[7] Chlamtac I and Fumagalli A, "Performance of reservation based (Quadro) WDM star networks", in *Proceedings IEEE INFOCOM'92*, Florence, Italy, vol. 3, May 1992.
[8] Chlamtac I and Fumagalli A, "QUADRO-Star: High performance optical WDM star networks", in *Proceedings IEEE Globecom'91*, Phoenix, Arizona, USA, December 1991.

[9] Chlamtac I and Ganz A, "A multibus train communication architecture (AMTRAC) for high-speed fibre optic networks", *IEEE Journal on Selected Areas in Communications*, vol. 6, pp. 903–912, July 1998.

[10] Chlamtac I and Ganz A, "Channel allocation protocols in frequency-time controlled high speed networks", *IEEE Transactions on Communications*, vol. 36, pp. 430–440, April 1988.

[11] Choi JS, Golmie N and Su D, "A bandwidth guaranteed multi-access protocol for WDM local networks", in *Proceedings ICC'00*, vol. 3, pp. 1270–1276, 2000.

[12] Diao J and Chu PL, "Packet rescheduling in WDM star networks with real-time service differentiation", *IEEE Journal of Lightwave Technology*, vol. 19, no. 12, pp. 1818–1828, December 2001.

[13] Dowd PW, "Random access protocols for high speed interprocessor communication based on an optical passive star topology", *IEEE Journal of Lightwave Technology*, vol. 9, pp. 799–808, June 1991.

[14] Ganz A, "End-to-end protocols for WDM star networks", in *IFIP/WG6.1-WG6.4 Workshop on Protocols for High-Speed Networks*, Zurich, Switzerland, May 1989.

[15] Ganz A and Koren Z, "WDM passive star—protocols and performance analysis", in *Proceedings IEEE INFOCOM'91*, Bal Harbour, Florida, USA, vol. 3, pp. 991–1000, April 1991.

[16] Habbab IMI, Kavehrad M and CW Sundberg, "Protocols for very high-speed optical fibre local area networks using a passive star topology", *IEEE Journal of Lightwave Technology*, vol. LT-5, no. 12, December 1987.

[17] Hall E, Kravitz J, Ramaswami R, Halvorson M, Tenbrink S and Thomsen R, "The Rainbow-II gigabit optical network", *IEEE Journal on Selected Areas in Communications/Journal of Lightwave Technology*, special issue on Optical Networks, vol. 14, no. 6, pp. 814–823, June 1996.

[18] Humblet PA, Ramaswami R and Sivarajan KN, "An efficient communication protocol for high-speed packet switched multichannel networks", *IEEE Journal on Selected Areas in Communications*, vol. 11, no. 4, pp. 568–578, May 1993.

[19] Janniello FJ, Ramaswami R and Steinberg DG, "A prototype circuit-switched multi-wavelength optical metropolitan-area network", *IEEE/OSA Journal of Lightwave Technology*, vol. 11, pp. 777–782, May/June 1993.

[20] Jeon WS and Jeong DG, "Contention-based reservation protocol for WDM local lightwave networks with nonuniform traffic pattern", *IEICE Transactions on Communications*, vol. E82-B, no. 3, March 1999.

[21] Jia F and Mukherjee B, "The receiver collision avoidance (RCA) protocol for a single-hop WDM lightwave network", *IEEE Journal of Lightwave Technology*, vol. 11, no. 5/6, pp. 1053–1065, May/June 1993.

[22] Jia F, Mukherjee B and Iness J, "Scheduling variable-length messages in a single-hop multichannel local lightwave network", *IEEE/ACM Transactions on Networking*, vol. 3, no. 4, pp. 477–488, August 1995.

[23] Johnson E, Mishra M and Sivalingam KM, "Scheduling in optical WDM networks using hidden Markov chain based traffic prediction", *Kluwer Journal of Photonic Network Communications*, vol. 3, no. 3, pp. 271–286, July 2001.

[24] Karol M, Hluchyj M and Morgan S, "Input verses output queuing on a space division packet switch", *IEEE Transactions on Communication*, vol. 35, pp. 1347–1356, December 1987.

[25] Kim DS and Un CK, "Performance of contention processing schemes for WDM networks with control channels", *Performance Evaluation*, vol. 25, pp. 85–104, 1996.

[26] Ma M, Hamidzadeh B and Hamdi M, "An efficient message scheduling algorithm for WDM lightwave networks", *Computer Networks*, vol. 31, pp. 2139–2152, 1999.

[27] Marsan MA and Roffinella D, "Multichannel local area network protocols", *IEEE Journal on Selected Areas in Communications*, vol. SAC-1, no. 5, November 1983.

[28] Mehravari N, "Performance and protocol improvements for very high speed optical fibre local area networks using a passive star topology", *IEEE Journal of Lightwave Technology*, vol. 8, no. 4, April 1990.

[29] Modiano E, "Random algorithms for scheduling multicast traffic in WDM broadcast-and-select networks", *IEEE Transactions on Networking*, vol. 7, no. 3, June 1993.

[30] Modiano E and Barry R, "A novel medium access control protocol for WDM-based LANs and access networks using a master/slave scheduler", *IEEE Journal of Lightwave Technology*, vol. 18, no. 4, April 2000.

[31] Modiano E, Barry R and Swanson E, "Design and analysis of an asynchronous WDM local area network using master/slave scheduler", in *Proceedings IEEE INFOCOM'99*, New York, USA, pp. 900–907, 1999.

[32] Mukherjee B, "WDM-based local lightwave networks part I: single-hop systems", *IEEE Network Magazine*, vol. 6, no. 3, pp. 12–27, May 1992.

[33] Pountourakis IE, "Asynchronous transmission protocols for WDM LANs using multichannel control architecture", *Elsevier Computer Communications*, vol. 24, nos. 7–8, pp. 610–621, April 2001.

[34] Shi H and Kavehrad M, "Aloha/slotted-CSMA protocol for a very high-speed optical fibre local area network using passive star topology", in *Proceedings IEEE INFOCOM'91*, Bal Harbour, Florida, USA, vol. 3, April 1991.

[35] Sivalingam KM, Bogineli K and Dowd PW, "Acknowledgement techniques of random access based media access protocols for a WDM photonic environment", *Elsevier Computer Communications*, vol. 16, pp. 458–471, August 1993.

[36] Sivalingam KM, Bogineli K and Dowd PW, "Design and performance analysis of pre-allocation protocols for WDM photonic networks", in *Proceeding SPIE (High-Speed Fibre Networks and Channels)*, vol. 1784, pp. 193–204, September 1992.

[37] Sivalingam KM, Bogineli K and Dowd PW, "Pre-allocation media access control protocols for multiple access WDM photonic networks", in *Proceedings ACM SIGCOMM'92*, pp. 235–246, August 1992.

[38] Sivalingam KM and Dowd PW, "A multi-level WDM access protocol for an optically interconnected multiprocessor system", *IEEE/OSA Journal of Lightwave Technology*, vol. 13, no. 11, pp. 2152–2167, November 1995.

[39] Sivalingam KM and Dowd PW, "A performance study of photonic local area network topologies", in *Proceedings International Workshop on Modelling and Simulation of Computer and Telecommunications Systems (MASCOTS)*, pp. 79–83, January 1994.

[40] Sivalingam KM and Wang J, "Media access protocols for WDM networks with on-line scheduling", *IEEE/OSA Journal of Lightwave Technology*, vol. 14, no. 6, pp. 1278–1286, June 1996.

[41] Sivalingam KM, Wang J, Wu X and Mishra M, "An interval-based scheduling algorithm for optical WDM star networks", *Kluwer Journal of Photonic Network Communications*, vol. 4, no. 1, pp. 73–87, 2002.

[42] Sudhakar GNM Georganas ND and M Kavehrad, "Slotted Aloha and reservation Aloha protocols for very high-speed optical fibre local area networks using passive star topology", *IEEE Journal of Lightwave Technology*, vol. 9, no. 10, October 1991.

[43] Sue CC and Kuo SY, "Design and analysis of accelerative preallocation protocol for WDM star-coupled networks", *IEEE Journal of Lightwave Technology*, vol. 20, no. 3, March 2002.

[44] Tseng C and Chen B, "D-Net, A new scheme for high data rate optical local area networks", *IEEE Journal on Selected Areas in Communications*, vol. 1, pp. 493–499, April 1983.

[45] Yu J, Lee M, Kim Y and Park J, "WDM/SCM MAC protocol suitable for passive double star optical networks", in *Proceedings CLEO/PR'01*, vol. 2, pp. 582–583, July 2001.

[46] Yu J and Park J, "WDM multiple access protocol using node grouping scheme for passive double star networks", in *Proceedings IEE Communications*, vol. 146, no. 5, October 1999.

4

Adaptive Protocols

In this chapter we are going to examine an individual category of relatively recent protocols, which are proposed for media access control (MAC) in the wavelength-division multiplexed (WDM) broadcast-and-select architecture for optical local area networks. The special characteristic that distinguishes this class of protocols and applies to all of the representative examples we are about to explore, is that during network operation (i.e. in real-time) some form of *feedback* depending on the individual scheme is received from the network and comprises the basis upon which behaviour of the various individual stations is determined, e.g. exactly which stations should transmit and on which particular channels at the next time slot.[1] In some of the protocols that will be examined last, this feedback could essentially degenerate to some control information about the source and destination nodes of the various packets; this information, as we will see, may be received and used by some centralized mechanism in order to decide which packets should be allowed to pass through the star coupler of the broadcast-and-select architecture at the next time slot. However, for most of the schemes that will be presented in this chapter, the above mentioned network feedback could actually be the state of the individual channels (wavelengths) of the network, e.g. idle or busy, which is determined by examining a very small percentage of the same overall signal broadcast to every receiving station.

Multiwavelength Optical LANs, G.I. Papadimitriou, P.A. Tsimoulas, M.S. Obaidat and A.S. Pomportsis.
 ISBN 0–470–85108-2.

In both cases described, we say that the protocols are adaptive in that they involve mechanisms enabling them to adapt (their behaviour) according to network feedback information they receive and process in a proper way, representative of the individual protocol. Such mechanisms provide a so-called learning capability to the protocols; thus 'adaptation' is the result of a learning process incorporated in the various protocols of this category.

This is especially true and more obvious (as we mentioned in the first paragraph) when the state of the individual channels is examined. In this case, as a rule, the operation of the protocols is based on the use of *stochastic learning automata*. Before generally identifying what a stochastic learning automaton is, let us first define the concept (and process) of learning. Learning can be defined as any permanent change in behaviour, as a result of past experience. A learning system should therefore have the ability to improve its behaviour with time, toward a final goal. In a purely mathematical context, the goal of a learning system is the optimization of a function not known explicitly [15]. The idea behind designing a learning system is to guarantee robust behaviour without complete, or even partial, knowledge of the system/environment to be controlled [38]. Such learning systems are required to be designed in the context of the so-called *stochastic control theory*, which is a differentiation of the *classical control theory*. In the latter the control of a process is based on *complete* knowledge of the process/system. The mathematical model is assumed to be known and the inputs to the process are deterministic functions of time. On the other hand, in stochastic control theory the uncertainties present in the system/environment are also considered. Assumptions on those uncertainties and/or input functions to the system may be insufficient to control the system in the presence of various changes. It is then necessary to observe the operation of the system in *real-time* and obtain further useful knowledge, since *a priori* assumptions are not sufficient [38].

In our case the system/environment considered is actually a WDM Broadcast-and-Select network where the traffic demands of the various nodes are not known *a priori*. For example, in a network considering tunable transmitters and fixed-tuned receivers (TT-FR) we simply could not know in advance the amount of traffic (in packets) that would demand service, i.e. transmission on the various channels; at a particular time slot, none, only one or more than one nodes might tune their transmitters on a certain wavelength implying a particular station was their desired destination (in fact the one having this wavelength as its 'home channel' for reception). Thus, in this case stochastic control theory may apply and it would be sensible to make use of stochastic learning automata.

An *automaton* is a machine or control mechanism designed to automatically follow a predetermined sequence of operations or respond to encoded instructions.

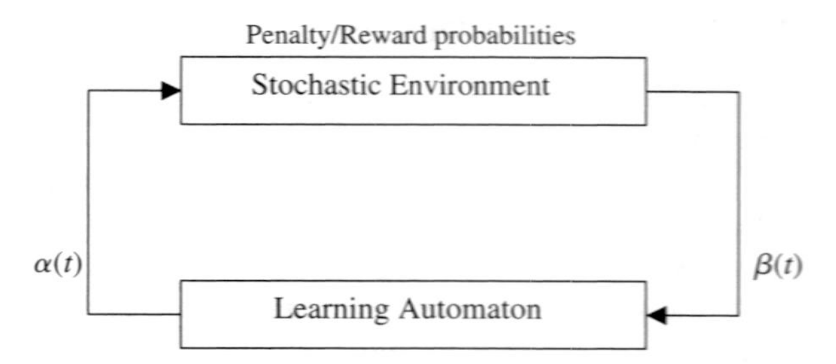

Figure 4.1 Learning automaton that interacts with stochastic environment, where $\beta(t)$ is the response of the environment that consists of the input to the automaton and $\alpha(t)$ is the action or output of the automaton at the time instant t.

The term *stochastic* emphasizes the adaptive nature of the automaton; that is to say, the automaton does not follow a set of predetermined rules but adapts to changes in its environment. It attempts a solution of a problem without any information on the optimal action and thus initially assigns equal probabilities to all actions. It then selects one action at random, observes the response from the environment and updates the action probabilities based on that response. This procedure, which obviously incorporates the concept of learning, is repeated and, consequently, the automaton can be characterized as a stochastic learning automaton [38]. A general depiction of the interaction between such a learning automaton and its environment is depicted in Figure 4.1. The response from the environment at time instant t, denoted by $\beta(t)$ in the figure consists of the input to the learning automaton. The output of the automaton is the action decided after examining the environment's response. This is denoted by $\alpha(t)$ in Figure 4.1. The set of actions or outputs of the automaton is finite, while the set of responses of the environment or inputs to the automaton may be either finite or infinite [38].

As we mentioned before, most of the protocols examined throughout this chapter make use of such stochastic learning automata. This way, they are capable of adapting better to the varying traffic demands of the nodes in an optical WDM local area network, thus generally achieving higher network performances than many other previously proposed solutions. It should be recalled, however, that some of the schemes, actually the ones proposed in Section 4.4, may still be characterized as adaptive without using learning automata in their internal operation mechanism. On the other hand, throughout Sections 4.1–4.3 several interesting Adaptive TDMA (Time Division Multiple Access), Adaptive Random Access and Adaptive Pretransmission Coordination protocols, respectively, will be described, which all make use of stochastic learning automata in order to improve network performance.

4.1 ADAPTIVE TDMA PROTOCOLS

We will begin our exploration of adaptive protocols by putting at the centre of our attention noteworthy schemes based on the Time Division Multiple Access (TDMA) method and, in parallel, making use of learning automata. Apparently, TDMA implies that we are talking about synchronous protocols (considering the discussion in the introduction of Chapter 3), where data channels are time-slotted and permission for transmission is generally assigned in a cyclic manner to all (or a proper subset) of the nodes, but always taking into account at the same time the network feedback information and the exact results of its processing by the learning automata used. Another common feature of these protocols is the hardware equipment used in each node in terms of transmitters and receivers, which actually comprises one tunable transmitter and one receiver fixed-tuned in a predetermined home channel. Thus we are generally considering WDM broadcast-and-select architectures that can be characterized as TT-FR, according to the discussion in Subsection 1.6.2. Moreover, transmitters are assumed to be capable of tuning over the entire range of wavelengths used in the network. Therefore, the optical local area networks considered in the context of the proposed protocols are single-hop.

4.1.1 SALP

In this subsection we present the *Self-Adaptive Learning Protocol* (*SALP*) which is capable of operating efficiently under *bursty* and *correlated* traffic [32–33]. By 'correlated' we imply that destinations of the various packets transmitted by a particular node within the network are not absolutely arbitrary, but have some kind of relationship between them.

Generally, it has been noticed that traffic in high-speed LANs (e.g. gigabit LANs as in the case of optical implementations) is highly bursty [11]. This is a straightforward consequence of the fact that most of the load in such networks is data traffic, which is known to be intrinsically bursty. Optical LANs offer the potential for higher speeds and therefore users' demand for service (i.e. transmission of their packets) may arrive at much higher peak rates in short time intervals, before ceasing again for a while (as usually happens with data traffic). In this way traffic is becoming even burstier in optical LANs and therefore only protocols that can handle this type of traffic well may achieve satisfactory performance. Moreover, the destinations of packets transmitted by a particular node in the network are highly correlated, since most of them are fragments of large messages [2]. Under these conditions, pure TDMA-based schemes are certain to suffer from low performance, since they would waste a considerable amount of time-slots for idle stations.

For example, let us consider R-TDMA and I-TDMA* which are compared to SALP in [33]. As described in Chapter 3, in R-TDMA [7] at each time slot the W available wavelengths are assigned to W different transmitting stations at random, while in I-TDMA* [1] the wavelengths are assigned to the transmitting stations in a round robin fashion. In a gigabit LAN with highly bursty and correlated traffic, the waste of a considerable amount of time slots would be inevitable (even though I-TDMA* would generally perform better than R-TDMA). This would be true even when W separate fixed length queues were used, each for one of the W available wavelengths, which has been shown [1] to favour the operation of these schemes in comparison to using just a single queue for buffering. The SALP protocol is proposed as an alternative, handling bursty and correlated traffic much better and leading to a significant improvement of network performance.

4.1.1.1 Network Model of SALP

The SALP protocol considers a network model comprising a WDM passive star broadcast-and-select network, like the one depicted in Figure 4.2. Let us define the set of stations as $U = \{u_1, u_2, \ldots, u_N\}$ and the set of wavelengths as $\Lambda = \{\lambda_1, \lambda_2, \ldots, \lambda_W\}$, where N and W are the number of stations and wavelengths respectively. Each station is provided with a transmitter, which is a tunable laser that can be tuned over the whole range of the W available wavelengths. Optical fibres are used to connect the outputs of the lasers to the network hub. The $N \times N$ passive star coupler included in the hub broadcasts all incoming optical signals to all output ports. Each output port of the star coupler is connected to the corresponding

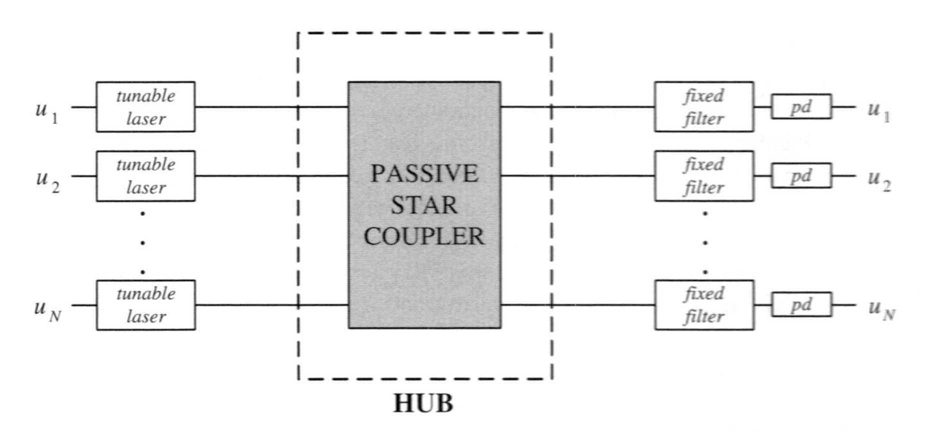

Figure 4.2 The WDM passive star network considered in SALP.

station by means of an optical fibre (other than the one used for transmission-input to the hub). Each station is also provided with a receiver. Actually this is a fixed-tuned optical filter that passes only one wavelength λ_{i_k}, where λ_{i_k} is an element of set Λ and is called the 'home channel' (or 'home wavelength') of station $u_k (k = 1, \ldots, N)$. The scheme for allocation of wavelengths-home channels to users considered in SALP is the so-called *interleaved allocation*[2] [1], where generally $i_k = ((k - 1) \text{ MOD } W) + 1$ for home channel λ_{i_k} of user u_k. Since $N \geq W$, it may be the case that more than one user is assigned the same wavelength as the home channel for reception. However, it should be noted that with interleaved allocation, sharing of same wavelengths between various stations is minimized. Apparently the system considered in SALP is TT-FR, therefore each station is capable of receiving only those packets transmitted on its home channel. The output of the optical filter is connected to a photodetector, which performs O/E (optoelectronic) translation of the incoming signal, before it reaches (the electronics of) the corresponding user. In Figure 4.2 the input ports of the hub (connected to the lasers) are shown on its left, while the output ports (connected to the fixed-tuned filters) on its right. However, the physical placement of the tunable laser and the fixed-tuned filter of a particular user must be considered 'inside' the same station.[3]

4.1.1.2 Detailed Description of SALP

As was mentioned in the introduction to this section, the operation of the SALP protocol is based on the use of learning automata. That is to say the WDM broadcast-and-select network, and more specifically the varying traffic demands of its N nodes, can be considered as a stochastic environment with which the learning automata are requested to interact. The actions of the automata will essentially be decisions determining which station is granted permission to transmit on each of the available wavelengths, at each time slot. That is why a set of exactly W learning automata $LA_i (i = 1, \ldots, W)$ are needed for each station in order to enable it (based on their decisions) to identify which stations are granted permission to transmit at the next time slot, on each of the W wavelengths. If the station itself is included in the transmitting nodes then it will have to prepare by tuning its laser appropriately. Each automaton LA_i corresponds to a specific wavelength λ_i and determines which station is granted permission to transmit on this wavelength. Note that for the specific wavelength λ_i, all in all there are exactly N automata LA_i, one for each station. As we will see, the N automata corresponding to the same wavelength (one automaton for each station) make the same decision (choice of a unique transmitter for the specific wavelength) at each time slot and thus the protocol becomes *fully distributed* and *collision-free*. But let us examine in more detail how the transmitting stations are selected every time.

Each learning automaton LA_i contains a probability distribution $P_i(t)$ over the set of stations. Thus, $P_i(t) = \{P_{i,1}(t), P_{i,2}(t), \ldots, P_{i,N}(t)\}$, where $P_{i,j}(t)$ is the basic choice probability of station u_j for wavelength λ_i at time slot t. The selection of the single station that is granted permission to transmit on each wavelength is based on these basic choice probabilities, i.e. the station with the biggest value of the relevant choice probability is selected. The selection of transmitting stations is described in the following algorithm.[4]

```
{                        *** Basic Notations ***
  U     : set of nodes = (u1, u2, ..., uN) - constant.
  Λ     : set of wavelengths = (λ1, λ2, ..., λW) - constant.
  G     : set of stations that have not been granted
          permission to transmit - variable.
  R     : set of wavelengths for which no source station has
          been selected - variable.
  Ci(t) : station selected to transmit on λi at time slot t -
          variable.
                                                              }
PROCEDURE SALP_STATION_SELECTION;
BEGIN
  G := U;
  R := Λ;
  WHILE R ≠ Ø DO
  BEGIN
        Choose at random one wavelength λi ∈ R;
        Choose as source station the station uj ∈ G with the
        biggest value of normalized choice probability
```

$$\Pi_{i,j} = \frac{P_{i,j}(t)}{\sum_{u_m \in G} P_{i,m}(t)};$$

```
        Ci(t) := uj;
        G  := G - {uj};
        R  := R - {λi};
  END;
END;
```

After presenting the station selection algorithm, it is time to discuss the way the choice probabilities are updated every time. The *probability updating scheme* is very important, since it is in fact the heart of the learning process incorporated in the protocol.

As was mentioned in the introduction to SALP, this protocol is destined to handle the bursty and correlated traffic that is typically present in gigabit LANs better than other non-learning-based schemes. Taking burstiness of traffic and correlation of destinations into account, it can be said that if a station transmitted a packet during the last time slot, it is probable that it will have packets to transmit to the same destination (same wavelength accordingly) at the next time slot too. Thus, the probability updating method used in SALP increases the basic choice probability of such a station. On the contrary, when a selected station was idle in the last time slot, it is likely to remain so at the next slot too. So, its choice probability has to be decreased.

All stations have (the same) knowledge of exactly which stations were selected to transmit during the last time slot. In order to decide how to update the choice probabilities for the next time slot (increase or decrease), the stations have to be informed about the state of each wavelength during the last time slot. That is to say, they have to know if the selected stations actually transmitted or not, or, equivalently, whether the corresponding wavelengths were *busy* or *idle*. This information actually comprises the input from the stochastic environment to the W learning automata of each station.

The mechanism of SALP that provides stations with this necessary network feedback information is illustrated in Figure 4.3. A small fraction ϵ of the optical signal coming from the passive star coupler to each station is fed to a WDM Demultiplexer, which separates the different wavelengths. The output ports of the Demultiplexer are fed to an array of W photodetectors that detect whether the corresponding wavelength is idle (no packet transmission) or busy (a packet transmission is taking place). All this information determines the appropriate inputs to the array of W learning automata LA_i, $i = 1, \ldots, W$ of the station. It should be noted that no full reception of the incoming signals is performed in the feedback mechanism. Only a detection of the presence or absence of optical signal power on each wavelength is required. Accordingly, the splitting ratio ϵ can be very small. Therefore, the power of the incoming signal that is guided to the receiver (fixed-tuned filter) of the station (equal to $(1 - \epsilon) \times P_{in}$) remains practically unaffected, i.e. approximately equal to P_{in}.

Let $\text{SLOT}_i(t) \in \{\text{busy, idle}\}$ be the state of wavelength λ_i at time slot t. Obviously, considering the discussion in the previous paragraph, the feedback mechanism of Figure 4.3 provides each station exactly with this value of $\text{SLOT}_i(t)$ at each time slot t. Now let us see next just how this information is used by the W learning automata LA_i, $i = 1, \ldots, W$ of each station for updating the choice probabilities. The probability updating scheme is the following:

$$P_{i,j}(t+1) = P_{i,j}(t) + L(1 - P_{i,j}(t)),$$
$$\text{if } C_i(t) = u_j \text{ and } \text{SLOT}_i(t) = \text{busy}.$$

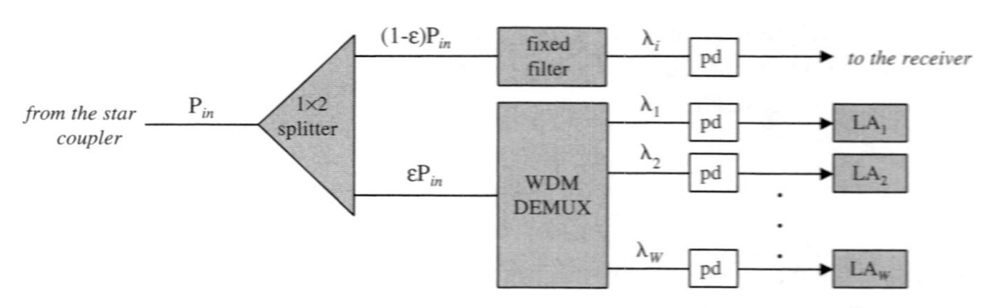

Figure 4.3 The feedback mechanism of SALP located in each station [29].(© 2001 Wiley)

$$P_{i,j}(t+1) = P_{i,j}(t) - L(P_{i,j}(t) - a),$$
$$\text{if } C_i(t) = u_j \text{ and } SLOT_i(t) = \text{ idle.}$$

In the above equations, $P_{i,j}(t)$ is the basic choice probability of station u_j, for wavelength λ_i at time slot t. By $C_i(t)$ we denote the station that is selected (granted permission to transmit) on wavelength λ_i at time slot t. Furthermore, α and L are special parameters of the protocol, with $L, \alpha \in (0, 1)$ and $P_{i,j}(t) \in (\alpha, 1)$ for all t. The role of parameter α is to prevent the basic choice probabilities of stations from taking values in the neighbourhood of zero, in order to increase the adaptivity of the protocol. The exact values of parameters α and L depend upon the environment, where the automata operate. When the environment is slowly switching or when the environmental responses have a high variance, α and L must be very close to zero in order to guarantee *high accuracy*. On the contrary, in a rapidly switching environment or when the variance of the environmental responses is low, higher values of α and L may be used so as to increase the *adaptivity* of the protocol. Thus when the burst length is high or when the queue length is low, small values of α and L must be selected. On the other hand, when the burst length is low or when the queue length is high, α and L can be much higher.

4.1.1.3 Overview of Analytical and Simulation Results for SALP

One of the most important characteristics of SALP protocol is derived by analysis, as provided in [33]. According to this characteristic, if we suppose that $d_{i,j}$ is the probability that station u_j has one or more packets waiting to be transmitted on wavelength λ_i, then, by using the probability updating scheme described in the previous subsection, SALP ensures that the probability of a station being granted permission to transmit asymptotically tends to be equal to $d_{i,j}$. That is to say, it can be proved that for any station $u_j \in \mathrm{U}$ and any wavelength $\lambda_i \in \Lambda$ we have:

$$\lim_{t\to\infty, L\to 0, \alpha\to 0} P_{i,j}(t) = d_{i,j}$$

Details of the proof of the above limit can be found in [33] (proof of Theorem 1). Considering this argument, for any two stations u_j, u_k and any wavelength λ_k (with $d_{k,j} \neq 0$), SALP asymptotically tends to satisfy the relation:

$$\frac{P_{k,i}}{P_{k,j}} = \frac{d_{k,i}}{d_{k,j}}$$

That is to say the ratios of the basic choice probabilities asymptotically tend to be equal to the ratios of the probabilities that there are packets for transmission. An analogous relation also holds for the normalized choice probabilities:

$$\frac{\Pi_{k,i}}{\Pi_{k,j}} = \frac{\frac{P_{k,i}}{\sum_{um \in G} P_{k,m}}}{\frac{P_{k,i}}{\sum_{um \in G} P_{k,m}}} = \frac{P_{k,i}}{P_{k,j}} = \frac{d_{k,i}}{d_{k,j}}$$

This key characteristic of SALP suggests that, although the traffic parameters are unknown and time-variable, the bandwidth of each wavelength is allocated by SALP to the stations according to their needs. In this way the number of idle slots is reduced, resulting in a significant improvement of network throughout.

The improvement of network throughput gained by using SALP has also been proved by simulation and comparison of SALP with I-TDMA* and R-TDMA (as mentioned in the introduction to this protocol) [33]. The same queuing model was used for all three protocols, so that we have a fair comparison of them. It was assumed that we have W separate queues of fixed length, one for each wavelength. As was mentioned in the introduction, this queuing approach favours the operation of R-TDMA and I-TDMA* in comparison to using just a single queue for all wavelengths.[5] If we denote the fixed length of each queue by Q, we observe that the total buffer size of each station rises up to $W \times Q$, since there is one queue for each channel. The performance metrics used for the comparison are the popular *delay versus throughput* and *throughput versus offered load* characteristics.

For the corresponding graphs the reader may consult [32–33]; we selectively present diagrams with performance results of SALP later on, when this is compared to protocol APS in Subsection 4.1.3. Here, we only outline the main conclusions from the comparison of SALP with R-TDMA and I-TDMA*:

- The use of the learning automata-based mechanism incorporated by SALP leads to a really notable improvement of the network performance. SALP outperforms the other two schemes when operating both under bursty and correlated traffic.

- SALP outperforms I-TDMA* and R-TDMA at a greater extent as the traffic gets burstier. In fact the operation of SALP, supported by the learning mechanism, is practically unaffected in terms of performance by the extent of traffic burstiness and correlation of destinations, while the performance degradation suffered by I-TDMA* and R-TDMA becomes much higher when burstier and correlated traffic is considered, since the number of idle slots increases dramatically due to the inflexible allocation of time slots to stations used by these schemes.
- By comparing the delay versus throughput and throughput versus offered load of the protocols for networks with different queue lengths (Q), we observe that when the queue length is low, each station has packets only for a small number of wavelengths and, apparently, this favours an intelligent assignment of channels (wavelengths) to transmitting stations like the one incorporated by SALP, rather than the inflexible ('blind') assignment of wavelengths to stations applied in the context of I-TDMA* and R-TDMA.

4.1.1.4 Concluding Remarks about SALP

Thus, we have seen that SALP is a relatively recent adaptive protocol that improves network performance in comparison to other non-adaptive schemes like I-TDMA* and R-TDMA, because of the learning-automata based mechanism of each station that adds intelligence to the assignment of wavelengths to transmitting stations; i.e. SALP is capable of adapting better to unknown and time-variable traffic conditions. The performance improvement it introduces is even greater when the traffic becomes burstier, the destinations of the packets transmitted by a single station more correlated and when the length of the queues used for each wavelength is short. In SALP there is no centralized control, but W learning automata controlling the W wavelengths are located in each station. The learning automata of all stations always select the same station as a source (transmitting station) for each wavelength at each time slot, i.e. they make exactly the same decisions all the time. Therefore the protocol is fully distributed and consequently it is also fault-tolerant, since the failure of one station would not affect its operation. Also, it should be noted that the only additional hardware required is the feedback mechanism of Figure 4.3 for each station, which cannot be considered to increase significantly the implementation cost of the SALP protocol.

4.1.2 LABP

In this subsection another self-adaptive learning automata-based protocol for WDM passive star networks of the broadcast-and-select architecture is presented.

This protocol is called *Learning-Automata-Based Protocol* (*LABP*) [34] and has several similarities with the previously examined SALP protocol. That is to say LABP is also fully distributed, collision-free and destined to handle better than other non-learning-based schemes the gigabit LANs traffic, considering its particularity of being bursty and correlated (as far as the destinations of a single transmitter are concerned). The burstiness and correlation of gigabit LANs traffic were discussed in the previous subsection about SALP, so we shall not extend the discussion about this subject here. LABP uses a probability updating scheme (different from the one used by SALP), which enables it to allocate the bandwidth of each channel to transmitting stations according to their needs and, thus, lead network operation in the desirable state where the number of idle slots tends to be minimized.

4.1.2.1 Network Model of LABP

The LABP protocol assumes a WDM passive star network where each station is equipped with one tunable laser and one filter for reception, fixed-tuned to a predetermined 'home wavelength'. Thus it is applied to a TT-FR system according to the relevant discussion of Chapter 1. Such a network was also assumed in the case of SALP and is illustrated in Figure 4.2.

Let us define here as well the set of stations to be $U = \{u_1, u_2, \ldots, u_N\}$ and the set of wavelengths as $\Lambda = \{\lambda_1, \lambda_2, \ldots, \lambda_W\}$, where N and W are the number of stations and wavelengths respectively. The lasers are assumed to be capable of tuning over the whole range of Λ. They are connected to the hub by means of a separate optical fibre which is the case as well with fixed-tuned filters.

4.1.2.2 Detailed Description of LABP

The operation of LABP is based on the use of 'discretized' learning automata [17]. Specifically, each station has to be provided with a set of W discretized learning automata LA_i $(i = 1, \ldots, W)$ that respectively control each one of the W available wavelengths; automaton LA_i is supposed to control wavelength λ_i and allocate the bandwidth of this wavelength to the appropriate transmitting station at each time slot. Thus, actions of the automata are in fact decisions determining which station is granted permission to transmit on each of the available wavelengths, at each time slot. The N automata corresponding to the same wavelength (one automaton for each station) make exactly the same decision (choice of a unique transmitter for the specific wavelength) at each time slot and thus LABP becomes fully distributed and collision-free.

Each learning automaton LA_i contains a probability distribution $P_i(t)$ over the set of stations. Thus, $P_i(t) = \{P_{i,1}(t), P_{i,2}(t), \ldots, P_{i,N}(t)\}$, where $P_{i,j}(t)$ is the basic choice probability of station u_j for wavelength λ_i at time slot t. The selection of the single station that is granted permission to transmit on each wavelength is based on these basic choice probabilities. If we denote by $C_i(t)$ the station that is granted permission to transmit on wavelength λ_i at time slot t, then its selection by LA_i is generally carried out as follows:

If $\sum_{m=1}^{N} P_{i,m}(t) = 0$ then $C_i(t)$ is selected at random.

If $\sum_{m=1}^{N} P_{i,m}(t) > 0$ then $C_i(t)$ is selected according to the normalized choice probabilities $\Pi_{i,j}(t) = \dfrac{P_{i,j}(t)}{\sum_{m=1}^{N} P_{i,m}(t)}, \; j = 1, \ldots, N.$

As in the case of SALP, considering that gigabit LANs traffic is bursty and there is correlation among the destinations of a single transmitter, it can be supported that if a station transmitted a packet during the last time slot, it would be probable that it will have packets to transmit to the same destination (thus wavelength) at the next time slot as well. Therefore the probability updating method used in LAPB increases the basic choice probability of such a station. On the other hand, if a selected station was idle in the last time slot, it would be probable to remain idle at the next time slot too. Consequently, LAPB decreases its basic choice probability. Since the learning automata of all stations make the same decisions, each station is aware of exactly which stations were selected as transmitters for each wavelength at each time slot. It remains to know whether they actually transmitted or not, i.e. accordingly whether the corresponding wavelength was busy or idle at each specific time slot. As in the case of SALP, let us denote by $\mathrm{SLOT}_i(t) \in \{\text{busy, idle}\}$ this information regarding wavelength λ_i. The mechanism of LAPB that provides stations with this necessary network feedback information is exactly the same as the one used in SALP and is shown in Figure 4.3. The probability updating scheme of LAPB that uses this information is quite different from SALP and is described by the following equations:

$$P_{i,j}(t+1) = P_{i,j}(t) + \frac{1}{k},$$

$$\text{if } C_i(t) = u_j,\ \mathrm{SLOT}_i(t) = \text{busy and } P_{i,j}(t) \leq 1 - \frac{1}{k}.$$

$$P_{i,j}(t+1) = P_{i,j}(t) - \frac{1}{k},$$
$$\text{if } C_i(t) = u_j,\ \text{SLOT}_i(t) = \text{idle and } P_{i,j}(t) \geq \frac{1}{k}.$$

It should be noted here that initially the basic choice probabilities of every station, regarding choice for transmission on every wavelength, are set to $1/k$, i.e. $P_{i,j}(t) = 1/k, \forall i, j$. Parameter k is a parameter of the protocol and determines the size of the probability quantity that comprises the step of increment or decrement every time. Obviously, a large value of k would result in small changes of the selected stations' basic choice probabilities at each time slot, thus high accuracy of LAPB, and, conversely, a small k would make the step of probability update big, thus LAPB would adapt more rapidly to changes in the environment. By carefully looking at the above probability updating scheme, it is obvious that the basic choice probabilities take values from a finite set, specifically we have: $P_{i,j}(t) \in \{0, 1/k, 2/k, \ldots, 1\}$.

Both the proper update of the basic choice probabilities and the selection of a transmitter for every wavelength at each time slot are tasks performed by the learning automata that are located in each station.[6] Actually, in the beginning of each time slot the former task is carried out first and the latter second. In the case of SALP we presented only the implementation (in form of pseudocode) for the procedure of station selection and omitted the pseudocode for the probability updating scheme. Here we shall present in detail an interesting implementation of both the previously mentioned tasks performed by the learning automata in the context of LAPB.

The algorithm is based on keeping a variable-sized array A, which contains 'probability cells', with each cell representing a probability mass equal to $1/k$. Each cell belongs to a specific station. The actual content of the array in each position is a station number that is actually an identifier of the station. Thus, if for example the number of a certain station is present three times in the array, this implies that its choice probability is equal to $3/k$. Therefore, the choice probability of each station is proportional to the number of cells it has in array A. Consequently, in order to select a station according to the present probability distribution, it suffices to select at random the content (station identifier) of one of the array's positions (with uniform distribution). This would mean that the corresponding station is granted permission to transmit. Accordingly, when the learning automata wish to increase the basic choice probability of a specific station, they just have to add one more 'cell' for this station in the array, i.e. the number of the station in one additional position. On the other hand, decreasing the choice probability of a station would just require removing one of the station's cells from the array. The implementation of the algorithm that each learning automaton must run is presented next in detail:

```
{                        *** Basic Notations ***
  A[]                   : array which contains 'probability cells'
                          for stations - variable.
  asize                 : current size of array A - variable
  index                 : index of currently selected cell (and
                          thus station) in array A - variable.
  t                     : current time slot - variable.
  slot_i(t)             : can be either busy or idle according to
                          the network feedback information -
                          variable flag.
  number_of_cells[N]    : array containing the current number of
                          probability cells for each station -
                          variable.
  C_i(t)                : station selected to transmit on λ_i at
                          time slot t variable.
                                                                  }
PROCEDURE LABP_LA_i;
BEGIN
 REPEAT
       t := t+1;
       j := C_i(t);
  (* Update the choice probability of the selected station *)
       IF (slot_i(t) = busy) AND (number_of_cells[j] < k) THEN
       BEGIN
            inc (asize);
            A[asize] := j;
            inc (number_of_cells[j]);
       END
       ELSE IF (slot_i(t) = idle) AND (number_of_cells[j] > 0)
       THEN
       BEGIN
            A[index]:= A[asize]; (* overwrite last value
            of A *)
            dec (asize);
            dec (number_of_cells[j]);
       END
  (* Select a station for the next time slot *)
       IF (asize > 0) THEN
       BEGIN
            index := ⌈random * asize⌉;
            C_i(t+1) := A[index];
       END
       ELSE C_i(t+1) := ⌈random * N⌉;
 FOREVER;
END;
```

The network feedback information regarding the state (busy or idle) of each wavelength is provided by the mechanism of Figure 4.3 to the learning automata at the beginning of each time slot. Each learning automaton has to update the probability distribution $P_i(t)$ based on this information and then select a station as a transmitter for the channel it controls, according to the new probability distribution; however, it is crucial that both of these tasks are performed within the interval of one time slot. Otherwise, a longer computational time would result in considerable performance degradation, since it would cause time gaps between packets. The implementation presented above is indeed sufficiently fast. In [34] it is assumed, for example, that a really slow, according to today's standards, Pentium/166MHz microprocessor is used for the implementation of the W learning automata in assembly language. The required clock cycles for execution are 110. All the variables of the above algorithm are stored in the internal cache of the microprocessor, due to the spatial and temporal locality of reference. As a result, there are no wait states and $0.663 \times W\mu s$ are enough for the execution of the algorithm. Considering that each station has a bit rate of 2 Gb/s and that the packet length is equal to 4 KB [4], the slot duration becomes 16 μs. Consequently, even with such a slow microprocessor, the network is capable of supporting up to 24 wavelengths and an overall network capacity of 48 Gb/s.

4.1.2.3 Overview of Simulation Results for LABP

In [34] the LABP protocol is simulated and compared to the well-known R-TDMA protocol for bursty and correlated traffic.

References [6] and [14] were used as exemplars for modelling bursty traffic. According to them, each station may either have no packet arrivals (i.e. packets having 'arrived' for transmission) thus be in state X_0 or may have one packet arrival at each time slot, thus be in state X_1 with probability Z. The transition probabilities are P_{01} for transition from state X_0 to state X_1 and P_{10} for the opposite transition. Assuming a network load equal to L packets/slot and a mean burst length of B slots, it can be shown that $P_{01} = L/B(NZ - L)$ and $P_{10} = 1/B$.

The correlation of packet destinations for a single transmitter was modelled with reference [23] as an exemplar. Specifically, it is assumed that the destination of a newly arriving packet is selected at random among all destinations with probability R and it is the same as the previous packet's destination with the remaining probability $(1 - R)$.

We shall present from [34] simulation results of the comparison of the LABP and R-TDMA protocols when applied to the following two network configurations:

1. Network N_1: $N = 10$, $W = 5$, $Q = 10$, $B = 20$, $Z = 1.0$, $R = 0.05$.
2. Network N_2: $N = 50$, $W = 10$, $Q = 10$, $B = 5$, $Z = 1.0$, $R = 0.2$.

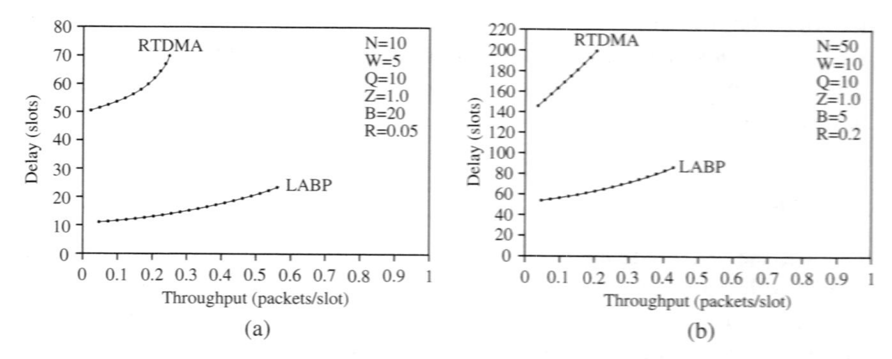

Figure 4.4 Results taken from simulation and comparison of LABP with R-TDMA [34] for two network configurations $N_1(N = 10$ nodes, $W = 5$ wavelengths, W queues with length $Q = 10$, mean burst size $B = 20$, traffic parameters $Z = 1.0$ and $R = 0.05)$ and N_2 (with $N = 50$, $W = 10$, $Q = 10$, $B = 5$, $Z = 1.0$ and $R = 0.2$). Figure shows delay versus throughput characteristic of protocols (a) for N_1 and (b) for N_2. (© 1999 IEEE)

Parameters B, Z and R are traffic parameters as described above. N, W and Q are the number of stations, wavelengths and the queue size, respectively. The delay versus throughput characteristic of the two compared protocols for network configurations N_1 and N_2 is depicted in Figure 4.4(a) and Figure 4.4(b) respectively.

It is obvious that the LABP protocol significantly outperforms R-TDMA especially for bursty and correlated traffic.

4.1.2.4 Concluding Remarks about LABP

Concluding LABP, we have seen that it is also a relatively recent adaptive protocol that improves network performance in comparison with other non-adaptive schemes like R-TDMA, because of the learning-automata based mechanism of each station that adds intelligence to the assignment of wavelengths to transmitting stations. As in SALP, there is no centralized control, but W learning automata controlling the W wavelengths are located in each station. The learning automata of all stations always select the same station as a transmitter for each channel during each time slot, i.e. they make exactly the same decisions all the time. That is because they always contain the same choice probabilities and, moreover, because the same random number generator and the same seed is used by all stations (in the part of the implementation of the algorithm executed by each automaton that calls for a random number, please refer to the pseudocode of Section 4.1.2.2). Taking all these into account, it is obvious that the protocol is collision-free, fully distributed and, by extension, it is also fault-tolerant, since the failure of one station would not affect the overall network operation. As far as LABP's implementation cost is concerned, it should be noted that the only additional hardware required is the feedback mechanism of Figure 4.3 for each station, which does not increase the implementation cost really much.

4.1.3 APS

We close the discussion about adaptive TDMA protocols by examining here another learning automata-based protocol for WDM passive star networks of the broadcast-and-select architecture. The new protocol seems to bring about further improvements in terms of performance, as compared to the first studied SALP protocol. It is called *Adaptive Passive Star* (*APS*) [29] protocol and has quite a few similarities with the two previously examined protocols. That is to say, APS is also fully distributed, collision-free and destined to handle better than other non-learning-based schemes the bursty and correlated gigabit LANs traffic. APS, however, is quite innovative in the way it selects stations as transmitters on each wavelength at each time slot. As we will see, a set of stations which are expected to have packets for each wavelength is maintained by the protocol (one set for each wavelength) and the transmitter is commonly (but not always as we will see later) selected in a round robin fashion among the stations in the set. Thus, APS is capable of allocating the bandwidth of each channel to transmitting stations according to their needs and, therefore, tends to minimize the number of idle slots. This implies that a notable improvement in network throughput is brought about by the use of the protocol in question.

4.1.3.1 Network Model of APS

The APS protocol assumes a WDM passive star network where each station is equipped with one tunable laser and one filter for reception, fixed-tuned to a predetermined 'home wavelength'. Thus, it is applied to a TT-FR system according to the relevant discussion of Chapter 1. Such a network was also assumed in the case of SALP and LABP and is illustrated in Figure 4.2.

Let us define here, as well, the set of stations to be $U = \{u_1, u_2, \ldots, u_N\}$ where N is the number of stations in the network and the set of wavelengths as $\Lambda = \{\lambda_1, \lambda_2, \ldots, \lambda_W\}$ where W is the number of available wavelengths. The lasers are assumed to be capable of tuning over the whole range of Λ. They are connected to the hub by means of a separate optical fibre which is the case as well with fixed-tuned filters. It should be noted that each station is assumed to have a waiting queue of length Q, where all the arriving packets are buffered before actual transmission.

4.1.3.2 Detailed Description of APS

The operation of APS is based on the use of learning automata, exactly as is the case for SALP and LABP. The WDM broadcast-and-select network, and more specifically the varying traffic demands of its N nodes, can be considered as a stochastic environment with which the learning automata are requested to interact. The actions of the automata are in essence decisions about which station is granted permission to transmit on each of the available wavelengths, at each time slot. Thus, a set of exactly W learning automata LA_i ($i = 1, \ldots, W$) is needed for each station in order to enable it (based on their decisions) to identify which stations are granted permission to transmit at the next time slot, on each of the W wavelengths. Each automaton LA_i controls a specific wavelength λ_i and determines which station is granted permission to transmit on this wavelength. As in the previous protocols, for the specific wavelength λ_i, all in all there are exactly N automata LA_i, one for each station. The N automata corresponding to the same wavelength (one automaton for each station) make the same decision, i.e. choose a unique transmitter for the specific wavelength at each time slot. Therefore, the APS protocol becomes fully distributed and collision-free. Traffic in the context of APS is considered to be bursty and the packet destinations of a single transmitter correlated; the packets included in a burst are actually considered fragments of larger messages with each message, of course, heading to the same destination.

In the context of APS each station can either be in *active* or *idle* state, i.e. it may respectively have one or no packet arrivals, at each time slot. More specifically, for a certain wavelength λ_i, we can characterize a station as active for λ_i, if it has one or more packets to transmit on this wavelength; otherwise, the station may be characterized as idle for λ_i. The fundamental characteristic of APS protocol is that the learning automata maintain and update (at each time slot according to network feedback information) a set of stations, which are *expected to be active* for the wavelength they control. To be exact, the learning automaton LA_i (as we saw in total there are N such identical automata, one located at each station) maintains a set A_i of stations that are expected to be active for wavelength λ_i and also updates this set at each time slot by adding or removing stations from it according to network feedback. The selection of the station that is granted permission to transmit on λ_i is carried out in a round robin fashion from set A_i. A central point in APS protocol is to determine exactly which stations should be considered applicable to enter the sets A_i, $i = 1, \ldots, W$. Since the offered traffic is assumed to be bursty and all packets in a burst are considered fragments of large messages, thus are typically destined to the same recipient, it follows that the packets arriving to a certain transmitter are waiting to be forwarded on the same wavelength towards the same destination node. Consequently, if a station given permission to transmit on a wavelength λ_i was active, it is highly probable that it will remain active for this wavelength in the near future. Therefore it is included in set A_i. On the contrary, if it was granted permission to transmit but was idle for wavelength λ_i, it is probable that it will remain idle for the specific wavelength in the near future, thus it is not included in set A_i.

However, it is easy to see that the APS protocol would be neither efficient nor fair if it was exactly as described so far. Stations that were idle for a specific wavelength for a long time would never have a chance to enter the corresponding set of 'expected-to-be' active stations for this wavelength (from which, only, transmitters are supposed to be chosen), even if they actually wanted to transmit something on this wavelength at some time, which is very natural to happen. For example, if a station $u_j \notin A_i$ was continuously idle for wavelength λ_i but at some point in time it actually wanted to transmit a burst of packets on it (i.e. it became active), the APS protocol would be incapable of sensing such a transition from idle to active state. In order to overcome this important deficiency, APS grants transmission permissions to stations outside the corresponding sets A_i after a certain timing threshold. Specifically, APS keeps track of the time $d_{i,j}$ that passed from the last time each station $u_j \in U$ was granted permission to transmit on wavelength λ_i, for $i = 1, \ldots, W$. In practice, APS maintains W sets of these times (one set per wavelength) which are defined as $D_i(t) = \{d_{i,1}(t), d_{i,2}(t), \ldots, d_{i,N}(t)\}, i = 1, \ldots,$ W, for time slot t. For each station u_j, when $d_{i,j}$ exceeds a predefined threshold Δ, the protocol grants to u_j permission to transmit on wavelength λ_i at the next time slot, even if $u_j \notin A_i$. This feature of APS ensures that no station has to wait more than Δ time slots before it is granted permission to transmit on any wavelength and, apparently, it makes the protocol capable of adapting better to varying traffic conditions, more effective and fair.

Before proceeding to the algorithmic description of the protocol it should be noted that, as in the case of SALP and LABP, the learning automata that perform all the previously described operations must be informed of the state of the wavelength they control at each time slot, i.e. for wavelength λ_i, LA_i must know the value of $\text{SLOT}_i(t) \in \{\text{busy, idle}\}$. This information is provided by a network feedback mechanism located at each station, which is identical to the two previously described protocols and is illustrated in Figure 4.3. Based on this network feedback information, the learning automata update the sets A_i in the proper way, as described above. In the following, the algorithm which the learning automaton LA_i runs in order to determine the transmitter for wavelength λ_i at each time slot is presented:

```
{                    *** Basic Notations ***
  t          : time slot - variable.
  Ai(t)      : set of stations that are expected to be active
               for wavelength λi at time slot t - variable.
  sloti(t)   : can be either busy or idle according to the
               network feedback information - variable flag.
  di,j(t)    : time slots passed since uj was granted
               permission to transmit on λi, at time slot t -
               variable.
  si(t)      : number of station that was granted permission to
               transmit on λi, at time slot t - variable.
  Usi(t)     : station that was granted permission to transmit
               on λi, at time slot t - variable.
  k          : auxiliary variable.
                                                              }
PROCEDURE APS_LAi;
 REPEAT
       t:= t+1;
       FOR each wavelength λi ∈ Λ DO
       BEGIN
            k :=si(t);
            (* Update vector Di *)
            FOR all j ≠ k DO di,j(t) := di,j(t−1)+1;
            di,k(t) := 0;
            (* Update the set of active users Ai *)
            IF (sloti(t) = busy) AND (uk ∉ Ai(t−1)) THEN
                 Ai(t) := Ai(t−1) ∪ {uk};
            IF (sloti(t) = idle) AND (uk ∈ Ai(t−1)) THEN
                 Ai(t) := Ai(t−1) − {uk};
 (* Select station usi(t + 1) which is granted permission to
   transmit *)
            Select k such that di,k(t) = maxj{di,j(t)};
            IF (di,k(t) ≥ Δ) THEN si(t+1) := k
            ELSE IF (Ai(t) = ∅) THEN si(t+1) := (k mod N)+1
            ELSE
            BEGIN
                 REPEAT
                      k := (k mod N) + 1;
                 UNTIL uk ∈ Ai(t);
                 si(t+1) := k;
            END
       END
FOREVER;
```

4.1.3.3 Overview of Simulation Results for APS

In [29] the operation of the APS protocol is simulated and its performance is evaluated. For the sake of this evaluation APS is compared to the SALP protocol (described in a previous subsection). These protocols have many similarities, e.g. they are both self-adaptive, learning automata-based, applied to the same network model and assume the same network feedback mechanism at each station. Thus a comparison would be very helpful in order to evaluate the possible performance improvements introduced by APS.

Traffic was assumed to be bursty and correlated as is the case for gigabit LANs. The burstiness of the offered traffic was modelled in a way identical to the one used for the simulation of SALP, i.e. using the *packet train model*, which is also applied in [6, 10, 13–14]. According to it, each station alternately produces a 'train' (i.e. burst) of packets with the same destination followed by one or more empty slots. The bursts contain a geometrically distributed number of packets. The mean burst length B is one of the parameters of the packet train model.

We shall present from [29] the results from the simulation and application of APS and SALP to the following three network configurations:

3. Network N_1: $N = 40$, $W = 20$, $Q = 20$, $B = 20$.
4. Network N_2: $N = 100$, $W = 20$, $Q = 20$, $B = 20$.
5. Network N_3: $N = 100$, $W = 20$, $Q = 40$, $B = 40$.

Parameters N, W, Q and B are the number of stations, available wavelengths, queue size and mean burst length, respectively. The performance metric used was the delay versus throughput characteristic. It should also be pointed out that the special parameter Δ of the APS protocol was taken to be equal to 500 time slots.

The delay versus throughput characteristic of protocols APS and SALP for networks N_1, N_2 and N_3 are shown in parts (a), (b) and (c) of Figure 4.5, respectively.

It is obvious that APS introduces a significant improvement in performance for networks with bursty traffic and correlated packet destinations, as far as the packets of a single transmitter are concerned. However, as was described extensively before, both protocols refer to the same network model and make use of the same network feedback mechanism. In an effort to explain why APS outperforms SALP we could note the following:

- In the context of APS once a station enters the set of active stations for a specific wavelength, it is guaranteed that it is going to be granted permission to transmit on it within the next few time slots. Thus APS is a *deterministic* protocol. On the other hand, in the context of SALP it is possible for a station having packets for

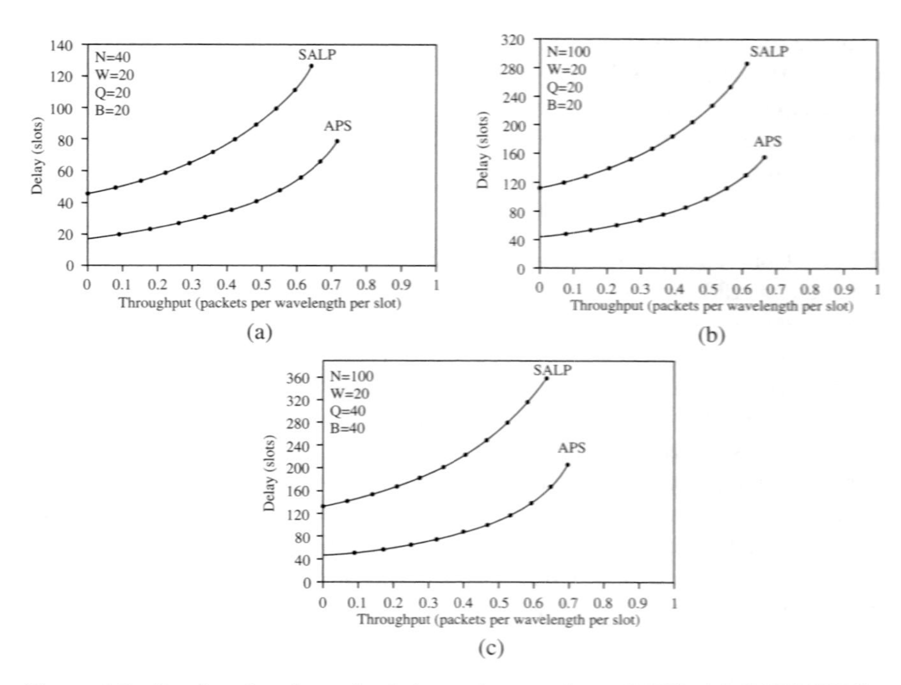

Figure 4.5 Results taken from simulation and comparison of APS with SALP [29] for three network configurations: $N_1(N = 40$ nodes, $W = 20$ wavelengths, W queues with length $Q = 20$, mean burst size $B = 20)$, N_2 (with $N = 100$, $W = 20$, $Q = 20$, $B = 20$) and N_3 ($N = 100$, $W = 20$, $Q = 40$, $B = 40$). Figure 4.5 shows the delay versus throughput characteristic of the protocols: (a) for N_1, (b) for N_2 and (c) for N_3. (© 2001 Wiley)

a certain wavelength to be forced to wait for an arbitrary number of time slots before being granted permission to transmit. This is because of the *stochastic* nature of SALP.

- APS provides better for the case where a station is continuously idle for a certain wavelength and suddenly forms a burst of packets destined for a station having this wavelength as home channel, i.e. transits from idle to active state for this wavelength. This situation is not an exception during typical operation of a WDM broadcast-and-select LAN, but may appear quite often if we consider the operation in the long run. As we saw, APS periodically grants transmission permissions to stations that were not selected for quite a long time (determined by parameter Δ). On the other hand, SALP does not sense efficiently enough (in comparison to APS) the transition of a station from idle to active state for a specific wavelength.

In the subsection where SALP was examined we saw that SALP brings about significant performance improvement in a network with bursty and correlated traffic in comparison to I-TDMA* and R-TDMA protocols. Consequently, APS also outperforms these protocols, since it clearly outperforms SALP, as is shown in Figure 4.5.

4.1.3.4 Concluding Remarks about APS

Concluding APS, we have seen that it is also a recent adaptive protocol that improves network performance in comparison to other both non-adaptive schemes, such as R-TDMA and I-TDMA*, and adaptive ones, such as the previously described SALP. As in SALP, there is no centralized control, but W learning automata controlling the W wavelengths are located in each station. The learning automata of all stations always select the same station as a transmitter for each channel during each time slot, i.e. they make exactly the same decisions all the time. This is because they always run the same algorithm presented in form of pseudocode before and always have common feedback information on account of the broadcast nature of the network. Consequently, APS protocol is collision-free, fully distributed and, by extension, it is also fault-tolerant, because a possible failure of one station would not affect the overall network operation. As far as the implementation cost of APS is concerned, as in the previous schemes, APS does not require significant additional cost, since the only additional hardware required is the feedback mechanism of Figure 4.3.

4.2 ADAPTIVE RANDOM ACCESS PROTOCOLS

We have already referred to random access MAC protocols designed for WDM local area networks of the broadcast-and-select architecture in Chapter 3. In random access protocols like slotted ALOHA, for example, each ready station transmits with a fixed transmission probability p and postpones transmission with the remaining probability $1 - p$ in the general case. Thus, it is obvious that such protocols do not provide any solution for the case where two or more stations choose to transmit on a specific wavelength at the same time; in other words, collisions are allowed to occur and exactly this is the cost of these protocols' comparative simplicity. In this section two random access schemes that make use of learning automata will be presented. By using learning automata, these protocols turn out to be adaptive in that they become capable of dynamically determining the transmission probability p for each wavelength, which is no longer fixed. In fact this probability is properly adapted close to an optimal value according to the current network traffic conditions with the help of the learning automata. The eventuality of collisions is not eliminated, however, since due to the random access nature of the protocols, two or more stations may still choose to transmit on a certain channel at the same time slot. Still, these new self-adaptive random access schemes introduce a significant improvement in network performance, as will be shown in more detail in the following subsections.

4.2.1 LABRA

We begin the presentation of adaptive random access protocols for WDM passive star networks of the broadcast-and-select architecture with the so-called *Learning Automata-Based Random Access* (*LABRA*) protocol [22, 24]. The LABRA protocol can be considered an extension of the well-known slotted ALOHA protocol which was examined in Chapter 3. Thus, LABRA considers the same network model as slotted ALOHA considers time-slotted data channels (synchronous) and is not collision-free, i.e. it does not eliminate the eventuality of concurrent transmissions on the same channel (collisions). The difference is that LABRA is capable of operating efficiently under any load conditions, because it is based on the use of learning automata.

We have already mentioned that, in the context of the slotted ALOHA protocol, each station which is ready to transmit on a certain wavelength chooses to transmit with a *fixed* transmission probability p and postpones transmission with probability $1 - p$. It is apparent that if two or more stations chose to transmit over the same channel at the same time slot, the transmitted packets would be destroyed, i.e. we would have a collision. Each of the transmitted stations that caused the collision would sense that a collision had occurred and, according to slotted ALOHA, would have to retransmit its packet at the next time slot with the same fixed transmission probability p. A very important deficiency of slotted ALOHA protocol, especially if it is considered to operate in gigabit local area networks like the multiwavelength optical LANs we consider, is that the assumption of a fixed transmission probability for each wavelength does not comply well at all with the variable traffic conditions that are encountered in practice.

This can easily be explained if we examine the probability of a successful transmission $P_{succ}(t)$ on a specific wavelength λ_i at time slot t under the slotted ALOHA protocol. Let M be the number of stations that are ready to transmit on λ_i at time slot t. This probability is equal to the probability that only the first station (out of the M) transmits and all others postpone transmission, or only the second (out of the M) transmits and all others postpone transmission, and so on, until the case only the Mth station (out of the M) transmits and all others postpone transmission. Thus, we have:

$$P_{succ}(t) = \sum_{k=1}^{M} p(1-p)^{M-1} = Mp(1-p)^{M-1}.$$

It is desirable to maximize this probability of successful transmission. By taking the first derivative of the right-hand side of the above relation and making it equal to zero we find that the probability of successful transmission would be maximized if we took $p = 1/M$, i.e. we have maximum throughput if the transmission probability is taken to be equal to the inverse of the number of ready stations for the specific wavelength λ_i at the specific time slot t. Generally, we notice that the optimum value of the transmission probability p on a certain wavelength depends on the number of users waiting to transmit on this wavelength at each time slot; in other words, it depends on the offered load to the specific wavelength. The following two observations can be made on the offered load of each specific wavelength of the networks we examine:

- Since we consider gigabit LANs, it has already been shown that traffic in such networks in highly bursty.[7] A high burstiness of offered traffic implies a high variability of the load offered to each wavelength.
- Even if we assume that generally the load offered to a WDM passive star network is stable, the load of each specific wavelength is time-variable. This is because at a certain time instant a large percentage of the (overall stable) offered load may be destined for the same wavelength. Under slotted ALOHA, this wavelength would suffer from a large number of collisions that would probably make the transmission rate much lower than the arrival rate of packets for this wavelength. As a result, the number of waiting packets would further increase and obviously the situation would form a vicious circle. In this case we say that the wavelength is overloaded. *Wavelength overloading* is a really undesirable situation that brings about a dramatic decrease in network throughput.

Since the load offered to each wavelength is variable, it follows that a choice of a fixed transmission probability for each wavelength does not comply well at all with this fact. In practice, a fixed transmission probability means that the protocol in question (e.g. slotted ALOHA) is incapable of adapting to the varying load conditions and thus it is highly probable that it will bring the network throughput down to extremely low levels, especially if we consider the highly bursty gigabit LAN traffic and the eventuality of wavelength overloading discussed above.

The operation of the examined LABRA protocol is based on the use of learning automata in order to properly adapt to the variable offered load conditions by dynamically determining the transmission probability p for each wavelength at each time slot. As we will see, in the context of LABRA the transmission probability asymptotically tends to take its optimum value presented above. Therefore, the protocol in question is capable of operating efficiently under any load conditions.

4.2.1.1 Network Model of LABRA

The LABRA protocol considers a network model comprising a WDM passive star broadcast-and-select network like the one depicted in Figure 4.2, which was also used in the previous adaptive schemes. Let us define here as well the set of stations as $U = \{u_1, u_2, \ldots, u_N\}$ and the set of wavelengths as $\Lambda = \{\lambda_1, \lambda_2, \ldots, \lambda_W\}$, where N and W are the number of stations and wavelengths respectively. Each station is provided with a transmitter which is a tunable laser capable of tuning over the whole range of the W available wavelengths. Optical fibres are used to connect the outputs of the lasers to the network hub. Each station is also provided with a receiver, which actually is a fixed-tuned optical filter that passes only one wavelength called the 'home wavelength' of the station, as we have seen. Obviously the system considered in LABRA is TT-FR; therefore, each station is capable of receiving only those packets transmitted on its home channel.

Of course, it should be noted that Figure 4.2 only illustrates the general network model used by LABRA. As we describe in the following subsection and the next figure, there is a slight variation in the receiving part of each station. Specifically, stations no longer need to have fixed optical filters for receiving the optical signals transmitted on their home wavelengths. This is because the operation of each filter is replaced in LABRA by directly connecting the corresponding photodetector to the appropriate output of a WDM demultiplexer, which has to be included in the internal architecture of each station. However, Figure 4.2 still generally depicts the network model used, as it shows a TT-FR system operating under the slotted ALOHA protocol, which is the base of LABRA as well. All the details of LABRA including the variation mentioned before are presented extensively in what follows.

4.2.1.2 Detailed Description of LABRA

As was stated in the introduction to LABRA, the optimum transmission probability varies with time and wavelength, since it depends on the number of packets waiting to be transmitted on each wavelength during each time slot. In order to dynamically determine, as optimally as possible, the transmission probabilities for each wavelength, an array of W learning automata LA_i where $i = 1, \ldots, W$, has to be placed at each station u_k. Each learning automaton corresponds to a specific wavelength. In general LA_i corresponds to wavelength λ_i and determines the transmission probability $P_i(t)$ at each time slot t for this wavelength. Apart from this important differentiation, the operation of LABRA remains similar to that of slotted ALOHA. Thus, at each time slot t, each station u_k which is ready to transmit on wavelength λ_i, transmits its packet with probability $P_i(t)$ and postpones transmission with the remaining probability $1 - P_i(t)$. Therefore, we still have a random access protocol, but with suitably varying transmission probabilities.

The central issue in LABRA is how to determine transmission probabilities. A high value of transmission probability $P_i(t)$ would commonly result in a large number of collisions on wavelength λ_i. On the other hand, a relatively small value of $P_i(t)$ would generally mean that the number of idle slots on λ_i is increased. Both cases are undesirable, since they lead to a decreased throughput of channel λ_i. As discussed in the introduction to LABRA, the ideal value of the transmission probability at each time slot is equal to $1/M$, where M denotes the number of users that are ready to transmit on this wavelength. The problem is that M varies with time and is generally unknown. Next, let us describe in detail just how LABRA protocol tries to deal with this situation, i.e. how the learning automata of LABRA update the transmission probabilities for each wavelength at each time slot.

Since the protocol is not collision-free, there are three possible events that may take place for wavelength λ_i at time slot t; namely, we may have a successful transmission of a single packet, no transmission at all (wavelength stays idle) or a transmission of at least two packets which inevitably leads to a collision. If we denote by M the number of users ready to transmit on wavelength λ_i at time slot t, the probabilities of these three events respectively are:

$$P_{succ}(t) = \sum_{k=1}^{M} P_i(t)(1 - P_i(t))^{M-1} = MP_i(t)(1 - P_i(t))^{M-1};$$

$$P_{id}(t) = (1 - P_i(t))^M;$$

$$P_{col}(t) = 1 - P_{id}(t) - P_{succ}(t) = 1 - (1 - P_i(t))^M - MP_i(t)(1 - P_i(t))^{M-1}.$$

The main objective of the learning automata used by LABRA is to determine the transmission probability as near as possible to the optimal $P_i(t) = 1/M$, so that $P_{succ}(t)$ always tends to be maximized. In this direction, if the wavelength was idle at the last time slot $(t - 1)$, the learning automata decide to increase probability $P_i(t)$, because an idle slot is probably due to a relatively small value of the transmission probability. Accordingly, if a successful transmission took place at the last time slot $(t - 1)$, it is concluded that this probably occurred due to the small number of packets waiting to be transmitted on wavelength λ_i. That is why the transmission probability $P_i(t)$ is also increased. Thus, in both the events of successful and no transmission at all, the N identical learning automata LA_i which are located at the N stations decide to increase the transmission probability $P_i(t)$ for wavelength λ_i. On the contrary, a collision during the last time slot $(t-1)$ can be ascribed to a high value of transmission probability; consequently, the learning automata LA_i choose to decrease the transmission probability $P_i(t)$ on λ_i for the next time slot.

Let us denote by $\mathrm{SLOT}_i(t) \in \{\mathrm{SUCCESS, IDLE, COLLISION}\}$ the state of wavelength λ_i during time slot t. The previous observations imply that the learning automata should generally update the transmission probabilities as follows:

$$\begin{aligned} &P_i(t+1) = P_i(t) + \Delta_1, \\ &\quad \text{if } \mathrm{SLOT}_i(t) = \mathrm{SUCCESS} \text{ or } \mathrm{SLOT}_i(t) = \mathrm{IDLE}. \\ &P_i(t+1) = P_i(t) - \Delta_2, \\ &\quad \text{if } \mathrm{SLOT}_i(t) = \mathrm{COLLISION}. \end{aligned}$$

In the above equations, we have $0 < \Delta_1 < 1 - P_i(t)$ and $0 < \Delta_2 < P_i(t)$. Thus, the main objective of LABRA reduces to an appropriate choice of Δ_1 and Δ_2, so that $P_i(t)$ tends to be equal to $1/M$.

In the direction of finding the appropriate values of Δ_1 and Δ_2, we could notice that if $P_i(t) = 1/M$, $\lim_{M\to\infty}\{P_{succ}(t) + P_{id}(t)\} = \lim_{M\to\infty} d_i(t) = 2e^{-1} = 0.736$, where $d_i(t)$ denotes the probability of either having a successful or no transmission at all; since, according to the general probability updating scheme, this implies that the transmission probability should be increased, probability $d_i(t)$ is also called 'reward probability'. Moreover, we notice that for small values of M $(M \geq 2)$, the reward probability $d_i(t)$ takes values very close to $2\mathrm{e}^{-1} = 0.736$. For example, considering the relations of $P_{succ}(t)$ and $P_{id}(t)$ presented before and bearing in mind that $d_i(t) = P_{succ}(t) + P_{id}(t)$, we have that $M = 2 \Rightarrow d_i(t) = 0.750$, $M = 3 \Rightarrow d_i(t) = 0.741$, $M = 4 \Rightarrow d_i(t) = 0.738$, $M = 5 \Rightarrow d_i(t) = 0.737$, etc. This important observation guarantees that the unknown optimal value of $P_i(t) = 1/M$ corresponds to a known value of $d_i(t) = 2\mathrm{e}^{-1} = 0.736 = \upsilon$.

Furthermore, we could notice that for any wavelength λ_i (with $M \geq 2$) the reward probability $d_i(t)$ is a monotonically decreasing function of the transmission probability $P_i(t)$. Indeed, if we consider the reward probability as a function of the transmission probability P_i and take its first derivative, for $P_i \in (0, 1)$ and $M \geq 2$ we have that:

$$d_i'(P_i) = \left[(1-P_i)M + MP_i(1-P_i)^{M-1}\right]' = -(M-1)MP_i(1-P_i)^{M-2} < 0.$$

This fact implies that the wanted value υ of the reward probability $d_i(t)$ can be achieved by increasing or decreasing the value of $P_i(t)$ appropriately. Specifically, if we had $d_i(t) > v$ and therefore wanted the reward probability to be decreased (i.e. $\delta d_i(t) = d_i(t+1) - d_i(t) < 0$) so that it gets closer to v, we would just have to increase the transmission probability i.e. $\delta P_i(t) = P_i(t+1) - P_i(t) > 0$. And inversely, if we wanted $d_i(t+1) - d_i(t) > 0$, i.e. a higher value for the reward probability, we would have to decrease the transmission probability thus get $P_i(t+1) - P_i(t) < 0$. If we further analyse the general updating scheme presented before, we get the following equation (4.1):

$$\begin{aligned} E[\delta P_i(t)] &= E[P_i(t+1) - P_i(t)] = d_i(t)\Delta_1 - (1 - d_i(t))\Delta_2 \\ &= \left(\Delta_1 + \Delta_2\right)\left(d_i(t) - \frac{\Delta_2}{\Delta_1 + \Delta_2}\right) \end{aligned} \tag{4.1}$$

In order to asymptotically converge to the desirable point where $d_i(t) = v$, the probability updating scheme used by the learning automata LA_i must satisfy the following three properties:

1. If $d_i(t) > v$ then the transmission probability must be increased, i.e. we should have $E[\delta P_i(t)] > 0$ and consequently $E[\delta d_i(t)] < 0$ (due to the fact that $d_i(t)$ is a monotonically decreasing function of the transmission probability $P_i(t)$).
2. If $d_i(t) < v$ then the transmission probability must be decreased, i.e. we should have $E[\delta P_i(t)] < 0$ and consequently $E[\delta d_i(t)] > 0$.
3. If $d_i(t) = v$ then the transmission probability must remain the same, i.e. we should have $E[\delta P_i(t)] = 0$ and consequently $E[\delta d_i(t)] = 0$.

Since from equation (4.1) $(\Delta_1 + \Delta_2)$ is always positive, the sign of $E[\delta P_i(t)]$ is determined by the difference $d_i(t) - \Delta_2/(\Delta_1 + \Delta_2)$. It is easy to see that all the three previous properties are satisfied if $\Delta_2/(\Delta_1 + \Delta_2) = v$ or equivalently if $\Delta_1 = ((1-v)/v)\Delta_2$. By denoting $((1-v)/v)$ as h, it would suffice to have $\Delta_1 = h\Delta_2$, where $h = ((1-v)/v) = (1 - 2\mathrm{e}^{-1})/2\mathrm{e}^{-1} = 0.359$. If we set $\Delta_2 = \Delta$, the general probability updating schemes changes to the following:

$$\begin{aligned} &P_i(t+1) = P_i(t) + h\Delta, \\ &\quad \text{if } \mathrm{SLOT}_i(t) = \mathrm{SUCCESS} \text{ or } \mathrm{SLOT}_i(t) = \mathrm{IDLE}. \\ &P_i(t+1) = P_i(t) - \Delta, \\ &\quad \text{if } \mathrm{SLOT}_i(t) = \mathrm{COLLISION}. \end{aligned}$$

In this probability updating scheme $0 < \Delta < (1 - P_i(t))/h$ and $0 < \Delta < P_i(t)$. Now, the problem of how the learning automata should update the transmission probabilities reduces to the appropriate choice of parameter Δ. In [22, 24] a first attempt for this is to select $\Delta = LP_i(t)(1 - P_i(t))$, where $0 < L < 1$. The probability updating scheme would then become:

$$P_i(t+1) = P_i(t) + hLP_i(t)(1 - P_i(t)),$$
$$\text{if } \mathrm{SLOT}_i(t) = \mathrm{SUCCESS} \text{ or } \mathrm{SLOT}_i(t) = \mathrm{IDLE}.$$
$$P_i(t+1) = P_i(t) - LP_i(t)(1 - P_i(t)),$$
$$\text{if } \mathrm{SLOT}_i(t) = \mathrm{COLLISION}.$$

However, a serious weakness of this scheme led the authors in [22, 24] to appropriately modify it; it can be noticed that when the transmission probability takes values very close to zero or one, the probability updating step becomes too small. This implies that the adaptivity of the corresponding learning automata would decrease dramatically and, accordingly, this would be the case as well for network throughput. Thus, the probability updating scheme is eventually proposed to be the following:

$$\begin{aligned} P_i(t+1) &= P_i(t) + hLP_i(t)(1 - P_i(t)) + aL^2(1 - P_i(t))^2, \\ &\quad \text{if } \mathrm{SLOT}_i(t) = \mathrm{SUCCESS} \text{ or } \mathrm{SLOT}_i(t) = \mathrm{IDLE}. \\ P_i(t+1) &= P_i(t) - LP_i(t)(1 - P_i(t)) - bL^2(P_i(t))^2, \\ &\quad \text{if } \mathrm{SLOT}_i(t) = \mathrm{COLLISION}. \end{aligned} \tag{4.2}$$

In this scheme parameter $L \in (0, 1)$ and, also, for the two new parameters a and b we have $a, b \in (0, 1/L)$. As we shall see when we report the main results from the analysis of LABRA, the above mentioned probability updating scheme tends to assign an optimal value to the transmission probability of each wavelength at each time slot.

As far as the actual implementation of LABRA is concerned, we notice that the W learning automata located at each station only require knowing the value of $\mathrm{SLOT}_i(t)$ at each time slot t, because in this case they would know exactly how to update the transmission probabilities according to the scheme above. It is easy to notice that the learning automata actually need to know whether there was a collision or not during the last time slot.[8] This information is provided to them by the network feedback mechanism included in the internal architecture of each LABRA station, as illustrated in Figure 4.6.

Each station operating under the LABRA protocol is provided with a WDM demultiplexer, which gets the optical signal from the star coupler as input and separates the individual wavelengths. Each one of the separated wavelengths is then detected for collision. The collision detection operation can be implemented either by computing the checksum of the packet's header or by measuring the

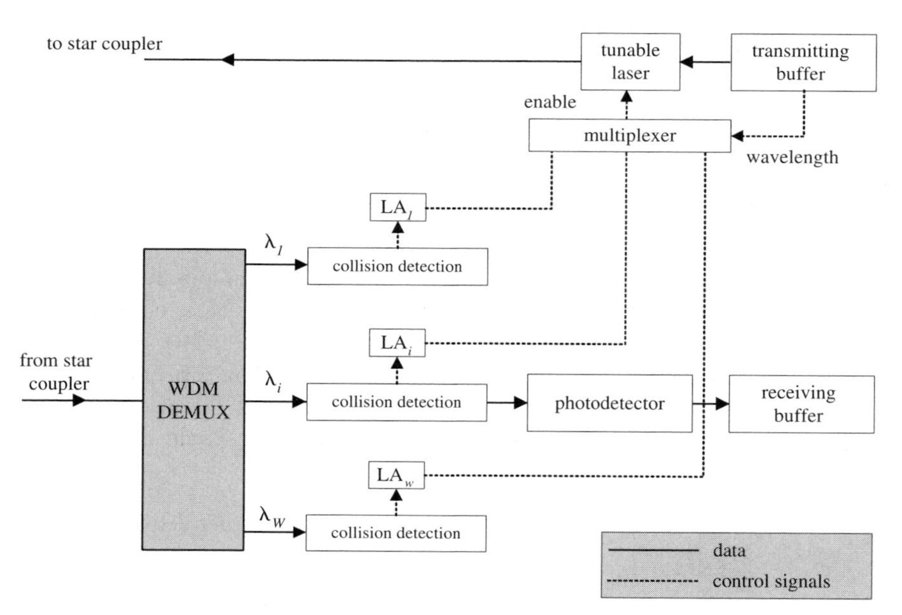

Figure 4.6 The internal architecture of a station with wavelength λ_i as its home channel in the context of LABRA protocol [22]. (© 1993 IEEE)

optical power of the signal by means of a photofet. As is shown in Figure 4.6 and described in the previous paragraph, the output of each collision detection mechanism is fed to the corresponding learning automaton, since it provides the learning automaton with all the necessary information needed in order to update the transmission probability of the wavelength it controls for the next time slot. It should be noted here that the end-to-end propagation delay is assumed to be negligible; thus the network feedback information concerning the presence or absence of a collision during the last time slot is immediately available to the corresponding learning automata.[9]

Figure 4.6 also shows why the stations using the LABRA protocol no longer need to have fixed optical filters to receive the optical signals transmitted on their home wavelengths. This statement was first made when we examined the network model of LABRA in the previous subsection. By looking at the figure, it is clear that the operation of each filter is replaced in LABRA by directly connecting the corresponding photodetector to the appropriate output of the WDM demultiplexer. Figure 4.6 depicts a station having wavelength λ_i as its home wavelength for reception. Thus it is obvious that the role of the WDM demultiplexer is twofold.

Given that all stations have the same architectural structure shown in Figure 4.6, that the network feedback is common for every one and that all learning automata use the same scheme for updating the transmission probability of the wavelength they control, it follows that the N learning automata LA_i always contain the same transmission probability for wavelength λ_i, for all i, i.e. for $i = 1, \ldots, W$. Therefore, since there is no central coordination of the learning automata, the LABRA protocol has the virtue of being fully distributed. This feature also implies that LABRA is fault-tolerant; the network feedback and the operation of the learning automata are not influenced by the failure of a single station, thus the overall operation of the LABRA protocol remains practically unaffected in such an event.

4.2.1.3 Overview of Analytical and Simulation Results for LABRA

In this subsection an overview of the main results taken from the analysis and simulation of the LABRA protocol is presented.

The final probability updating scheme represented by relations (4.2) of the previous subsection is analyzed in [22, 24] and it is proved that for small values of parameter L, the transmission probability $P_i(t)$ asymptotically tends to take its optimum value, i.e. it tends to be equal to $1/M$, where M represents the number of stations that are ready to transmit on wavelength λ_i. Thus, we have:

$$\lim_{L\to 0, t\to\infty} P_i(t) = \frac{1}{M},$$

if we suppose that the probability updating scheme given by relations (4.2) is used. Of course, this is the case for all wavelengths, where M has different values from wavelength to wavelength and is also slowly varying with time. Details of the proof of this result can be found in [24].

As far as simulation of the protocol's behaviour is concerned, LABRA is simulated and compared to the slotted ALOHA protocol in [24]. As described in the introduction, the LABRA protocol is a random access protocol that applies to exactly the same network model as the slotted ALOHA protocol. Also, its operation is identical besides the fact that the transmission probabilities are not fixed as slotted ALOHA assumes, but they are determined in a dynamic way according to the relevant decisions of learning automata. Hence, LABRA can be considered an extension of slotted ALOHA protocol and a comparison of these schemes would visibly reveal the performance improvement introduced by LABRA.

For each simulated network in [24], the slotted ALOHA protocol was simulated for certain different values of the fixed transmission probability p. At any rate, the objective of each selection was to ensure that there is no other value of p, which may lead to a considerably higher throughput of the slotted ALOHA protocol.

Next we only summarize the simulation results from the comparison between the two protocols; we defer the presentation of a diagram illustrating the performance difference between LABRA and slotted ALOHA until the next subsection. (Figure 4.8 there actually illustrates the performance comparison of slotted ALOHA with a protocol performing just like LABRA, namely LABON.) It is observed [22, 24] that LABRA achieves a lower delay and a higher throughput than the slotted ALOHA protocol. All in all, the simulation result can be summarized in that the throughput of the LABRA protocol is higher than that of slotted ALOHA *under any load conditions*. That is to say, LABRA outperforms slotted ALOHA regardless of considering high or low load conditions, or assuming the load is fixed or variable. Of course, it should be noted that under gigabit LAN traffic (with the attributes already described) the improvement in throughput introduced by LABRA becomes even more remarkable.

4.2.1.4 Concluding Remarks about LABRA

In conclusion, we have seen that it is a relatively recent random access self-adaptive protocol that improves network performance in comparison with other non-adaptive random access schemes. In the context of LABRA, there is no centralized control, but W learning automata dynamically determining the transmission probabilities of the W wavelengths are located in each station. The learning automata of all stations in the TT-FR system assumed by LABRA, always contain the same transmission probabilities; hence, LABRA is fully distributed and, by extension, it is also fault-tolerant as explained before. On the other hand, due to its random access nature, LABRA is not collision-free, like the schemes that had been described so far. However, the LABRA protocol still significantly improves the network throughput on account of the learning mechanism of each station, which adapts the values of the transmission probabilities for each wavelength in such a way that they asymptotically tend to take their optimum values.

4.2.2 LABON

In this section another relatively recent random access MAC protocol that is based on the use of learning automata is presented. The proposed protocol is also destined to be applied in WDM passive star networks of the broadcast-and-select architecture. As we mentioned before, the following two observations can be made about gigabit LAN traffic which is considered to be the case in such networks; these observations mainly relate to schemes suggesting that the various stations attached to the network should access the shared medium in a random access fashion:

- Gigabit LAN traffic is highly bursty, thus the offered load to the network is intrinsically variable with time.
- Even if the overall offered load was considered stable, this would not imply that the load offered to the individual wavelengths would be stable as well. Indeed, as we described in the section about LABRA protocol, it is possible to have many stations wanting to transmit over the same wavelength at the same time. Random access schemes would suffer from a high number of collisions in this case and, since collided packets typically have to be retransmitted, it follows that the packets waiting to be transmitted over the specific wavelength would grow even more; hence we say that the wavelength would be overloaded in such a case.

In traditional random access protocols like slotted ALOHA, the selection of a fixed transmission probability p according to which stations should transmit over a wavelength is in full contradiction to the variability of the offered load (at least for individual wavelengths) described in the above observations. Therefore, wavelength overloading turns out to be really disadvantageous for the throughput of the network and generally such schemes are not recommended for high speed (gigabit) local area networks.

In this context, random access protocols would benefit a lot if they could somehow adjust the transmission probabilities of individual wavelengths according to the variations of the offered load. As we saw, LABRA is a protocol that achieves such an adaptivity of the transmission probabilities by incorporating a learning mechanism to the internal architecture of each station attached to the network. Thus, the adaptivity is included in the hardware implementation of the protocol in a distributed fashion.

The protocol proposed here attempts to achieve analogous results in a somewhat more centralized fashion, i.e. by incorporating a learning mechanism into the network hub. As a consequence the network hub is no longer passive and the protocol should be supported by a fairly modified architecture, especially as far as the hub is concerned. The protocol is called *Learning Automata-Based Optical Network* (*LABON*) protocol [21, 26] and in fact, apart from the learning algorithm, it includes the design of a specific architecture supporting its operation, the so-called *LABON architecture*. It should be noted, however, that LABON is still based on the same network model (as we will see) that is used by LABRA or slotted ALOHA. Actually, given that LABON is a random access protocol which makes use of learning automata in order to dynamically determine the transmission probabilities for each wavelength, we could also consider it an extension of the slotted ALOHA scheme, as we did LABRA. There are several similarities of the proposed LABON protocol with the previously described LABRA protocol, especially as far as the learning algorithm is concerned. Both the similarities with previous schemes and the innovations of LABON will become clear when we examine in more detail the LABON architecture and learning protocol. But before that, let us first describe the network model comprising the basis of LABON.

4.2.2.1 Network Model of LABON

The general network model considered by LABON can be described by the WDM passive star broadcast-and-select network shown in Figure 4.2, which was also used by the previous adaptive protocols. The set of stations may also be defined as $U = \{u_1, u_2, \ldots, u_N\}$ and the set of wavelengths as $\Lambda = \{\lambda_1, \lambda_2, \ldots, \lambda_W\}$, where N and W are the number of stations and wavelengths respectively. Each station is provided with a transmitter which is a tunable laser capable of tuning over the whole range of the W available wavelengths. Optical fibres are used to connect the outputs of the lasers to the network hub. Each station is also provided with a receiver, which actually is a fixed-tuned optical filter that passes only one wavelength called the 'home wavelength' of the station, as we have seen. As a result, the stations are capable of receiving only those packets transmitted on their home wavelength. Clearly, the system considered in LABON is TT-FR too.

The network of Figure 4.2, however, only comprises a depiction of the general network model upon which LABON is based. As we mentioned in the introduction to LABON, the hub of the network is no longer a $N \times N$ passive star coupler. Actually, the hub includes a $N \times N$ passive star coupler, but it also incorporates a learning mechanism which determines in a dynamic way the transmission probabilities of each wavelength according to network feedback information. The internal architecture of the hub will be described further in the following subsection.

4.2.2.2 Detailed Description of LABON Architecture and Protocol

As we saw, under the slotted ALOHA scheme the probability of a successful transmission $P_{succ}(t)$ on a specific wavelength λ_i at time slot t depends on the number of stations waiting to transmit on this wavelength at this particular instant. Let us denote the number of ready stations for wavelength λ_i at time slot t as M and the fixed transmission probability of slotted ALOHA as p. The probability of successful transmission was explained to be equal to the probability that only the first station (of the M) transmits and all others postpone transmission or only the second (of the M) transmits and all others postpone transmission and so on up to the case where only the Mth station (of the M) transmits and all others postpone transmission. Thus, we have:

$$P_{succ}(t) = \sum_{k=1}^{M} p(1-p)^{M-1} = Mp(1-p)^{M-1}.$$

We have shown that the probability of successful transmission would be maximized if we took $p = 1/M$. Since the offered load in gigabit local area networks was shown to be variable with time and wavelength, it follows that M varies accordingly; thus the optimal transmission probability cannot be effectively captured by a fixed value.

In order to dynamically determine the transmission probabilities for each wavelength, the LABON protocol assumes that additional functionality is placed at the hub of the network. First of all, an array of W learning automata LA_i, $i = 1, \ldots, W$, is placed at the network hub. In the context of LABON, unlike slotted ALOHA and LABRA, when a station has a packet to transmit on a certain wavelength, it transmits this packet with probability one immediately in the next time slot. The learning automata control the *passing* of the transmitted packets to the star coupler also included in the hub. Specifically, learning automaton LA_i controls wavelength λ_i and determines in a random access manner (that will be explained in a short while) exactly which packets transmitted on this wavelength will make it to the star coupler. Obviously, if two or more packets are allowed to pass on the same wavelength at the same time we will have a collision, i.e. the random access nature of the protocol implies that the eventuality of collisions is not eliminated. Given that each packet is transmitted with probability one and its passing to the star coupler is probabilistically decided at the network hub, it would be more appropriate to use the term ‘passing probability’ instead of the term ‘transmission probability’ in the context of this protocol. At each time slot t the learning automaton LA_i contains the passing probability $P_i(t)$ for wavelength λ_i. Therefore a packet transmitted on this wavelength will either pass to the star coupler with probability $P_i(t)$ or be blocked with the remaining probability $1 - P_i(t)$.

The passing and blocking of packets is a task carried out in practice by *Acousto-Optic Tunable Filters* (*AOTFs*), which are known to have multiple wavelength selection capability [4]. In point of fact, the number of filters used in the LABON architecture is equal to the number of stations. Thus, there is an array of N acousto-optic tunable filters F_k ($k = 1, \ldots, N$) placed at the network hub as well, where filter F_k corresponds to user u_k. Actually, the output of the tunable laser of user u_k is connected to filter F_k by means of an optical fibre, for all k. An example of the LABON architecture for a simple network with four stations and three

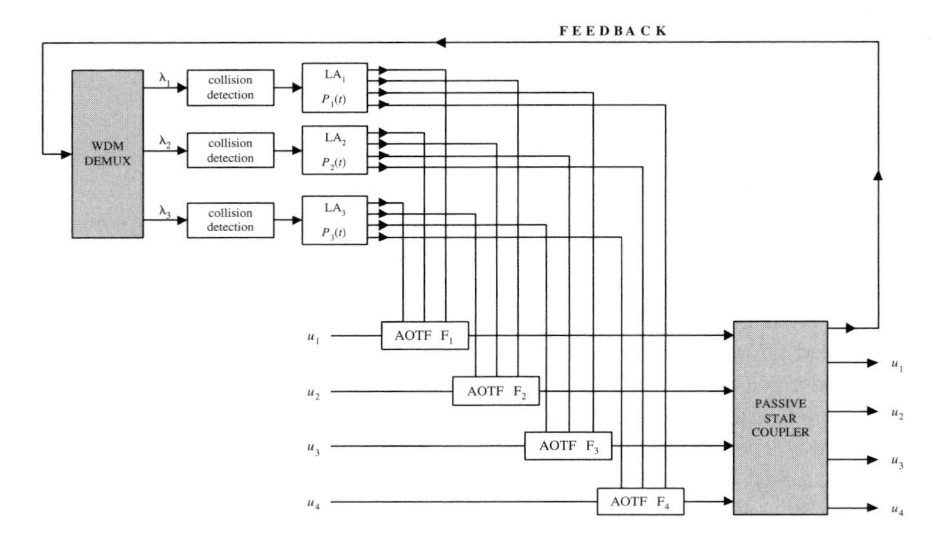

Figure 4.7 Hub of a network with four stations and three wavelengths for the LABON protocol. (Reprinted from [26], copyright 1996, with permission from Elsevier)

wavelengths is illustrated in Figure 4.7. The architecture will be further explained in the following.

At any time slot t, each filter F_k ($k = 1,\ldots, N$) is tuned by the learning automata to pass only a subset of wavelengths $\Lambda_k(t)$ (with $\Lambda_k(t) \subseteq \Lambda$). Let us assume that user u_k transmits a packet on wavelength λ_i at time slot t. If $\lambda_i \in \Lambda_k(t)$, the packet will be allowed by filter F_k to pass to the star coupler. On the other hand, if $\lambda_i \notin \Lambda_k(t)$, the packet will be blocked by the filter. A key issue in the LABON architecture and protocol is how the learning automata tune each optical filter in the beginning of each time slot, i.e. how they determine the subsets $\Lambda_k(t), k = 1, \ldots, N$. Since each wavelength is controlled by one separate learning automaton, it follows that the determination of the particular subset $\Lambda_k(t)$ requires control information by all the W learning automata. Hence, appropriate control information must be provided by each LA_i, ($i = 1, \ldots, W$) to the single filter F_k at each time slot. This is also depicted in Figure 4.7, where outputs of each learning automaton comprise inputs of control information for each filter.

The control information provided by the learning automata LA_i ($i = 1, \ldots, W$) to the filter F_k at the beginning of each time slot is actually a decision on whether wavelength λ_i should be included in subset $\Lambda_k(t)$ or not, or accordingly whether the filter should select wavelength λ_i or block it. In the former case, if the corresponding station u_k actually transmits a packet on λ_i at this time slot, the packet will pass to the star coupler, while in the latter case the packet will be blocked. Of course, if the station does not actually transmit on this wavelength, the specific decision about wavelength λ_i would be of no interest to it. In order to see how the decisions are made by the learning automata, let us assume that each automaton LA_i ($i = 1, \ldots, W$) contains the passing probability $P_i(t)$ of wavelength λ_i at each time slot t, without worrying now about the way this is determined.[11] At the beginning of each time slot, the automaton LA_i chooses N random numbers $r_{1,i}(t)$, $r_{2,i}(t)$,..., $r_{N,i}(t)$ uniformly distributed in the range of (0, 1). For $k = 1, \ldots, N$, the decision of LA_i that regards filter F_k is the following:

- If $r_{k,i}(t) < P_i(t)$ then $\lambda_i \in \Lambda_k(t)$, i.e. the filter F_k allows the wavelength λ_i to pass to the star coupler.
- Otherwise, if $r_{k,i}(t) \geq P_i(t)$ then $\lambda_i \notin \Lambda_k(t)$, i.e. the filter F_k blocks the wavelength λ_i.

Obviously, the way these decisions are made explains the random access character of the LABON protocol. According to the above decision scheme at a specific time slot t, the higher the value of $P_i(t)$, the more filters are probable to be tuned to pass wavelength λ_i; and conversely, the lower the value of $P_i(t)$, the less filters are likely to be tuned to pass λ_i. Hence, this decision scheme complies well with the definition of $P_i(t)$ as the 'passing probability' of wavelength λ_i.

The final fundamental issue of the LABON protocol that remains to be discussed is the learning algorithm used by the learning automata LA_i in order to determine the passing probabilities $P_i(t)$ of wavelengths λ_i, $i = 1, \ldots, W$, at each time slot t. According to the discussion in the beginning of this subsection, the probability of a successful transmission on wavelength λ_i at time slot t, and therefore the throughput of wavelength λ_i, is maximized when the passing probability takes the optimum value $P_i(t) = 1/M$, where M denotes the number of stations that are ready to transmit on λ_i at time slot t. It would be ideal to always choose $P_i(t) = 1/M$, but, since the value of M is unknown, the objective of the learning process included in LABON reduces to determining passing probabilities as close as possible to the optimum value. The learning process used by the LABON protocol is identical to the one presented in the context of the previously described LABRA protocol. Thus, the probability updating scheme used is the following:

$$
\begin{aligned}
&P_i(t+1) = P_i(t) + hLP_i(t)(1 - P_i(t)) + aL^2(1 - P_i(t))^2, \\
&\quad \text{if } \text{SLOT}_i(t) = \text{SUCCESS or } \text{SLOT}_i(t) = \text{IDLE}. \\
&P_i(t+1) = P_i(t) - LP_i(t)(1 - P_i(t)) - bL^2(P_i(t))^2, \\
&\quad \text{if } \text{SLOT}_i(t) = \text{COLLISION},
\end{aligned}
$$

where $L \in (0, 1)$ and $a, b \in (0, 1/L)$.

From the probability updating scheme above, it becomes clear that in order to determine the passing probability for the next time slot $P_i(t+1)$ the learning automata need to know whether we had a success, an idle slot or a collision during the last time slot t, i.e. the value of $\text{SLOT}_i(t)$. What is enough, more specifically, is to know whether there was a collision or not. In the former case the passing probability has to be increased and in the latter it must be decreased, according to the scheme above. This update of the passing probabilities takes place at the end of each time slot. The feedback mechanism used to provide the automata with the necessary information is also illustrated in Figure 4.7. We notice that one of the output ports of the star coupler is connected to a WDM demultiplexer which separates the individual wavelengths. Each wavelength is then detected for collision by computing the checksum of the packet's header. The feedback information produced is fed to the corresponding learning automaton, which updates appropriately the passing probability of the wavelength it controls. Subsequently, i.e. at the beginning of the new time slot, each learning automaton generates N random numbers and compares them to the value of the passing probability in order to make a decision about how it should tune each filter with regard to the wavelength it controls. This process was described in detail before.

One last point that has to be made about the operation of LABON regards the way transmitting stations are informed about the results of their transmissions. Considering the fact that stations always transmit with probability one and decisions about the blocking or passing of packets to the star coupler are made in a centralized fashion (inside the network hub), it becomes clear that there has to be some way for the stations to discover the results of their transmissions. Apparently, this is necessary since a blocked or collided packet always has to be retransmitted; if not, the packet would get lost, which is absolutely undesirable if we wish reliable network operation. The following two methods might be used for this purpose:

1. The transmitter could 'hear' the wavelength on which it transmitted, by using a tunable filter that should be tuned at the transmitting wavelength each time.
2. A short acknowledgement subslot could follow the data transmission slot. This subslot should be accessed in a time-division multiplexing fashion by the receiving stations. Because of this, the acknowledgements sent back by the receiving stations to the corresponding transmitting ones would never collide.

By making use of one of the above methods, if a station u_k realizes that the packet it transmitted did not reach its destination (because it was blocked by filter F_k or collided with another packet), it retransmits the packet at the next time slot (again with probability one according to LABON).

4.2.2.3 Overview of Analytical and Simulation results for LABON

In this subsection an overview of the main results taken from the analysis and simulation of the LABON architecture and protocol is presented.

The probability updating scheme presented in the previous subsection is analyzed in [21, 26] and it is proved that for small values of parameter L, the passing probability $P_i(t)$ asymptotically tends to take its optimum value, i.e. it tends to be equal to $1/M$, where M represents the number of stations that are ready to transmit on wavelength λ_i at time slot t. Thus, we have:

$$\lim_{L\to 0, t\to\infty} P_i(t) = \frac{1}{M},$$

if the probability updating scheme of LABON is used. The analysis and proof of this claim are identical with the analogous ones for LABRA protocol and can be found in [26].

As far as simulation is concerned, LABON is simulated and compared to the slotted ALOHA protocol. As described in the introduction, the LABON protocol is a random access protocol that applies to the same network model as the slotted ALOHA one. Hence, LABON can be considered an extension of slotted ALOHA protocol and a comparison of these schemes would clearly demonstrate the performance improvement introduced by LABON.

Suggestively, we present from [26] the simulation results for a network with $N = 16, W = 8$. Details considering the exact values of the total bandwidth, packet size, queue size and the allocation scheme of home channels to stations assumed for the real network that was simulated, can be found in [26]. Parameters L, a and b of the LABON protocol were taken to be equal to 0.3, 0.1 and 3.0 respectively for the sake of simulation.

The slotted ALOHA protocol was simulated for three different values of the fixed transmission probability p; specifically, p was alternately chosen to be 0.15, 0.20 and 0.25. The objective of each selection was to ensure that there is no other value of p which may lead to a considerably higher throughput of the slotted ALOHA protocol. Details on how and why these values were selected (for this and other simulated network configurations) may also be found in [26].

The throughput versus offered load characteristic of the compared protocols for the considered network is illustrated in Figure 4.8. From Figure 4.8 it becomes clear that LABON achieves a higher throughput than the slotted ALOHA protocol. In [26] the delay versus throughput characteristic of these protocols for the above network (and other configurations) is also illustrated.[12]

The overall simulation result can be summarized in that the throughput of the LABON protocol is higher than that of slotted ALOHA under any load conditions.

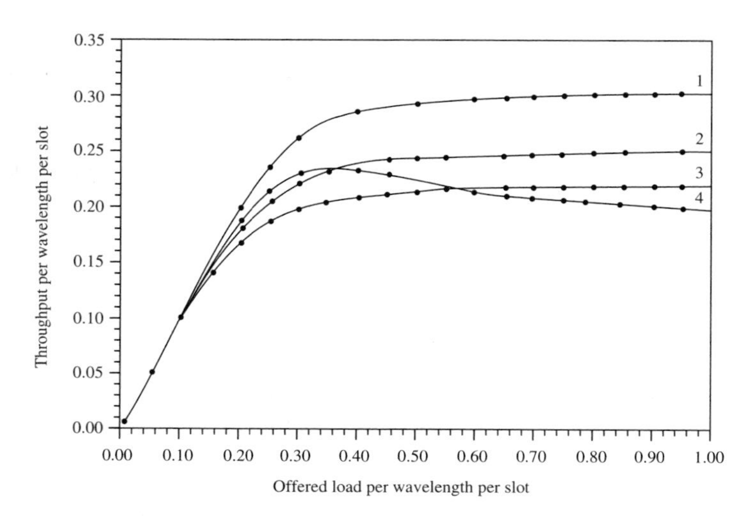

Figure 4.8 Results taken from simulation and comparison of LABON with slotted ALOHA, showing the throughput versus offered load characteristic of the protocols for a network with $N = 16$ nodes, $W = 8$ wavelengths. Curve 1: LABON, 2: s. ALOHA with $p = 0.20$, 3: s. ALOHA with $p = 0.15$ and 4: s. ALOHA with $p = 0.25$. (Reprinted from [26], copyright 1996, with permission from Elsevier)

That is to say, the LABON protocol outperforms slotted ALOHA regardless of considering high or low load conditions, or assuming the load is fixed or variable. Of course, it should be noted that under gigabit LAN traffic (with the attributes already described) the improvement in throughput introduced by LABON becomes even more remarkable. Note that the simulation results concerning LABON are identical to the ones for the (previous) LABRA protocol (thus, curve 1 of Figure 4.8 above represents the performance of both LABON and LABRA).

4.2.2.4 Concluding Remarks about LABON

In conclusion, we have seen that it is a relatively recent random access self-adaptive protocol that improves network performance in comparison to other non-adaptive random access schemes. In the context of LABON, W learning automata dynamically determining the passing probabilities of the W wavelengths to the star coupler are located in the network hub. The learning automata always contain the same passing probabilities and at the beginning of each time slot they choose N random numbers (one for each station), which they compare to the passing probabilities. Next, they tune N acousto-optic tunable filters (also located at the network hub and connected to the N stations) to allow or block the corresponding wavelengths, according to the results of the comparison mentioned before. The centralized nature of LABON results in the relevant simplicity and lower cost of the stations' implementation (e.g. a feedback mechanism is not required for each station as in the previous schemes, since it is included at the hub). The additional implementation cost required by LABON is the array of AOTFs and the feedback mechanism. However, both costs are only located at a single place (the hub) of the network. One disadvantage of LABON, like all centralized schemes, is that a possible failure inside the hub would bring about serious problems in the overall operation of the network. Furthermore, due to its random access nature LABON is not collision-free. Nonetheless, the LABON architecture and protocol still significantly improve the network throughput on account of the central learning mechanism, which adapts the values of the passing probabilities for each wavelength to the star coupler, in such a way that they asymptotically tend to take their optimum values.

4.3 ADAPTIVE PRETRANSMISSION COORDINATION PROTOCOLS

In this section we consider the use of learning mechanisms in protocols that are based on pretransmission coordination. In Chapter 3 we presented pretransmission coordination-based protocols for WDM passive star networks of the broadcast-and-select architecture. One of the first and more characteristic protocols of this category is the DT-WDMA protocol. As we saw, this is a 'tell-and-go' scheme, since each station first announces in a control channel that it is about to transmit to a specific destination and immediately after that (in the next slot) it begins transmission. In what follows we shall present an adaptive scheme which enhances the capabilities of DT-WDMA protocol. According to this adaptive scheme, each station is equipped with a learning automaton, which helps it to decide which of the packets waiting for transmission should be transmitted next, so that the network throughput is led to higher levels than several non-adaptive protocols. The protocol is distributed and it assumes that the network hub still remains passive, contrary to LABON (Subsection 4.2.2) and the adaptive protocols presented in Section 4.4.

4.3.1 RCALA

In this section a relatively recent adaptive pretransmission coordination protocol for multiwavelength optical LANs is presented. The main objective of this protocol is to reduce the number of *receiver conflicts* that may be present in networks, where each node has one fixed-tuned transmitter and one tunable receiver for data packets.[13] For this purpose, the protocol uses the so-called *Receiver Conflict Avoidance Learning Algorithm* (*RCALA*) [23] and thus will be referred as the RCALA protocol from now on. Before introducing the protocol let us recall what a receiver conflict is.

In the network considered, each station has its own unique home wavelength for transmission, as we will see. Since each destination node is provided with only one tunable receiver, it is impossible to receive two or more packets transmitted to it at the same time. This is because it can only tune to only one wavelength at each time slot, and thus receive only one packet; the other packets inevitably are lost. This situation is called a receiver conflict and contributes to the degradation of the network performance if present frequently. The RCALA protocol aims to prevent this from happening, i.e. it does not completely eliminate receiver conflicts, but it attempts to restrict their occurrence, so that network performance is kept at relatively high levels.

The RCALA protocol considers that the frequency of receiver conflicts strictly depends on the transmission strategy, i.e. the policy used by each station for choosing which of its waiting packets will be transmitted at each time slot. In the context of this protocol, a new transmission strategy is proposed, which is based on the use of learning automata. Specifically, each station is assumed to have one learning automaton, which decides which of its waiting packets should be transmitted every time according to network feedback information. This adaptive transmission strategy performs better (as we will see) than other more traditional strategies, such as first-in-first-out (FIFO) and random. The FIFO strategy [3] designates that each station should transmit the packet that first entered the waiting queue, i.e. the oldest. On the other hand, according to the random strategy [5] the station chooses to transmit one of the waiting packets at random. Note that the random strategy generally performs better than the FIFO one, as far as the number of receiver conflicts is concerned.

4.3.1.1 Network Model of RCALA

The network model considered by RCALA can be described by the WDM passive star broadcast-and-select network shown in Figure 4.9, which is also used by the DT-WDMA protocol. The set of stations may be defined as $U = \{u_1, u_2, \ldots, u_N\}$ and the set of available wavelengths for data as $\Lambda = \{\lambda_1, \lambda_2, \ldots, \lambda_N\}$; thus the number of data wavelengths is taken to be equal to the number of stations, as in DT-WDMA. Since the protocol is pretransmission coordination-based, a control

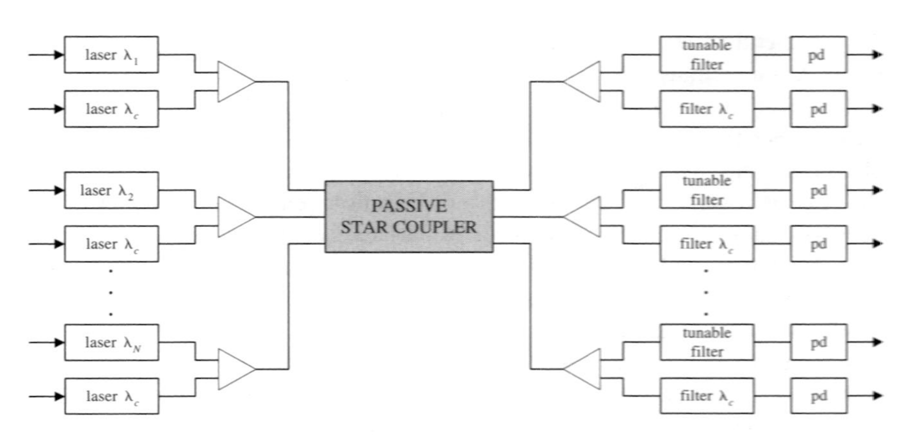

Figure 4.9 The network model of RCALA [23]. (© 1996 IEEE)

wavelength, denoted as λ_c in the figure, is also assumed to be used by all stations in order to preannounce their transmissions. Each station is provided with two fixed-tuned transmitters (lasers), one for data and one for control packets. These are connected to a 2×1 combiner before reaching the $N \times N$ passive star coupler via an optical fibre. The N outputs of the star coupler are connected via N separate fibres to the receiving parts of stations. These include 1×2 splitters that separate the data wavelength from the control wavelength. The former is led to a tunable receiver (filter) capable of tuning over the whole range of available data wavelengths, while the latter is connected to a receiver (filter) that is fixed-tuned to the control wavelength. Thus, the system considered in RCALA is a CC-FT^2-TTFR system, according to the relevant discussion of Chapter 1. Obviously, each station has a unique wavelength ('home wavelength') on which it always transmits its packets independently of destination.

As we described in the introduction to RCALA, receiver conflicts occur when two or more stations choose to transmit to the same destination at the same time slot. The transmitted packets are carried on separate wavelengths (the home wavelengths of the transmitters) and, since there is only one tunable filter available for data per station, the receiving station can only tune and collect one of the packets.

Each station has also one learning automaton, which determines which of the packets waiting for transmission will actually be transmitted at each time slot. The way these decisions are made is a key issue for RCALA protocol and will be described next.

4.3.1.2 Detailed Description of RCALA

RCALA is a pretransmission coordination-based adaptive protocol with the 'tell-and-go' feature. This means that a common control channel is used, where stations preannounce their transmissions before actually transmitting; transmissions take place right after the corresponding preannouncements, i.e. right in the next data slot. Let us assume that the round-trip propagation delay from a station to the network and back is exactly t_d slots for all stations. Let us also consider a specific time slot in which all transmitters preannounce their transmissions on the control wavelength (in N separate control minislots lasting in total as one data slot). After a delay of t_d slots, this control information will arrive at all stations due to the broadcast character of the network; that is to say, each station will be aware of the occurrence or not of receiver conflicts at each one of the destinations (conflicts concerning the transmissions that were announced before).

This information is used as network feedback information by the learning automaton of each station, in order to update the so-called 'basic transmission probability' $P_j(t)$ of each possible destination node $j (j = 1, \ldots, N)$. As we will describe later, the choice of a waiting packet is based on these basic transmission probabilities of destinations; this is because RCALA actually tries to *select the most appropriate destination node* each time, in an effort to reach its objective of limiting the frequency of receiver collisions. The basic transmission probabilities of possible destinations are updated by the learning automaton after each time slot on the basis of network feedback information, according to the following scheme $(j = 1, \ldots, N)$:

$$P_j(t+1) = P_j(t) - L \cdot P_j(t),$$

if a receiver conflict has occurred at destination node j during the $(t - t_d)^{th}$ time slot.

$$P_j(t+1) = P_j(t) + \epsilon \cdot L \cdot (1 - P_j(t)),$$

if no receiver conflict has occurred at destination node j during the $(t - t_d)^{th}$ time slot.

In the above scheme, $L \in (0, 1)$ is the 'step size parameter' (an internal parameter of the automaton) and ϵ is a very small positive real number $0 < \epsilon \ll 1$.

Parameter ϵ must be small enough in order to guarantee the accurate convergence of the automaton. The L parameter determines the step size of the probability update. A high value of L makes the automaton adapt rapidly to changes of the network state. However, it also results in a degradation of its accuracy. For example, an accidental receiver conflict at a relatively unloaded destination node j would probably decrease $P_j(t)$ more than it should. On the other hand, a relatively low

Table 4.1 RCALA performance for network N_1 for various values of parameter L and $\epsilon = 0.025$ [23]. 'Thr.' stands for throughput and 'del.' for delay. The column corresponding to the value of parameter L that maximizes performance has grey shading.

LOAD (p/slot)	L = 0.0 (dt-wdma, random tr.)		L = 0.1		L = 0.2		L = 0.3		L = 0.4		L = 0.5	
	thr.	del.	thr.	del.	thr.	del.	thr.	del.	thr.	del.	thr.	del.
1.0	0.559	18.9	0.604	17.6	0.608	17.4	0.609	17.4	0.607	17.5	0.604	17.6
0.8	0.557	16.9	0.595	15.7	0.598	15.6	0.599	15.5	0.597	15.6	0.596	15.6
0.6	0.526	11.8	0.544	10.8	0.545	10.6	0.545	10.6	0.545	10.6	0.544	10.6
0.4	0.396	5.3	0.397	4.9	0.397	4.9	0.397	4.9	0.397	4.9	0.397	4.9
0.2	0.200	2.3	0.200	2.3	0.200	2.3	0.200	2.3	0.200	2.3	0.200	2.3

value of parameter L results in high accuracy but low adaptation rate. Hence, there is a trade-off between adaptation speed and accuracy in the choice of parameter L, which calls for a careful selection that would maximize network performance each time.

Let us study how the choice of parameter L affects network performance through an example. Assume we want to choose L for a network[14] N_1 with 40 stations (and wavelengths), maximum queue length $Q = 10$ for each station, zero propagation delay and highly correlated destinations of the transmitted packets. The way the choice of parameter L affects network performance is shown in Table 4.1.

We notice that, initially, performance is gradually improving as L increases (due to the resulting increase in the adaptation speed) until it reaches a maximum, which corresponds to $L = 0.3$ in our example. After this value, network performance gradually decreases as L further increases (due to the parallel reduction of the automaton's accuracy). Generally, it should be pointed out that network performance is modified slowly with L, since it depends on both adaptation speed and accuracy of the automaton. Notice also that $L = 0$ implies that the protocol degenerates to DT-WDMA with the random transmission strategy, because probabilities of all destinations are always the same and never updated; thus RCALA is clearly an extension of the DT-WDMA protocol.

In order to see how the actual selection of a waiting packet for transmission is made by each learning automaton at each time slot, let us denote by $\mathrm{Q}_i(t)$ the queue with the waiting packets of source node i at time slot t. Also, let $Q_{i,j}(t)$ be the set of packets in $\mathrm{Q}_i(t)$ which are destined for node $j (j = 1, \ldots, N)$, $z_{i,j}(t)$ be the number of these packets (for node j) at time slot t i.e. $z_{i,j}(t) = |Q_{i,j}(t)|$ and $D_i(t)$ be the set of destination nodes for which source node i has at least one waiting packet at time slot t, i.e. $D_i(t) = \{\text{node k } | Q_{i,k}(t) \neq \emptyset\}$. Each waiting packet in $Q_i(t)$ corresponds to an action of the learning automaton. The probability of selecting a packet in $Q_i(t)$ depends on the destination node of this packet; therefore, each packet in $Q_{i,j}(t)$ is selected with the same probability $\pi_{i,j}(t)$. The choice probability $\pi_{i,j}(t)$ is computed by scaling [37] the basic transmission probability $P_j(t)$ in the following way:

$$\pi_{i,j}(t) = \frac{P_j(t)}{\sum_{k \in D_i(t)} z_{i,k}(t) \times P_k(t)} \quad \text{for } j \in D_i(t).$$

Thus at each time slot t, the learning automaton of each station i contains a probability distribution over the set $Q_i(t)$, i.e. a probability $\pi_{i,j}(t)$ for each of the waiting packets, with packets destined to the same destination (e.g. j) being assigned the same probability (i.e. $\pi_{i,j}(t)$). The packet to be transmitted is chosen according to these scaled probabilities $\pi_{i,j}(t)$. The probability $r_{i,j}(t)$ that station (source node) i transmits to station (destination node) j at time slot t is computed by multiplying the number of waiting packets for j times the same choice probability $\pi_{i,j}(t)$ of each one, i.e. it is the following:

$$r_{i,j}(t) = z_{i,j}(t) \times \pi_{i,j}(t) = \frac{z_{i,j}(t) \times P_j(t)}{\sum_{k \in D_i(t)} z_{i,k}(t) \times P_k(t)} \quad \text{for } j \in D_i(t).$$

4.3.1.3 Overview of Analytical and Simulation Results for RCALA

As far as the performance analysis of RCALA protocol is concerned, let us assume that a receiver conflict at station (destination) j at time slot t occurs with probability $c_j(t)$. Accordingly, the probability that there is no such conflict is $1 - c_j(t)$. It can be proved that $c_j(t)$ is a monotonically increasing function of the basic transmission probability $P_j(t)$ for this station. Furthermore, it can be proved that the basic transmission probability of each station asymptotically tends to be proportional to $(1 - c_j(t))$, i.e. to the probability that there is no conflict, and inversely proportional to the probability $c_j(t)$ that there is a conflict. This can be formally expressed as follows:

$$\frac{P_i(t)}{P_j(t)} = \frac{\frac{1-c_i(t)}{c_i(t)}}{\frac{1-c_j(t)}{c_j(t)}}.$$

This means that if a receiver conflict probability of a destination node is relatively low (because the number of packets destined for it is low), RCALA tends to increase the basic transmission probability for this destination, i.e. tends to select waiting packets that are destined for this node. The opposite would be decided by RCALA, if the receiver conflict probability of the node was assumed to be relatively high.

Next, we selectively present a part of the extensive simulation results from [23]. As discussed in the previous subsection, RCALA can be seen as an extension of DT-WDMA; that is why the latter was chosen for performance comparison with our protocol. In fact, separate simulations results about the FIFO and the random strategy are presented. The performance metrics used are the mean number of packets which are destroyed due to receiver conflicts and the delay versus throughput characteristic. We shall selectively present only the results for the following three network configurations (for more the reader may refer to [23]):

1. Network N_1: $N = W = 40$, $Q = 10$, $t_d = 0$, $L = 0.3$, $P_{12} = 1.0$ and $P_{21} = 0.1$.
2. Network N_2: $N = W = 40$, $Q = 15$, $t_d = 5$, $L = 0.2$, $P_{12} = 1.0$ and $P_{21} = 0.1$.
3. Network N_3: $N = W = 50$, $Q = 3$, $t_d = 0$, $L = 0.4$, $P_{12} = 0.5$ and $P_{21} = 0.5$.

Let us explain the symbols in the descriptions of the three configurations above: N and W are the number of stations and wavelengths respectively, taken to be equal in the context of both RCALA and DT-WDMA protocols; Q is the maximum length (in packets) of each station's queue for packets waiting to be transmitted, including transmitted but unacknowledged ones; t_d is the propagation delay in time slots; L is the step size parameter discussed before; and finally, P_{12} and P_{21} are the probabilities that a source node being in state S_1 will transit at state S_2 at the next time slot and vice versa, respectively. A station is said to be in state S_1, when the destination node of a newly arriving packet (for transmission) is selected at random among all other nodes; on the other hand, it is said to be in state S_2, when the destination node of a newly arriving packet is the same with the destination node of the previously arrived packet. Thus, probabilities P_{12} and P_{21} are expressions of the correlation of packet destinations. Accordingly, for the first two networks we assume high correlation of traffic (see values of probabilities above), while the correlation is taken to be significantly less for the third network. Finally, it should be noted that the choice of parameter L was made after testing a large number of possible values in a way similar to the one we presented in Table 4.1, while parameter ϵ (see probability updating

Table 4.2 Packet losses due to receiver conflicts for the three networks N_1, N_2 and N_3 [23]. (© 1996 IEEE)

Network	Packet losses due to receiver conflicts (node traffic = 1 packet / slot)		
	FIFO	RANDOM	RCALA
N_1	19.5	17.6	15.6
N_2	18.4	17.1	15.6
N_3	22.9	20.5	18.8

scheme of previous subsection) was assumed to be equal to 0.025 for all networks.

The results for the first metric, i.e. the mean number of packet losses due to receiver conflicts, for the three networks are shown in Table 4.2.

Clearly, there is a reduction in the number of lost packets, if the proposed adaptive protocol is used; this reduction is overall (taking into account all results from [23]) up to 20% as compared to DT-WDMA with the FIFO transmission strategy and up to 11% as compared to DT-WDMA with the random strategy.

The delay versus throughput characteristic of the protocols for networks N_1, N_2 and N_3 is presented in Figure 4.10.

From Figure 4.10 it becomes clear that RCALA outperforms DT-WDMA with either the FIFO or the random strategy. Specifically, the performance improvement in comparison to the random strategy of DT-WDMA is approximately equal to the one that the random strategy introduces in comparison to the FIFO strategy of DT-WDMA. Also, by studying both Figure 4.10 and the characteristics of each simulated network, it becomes clear that RCALA performs better in networks with low propagation delay and relatively high correlation of traffic; however, the performance is still improved in networks which do not satisfy one of the above conditions (or even both of them).

4.3.1.4 Concluding Remarks about RCALA

We close the discussion about RCALA by making some concluding and summarizing remarks about it. It is a new adaptive scheme that is based on the use of one learning automaton per station in order to reduce the number of receiver conflicts that occur in WDM broadcast-and-select star networks that assume tunable

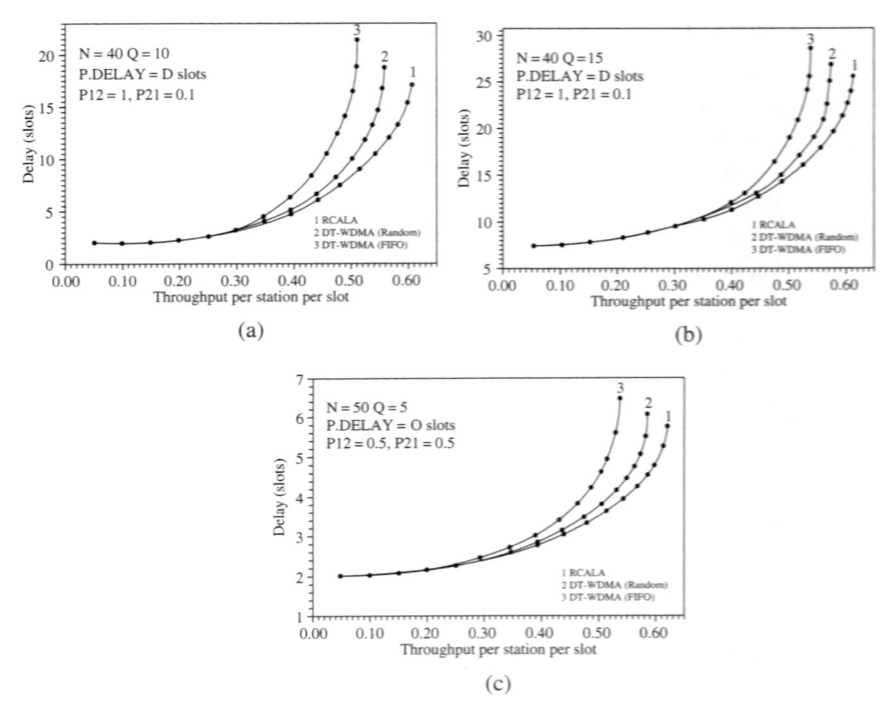

Figure 4.10 Delay versus throughput characteristic of the compared protocols [23]: (a) for network N_1, (b) for network N_2 and (c) for network N_3. In all three parts of the figure, curve number 1 is for RCALA, curve number 2 for DT-WDMA with the random strategy and curve number 3 for DT-WDMA with the FIFO strategy. (© 1996 IEEE)

receivers for nodes. More specifically, it is applied to the same network model as the well-established DT-WDMA protocol, i.e. to a CC-FT^2-TRFR system. Thus, it is a pretransmission coordination-based scheme, which in fact uses the control information transferred through the common control channel as network feedback information to update the basic choice probabilities of destinations for each source node. RCALA actually introduces a new strategy for choosing which of the waiting packets at each source node should be transmitted at the next time slot. For this purpose, it assigns selection probabilities to packets according to the basic choice probabilities of destinations and picks one packet for transmission (for each source at each time slot) according to these probabilities contained in each learning automaton. The result of the application of RCALA to multiwavelength optical LANs is a considerable performance improvement.

4.4 CENTRALIZED PACKET FILTERING PROTOCOLS

Throughout this section we will focus on some representative examples from a relatively new family of MAC protocols for WDM broadcast-and-select star networks. The base of these schemes is the *centralized filtering* of transmitted packets before they reach the star coupler, where they are broadcast back to every station. The term 'centralized' implies that the filtering function is carried out inside the network hub; thus, the hub is no longer considered to be passive.[15]

The benefits from filtering packets in a centralized fashion may include, for example, depending on the scheme, the elimination of collisions, the more efficient handling of bursty and correlated traffic and the balancing of the offered load between available wavelengths. In general, it has been proved that the protocols of this family achieve a significantly higher performance than the well-known contention-oriented or round-robin protocols. The control of the filtering function may be based on network feedback information from e.g. a common control channel; thus, the decided filtering function every time can be seen as an adaptation to the variable offered traffic conditions.

Four protocols of this category will be presented: the *Hybrid random Access and Reservation Protocol* (*HARP*), the *Centralized Packet Filtering Protocol* (*CPF*), the *Centralized Wavelength Conversion* (*CWC*) protocol and the *Optically Controlled Optical Network* (*OCON*) protocol. The first three assume an *electronic* centralized packet filtering mechanism, while the latter makes use of purely *optical* technology for this purpose.

4.4.1 HARP

The first proposed centralized packet filtering-based protocol is called *Hybrid random Access and Reservation Protocol* (*HARP*) [20, 25]. The operation of the protocol is based on the use of *polarization-independent acoustically tunable optical filters* (*PIATOFs*) inside the network hub. The result of the filtering is that at most one packet per wavelength is selected to pass to the star coupler each time, hence, HARP is collision-free. The passing packet (or equivalently station) for a wavelength, is selected in a random access fashion from those stations, which have tried to transmit in the last time slot, but whose transmissions were blocked by the filters. If there were no blocked transmissions, the transmitting station would be selected randomly among all stations. Thus, in the context of HARP the role of a packet transmission is twofold (one of the two applies for a specific transmission):

1. To transport data.
2. To make a *reservation* for the next time slot.

If a packet transmission fails to play the primary role (1), the secondary role (2) is activated. Hence, the name of the protocol can be explained, since HARP combines the characteristics and advantages of both random access and reservation-based schemes.

4.4.1.1 Network Model of HARP

The network model considered in HARP is generally depicted in Figure 4.2. However, as was already mentioned, the network hub is no longer passive, but includes a packet filtering mechanism that will be described later.

The set of stations can be defined as $U = \{u_1, \ldots, u_N\}$ and the set of wavelengths as $\Lambda = \{\lambda_1, \ldots, \lambda_W\}$, where N and W are the number of stations and wavelengths respectively. Each station is equipped with a tunable laser for transmission and a filter fixed-tuned to home wavelength for reception, as Figure 4.2 shows; thus the system considered is TT-FR. Next, the HARP protocol and the proposed internal architecture of the hub are described in more detail.

4.4.1.2 Detailed Description of HARP

According to the HARP protocol, an array of N PIATOFs F_k ($k = 1, \ldots, N$), one for each station, must be placed at the network hub. At each time slot t, each filter F_k is tuned to pass only a subset of wavelengths $\Lambda_k(t) \subseteq \Lambda$. Note that analogous filters were used in the LABON protocol and it was mentioned that they are indeed capable of selecting multiple wavelengths from an optical signal. If the corresponding user u_k transmits on one of the passing wavelengths, i.e. $\lambda_i \in \Lambda_k(t)$ where λ_i is the wavelength used by u_k, the packet passes to the star coupler. Otherwise, i.e. if $\lambda_i \notin \Lambda_k(t)$, the packet is blocked.

The set of passing wavelengths Λ_k for each filter F_k is generally constructed by adding to it every wavelength on which the station u_k was chosen to be the passing station at a specific time slot. Note that the same station may be selected as the passing station for more than one wavelength, if e.g. it makes a successful reservation through a previously blocked transmission on a certain wavelength and it is also selected for other wavelengths on which there were no previous blocked transmissions.[16] Thus, in order to see how Λ_k is constructed (for all possible k), it suffices to examine the way the passing packets (or stations equivalently) are selected for each wavelength at each time slot. This task is performed by a control circuit, which is used for each wavelength and will be described in the following.

Within each time slot t, for each wavelength λ_i ($i = 1, \ldots, W$), a separate control circuit determines the passing station $ps_i(t + 1)$ for the next slot according to the following algorithm:

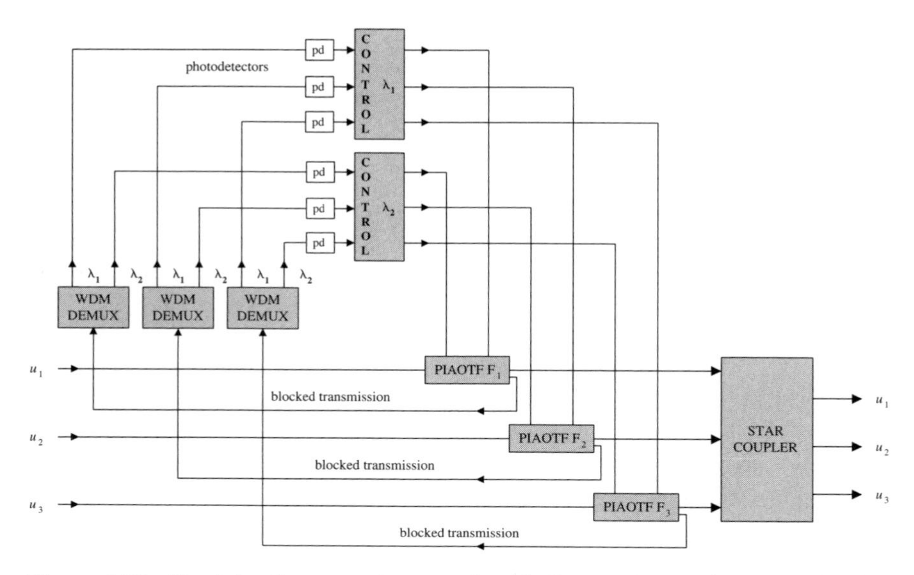

Figure 4.11 The hub of an example network with three stations and two wavelengths operating under the HARP protocol. (Reprinted from [25], copyright 1996, with permission from Elsevier)

```
1. Determine the set Bi(t) of stations which transmitted
   a packet on wavelength λi, but their transmissions were
   blocked by filters at time slot t.
2. IF (Bi(t) ≠ Ø) THEN
     Choose psi(t+1) at random among stations in Bi(t)
   ELSE (* i.e. if Bi(t) = Ø *)
     Choose psi(t+1) at random among all stations;
```

After the various control circuits determine the passing stations $ps_i(t+1)(i = 1, \ldots, W)$, the sets $\Lambda_k(t+1)$ of passing wavelengths for each filter $F_k(k = 1, \ldots, N)$ are constructed as follows:

$$\Lambda_k(t+1) = \{\lambda_i\text{: } ps_i(t+1) = u_k \text{ for } i = 1, \ldots, W\}.$$

One key issue in the HARP protocol is how the first step of the above algorithm is actually performed, i.e. how the set of stations $B_i(t)$, which transmitted a packet on each wavelength $\lambda_i(i = 1, \ldots, W)$ but had their transmission blocked by filters, is determined. In order to see this more clearly, let us consider the hub of an example network with three stations and two wavelengths operating under the HARP protocol and shown in Figure 4.11.

An important feature of the PIATOFs is that they have a second output port where the blocked wavelengths are sent.[17] The second port of each filter is connected to a WDM demultiplexer (there are in total N WDM demultiplexers, one for each filter), which separates the individual wavelengths. Each wavelength is detected for transmission by means of a photodetector, the output of which is connected to the control circuit corresponding to the specific wavelength. Thus, $N \times W$ photodetectors are required in total, six in our case as shown in the figure. In this way, the control circuit of each wavelength λ_i is informed about which stations have tried to transmit on this wavelength at each time slot t without success, hence the set $B_i(t)$ is determined. This feedback information is enough to choose the passing stations for the next time slot, as we described (previous algorithm).

Note that in the above description of HARP, the round-trip propagation delay from a station to the hub and back was taken to be negligible. Therefore, it is assumed that all transmitters can ascertain the result of their transmissions (by using e.g. one of the methods in [36]) very quickly, i.e. quickly enough to know whether they should retransmit during the next time slot or not. Generally, the stations operate in a slotted ALOHA fashion with $p = 1$, that is to say they transmit a packet immediately in the next time slot with probability one. From then on, the HARP protocol undertakes the task of allowing at most one packet per wavelength to pass to the star coupler.

If the round-trip propagation delay was assumed to be non-negligible in comparison to the data slot, HARP would have to be slightly modified and applied to (symmetric) networks by making use of *pipelining*. The above algorithm determining the passing stations should be modified so that the passing stations for slot $t + 1$ are determined according to the set of blocked stations of time slot $t - t_d$ (and not t), where t_d is the round-trip propagation delay. Also, the use of pipelining calls for additional buffers. More details of the extension of HARP to networks with large propagation delays can be found in [25].

4.4.1.3 Overview of Simulation Results for HARP

The HARP protocol is simulated and compared to the slotted ALOHA and the TDMA protocols in [25]. An overview of the simulation results is presented in this section.

The slotted ALOHA protocol was chosen to test the performance of HARP protocol under low-load conditions, where slotted ALOHA is known to perform well due to its random access nature. On the other hand, TDMA was chosen to test the performance of HARP under high-load conditions, where it is known to achieve a good performance owing to the fact that it is collision-free.

It can be understood that the performance of HARP depends on the tuning time of the PIATOFs and more specifically on the ratio of the tuning time to the packet transmission time (or data slot). The normalized tuning time of the PIATOFs was given values from 0 slots up to 0.4 slots, i.e. 0, 0.1, 0.2, 0.3 and 0.4 slots, in order to study the difference in performance [25]. Note that for uniformity among the protocols, by 'slot' we mean the packet transmission time in all protocols, i.e. in the HARP protocol the tuning time for the PIATOFs is not included in it.

Also, it should be noted that the values of the transmission probability for the slotted ALOHA protocol were selected carefully so that it achieves a high performance. For more details about this choice and about other characteristics of the simulation (such as the transmission rate at each wavelength, packet size, allocation of home wavelengths to stations and queue size of each station), the reader may refer to [25].

From the simulation [25] it becomes clear that the HARP protocol achieves a lower delay than slotted ALOHA and TDMA under low load conditions, since only one or two attempts would be enough for successful transmission in the context of HARP. Furthermore, HARP achieves a higher throughput-delay performance than the other protocols under high load conditions, because of the elimination of collisions and the incorporated intelligent reservation mechanism based on the centralized filtering module. We also notice that the performance of HARP remains significantly higher than slotted ALOHA and TDMA, even when the tuning time of the PIATOFs is taken to be relatively high.

A diagram illustrating the throughput versus offered load performance characteristic of HARP for an example network (and considering two values for the tuning time) will be presented in Subsection 4.4.3 about OCON protocol, where the two protocols are compared.

4.4.1.4 Concluding Remarks about HARP

Finally, let us make some concluding and summarizing remarks about the HARP protocol. We have seen that HARP is a protocol that uses a packet filtering mechanism in the network hub in order to allow at most one packet per wavelength to pass to the star coupler during each time slot, thus eliminating collisions. The selection of packets at each time slot is carried out by using principles of both random access and reservation based schemes. That is to say the selection of a passing station for each wavelength at a specific time slot, is done, as we saw, in a random access manner among those stations that failed (were blocked by filters) to transmit on the specific wavelength at the previous time slot; hence, blocked transmissions are always used as reservations for the near future.

Since HARP stands between random access and reservation-based schemes, it can achieve a very high throughput-delay performance under both high and low load conditions. Furthermore, we could notice that the stations transmit in a slotted ALOHA fashion, which implies that HARP can be applied in existing WDM passive star networks operating under slotted ALOHA protocol by adding the filtering mechanism at the network hub. Since the hub has almost all the responsibility for implementing HARP, it follows that the operation of the various stations attached to the network can be very simple; the latter is an important advantage of HARP that is also encountered in most of the centralized schemes.

The drawbacks of HARP, on the other hand, include the common disadvantage of centralized schemes that the hub of the network is no longer a passive device. A possible failure at the hub would affect the operation of the entire network. Moreover, HARP requires additional implementation cost at the network hub, namely the cost of adding N PIATOFs, N WDM demultiplexers, $N \times W$ photodetectors, W control circuits and all the necessary connections as illustrated in the example of Figure 4.11. However, the cost is limited by the fact that all the PIATOFs are placed at the same site and, consequently, they can be arrayed in a common device [4].

4.4.2 CPF

A new protocol for WDM star networks of the broadcast-and-select architecture, which is based on the centralized filtering of packets as well, is presented here. The proposed *Centralized Packet Filtering* (*CPF*) protocol [18–19, 30] achieves a high network performance owing to a packet filtering mechanism placed at the network hub, which allows at most one packet per wavelength to pass to the star coupler at each time slot. The basis of the filtering mechanism is an array of *electro-optic tunable filters*. Furthermore, according to CPF, the selection of the passing packets is performed in such a way that receiver conflicts are avoided; note that, as in the case of RCALA protocol (which aims at reducing the number of receiver conflicts), tunability is assumed at the receiving part of stations. Thus, overall, the CPF protocol eliminates both channel collisions and receiver conflicts, which explains the significant increase in network performance.

Another important feature of the CPF protocol is that, unlike RCALA and other important schemes proposed for an analogous network model[18] (as the DT-WDMA protocol), CPF removes the important limitation that the number of wavelengths should be equal to the number of stations. Thus, it allows the sharing of a specific wavelength among two or more stations. Next, let us present the network model on which the proposed CPF protocol is based.

4.4.2.1 Network Model of CPF

The network model considered by CPF can be described by the WDM passive star broadcast-and-select network shown in Figure 4.9, which is also used by the RCALA and DT-WDMA protocols. Let us define the set of stations as $U = \{u_1, u_2, \ldots, u_N\}$ and the set of available wavelengths for data as $\Lambda = \{\lambda_1, \lambda_2, \ldots, \lambda_W\}$, where N and W are the number of stations and wavelengths respectively. We notice that actually there is a small modification of the model considered in Figure 4.9 by DT-WDMA and RCALA, in that the number of available wavelengths may not necessarily be equal to the number of stations; in other words, a limited number of wavelengths may be shared by an arbitrary number of stations. This is an important advantage of the CPF protocol, as we mentioned in the introduction.

Since the protocol is pretransmission coordination-based, a control wavelength, denoted as λ_c in the figure, is also assumed to be used by all stations in order to preannounce their transmissions. Each station is provided with two fixed-tuned transmitters (lasers), one for data and one for control packets. These are connected to a 2×1 combiner before reaching the $N \times N$ passive star coupler via an optical fibre. The N outputs of the star coupler are connected via N separate fibres to the receiving parts of stations. These include 1×2 splitters that separate the data wavelength from the control wavelength. The former is led to a tunable receiver (filter) capable of tuning over the whole range of the W available data wavelengths, while the latter is connected to a receiver (filter) that is fixed-tuned to the control wavelength. Thus, the system considered in CPF is CC-FT^2-TTFR, according to the relevant discussion of Chapter 1. Like RCALA and DT-WDMA, the CPF protocol has the tell-and-go feature, i.e. each station having a packet to transmit first sends a control message on the common control wavelength containing the source and destination node of the packet, and then it transmits the data packet immediately in the next data slot.

Obviously, each station has a home wavelength on which it always transmits its packets independently of destination, which, however, is not unique, but may be used by other stations as well. The allocation of home wavelengths to transmitters is done in the following way: for each source node $u_k (k = 1, \ldots, N)$ data wavelength λ_{i_k} is allocated, where $i_k = \lceil kW/N \rceil$; thus $1 \leq i_k \leq W$. We shall assume that N/W is an integer in order to simplify the presentation of the protocol. Accordingly, each data wavelength is shared by exactly N/W stations.

The round-trip propagation delay from a station to the hub and back is assumed to be equal to D slots for all stations, where D is taken to be an even integer number, so that the time required for a packet to reach the hub is exactly $D/2$ slots.

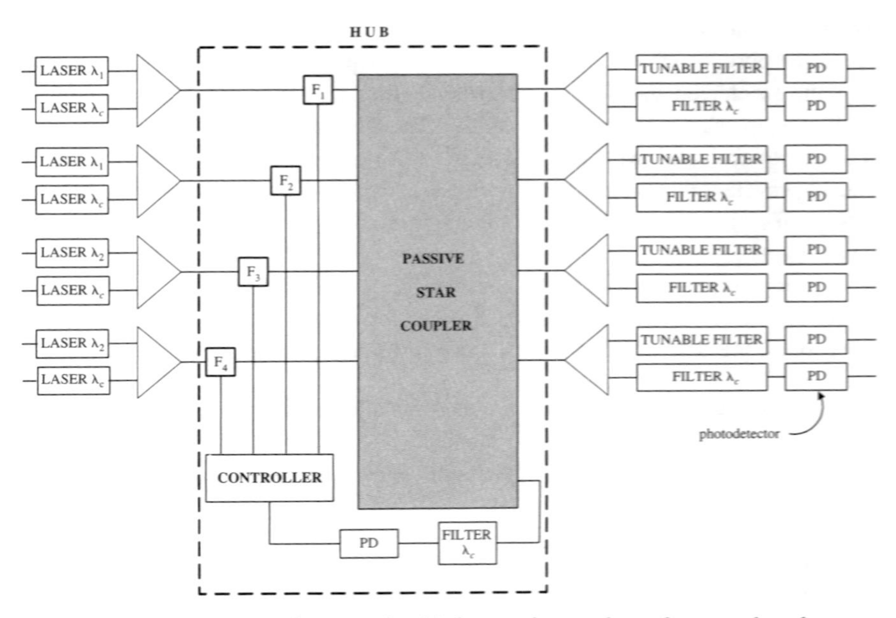

Figure 4.12 An example of a network with four stations and two data wavelengths operating under the CPF protocol [30]. (© 1999 Wiley)

4.4.2.2 Detailed Description of CPF

The hub of a network operating under the CPF protocol is embellished with a packet filtering mechanism. Specifically, an array of N electrooptic tunable filters, which control the passing of packets to the star coupler, are placed at the network hub. Each filter $F_k (k = 1, \ldots, N)$ corresponds to a specific station u_k and controls its transmitted packets. The output of the 2×1 combiner of u_k, which is an optical signal containing both the control and data wavelengths (λ_c and λ_{i_k} respectively), is connected via an optical fibre to filter F_k. An example of a network with four stations and two data wavelengths operating under the CPF protocol is depicted in Figure 4.12.

At each time slot t, each filter can be in one of two states, namely either in the ON or in the OFF state. If $F_k =$ ON, both wavelengths are allowed to pass; otherwise, if $F_k =$ OFF, only the control wavelength is allowed to pass. Notice that the common control wavelength is always allowed to pass to the star coupler by each filter of this mechanism.

Electro-optic tunable filters were chosen to implement the protocol, because they provide for a separate output port, where the unselected wavelengths are driven [8].

The OFF state is implemented by tuning the filter at wavelength λ_{i_k} and using as an output port (which is connected to the star coupler via an optical fibre) the port with the unselected wavelengths described before; thus, the control wavelength λ_c is allowed to pass, but not the selected λ_{i_k}. The ON state is implemented by tuning the filter to any other wavelength except for λ_{ik} and λ_c, in order to have both λ_{i_k} and λ_c wavelengths in the output port used. Thus, the filters can be tuned to the ON or OFF state by appropriately tuning them to (i.e. selecting) a wavelength as described before.

A *controller*, also placed at the network hub, undertakes the task of tuning the filters to the ON or OFF state at each time slot. Note that a unique controller is used for all the N available filters. Tuning of filters should ensure the following:

1. At most one packet per data wavelength is allowed to pass to the star coupler so that channel collisions are eliminated.
2. All the passing packets have different destination nodes, so that receiver conflicts are eliminated too.

In addition, the tuning operation must be implemented by a relatively simple algorithm so as to introduce the less possible computational overhead to the controller and not affect negatively the overall implementation and performance of CPF. The algorithm for tuning the filters (hence selecting the passing packets) that is used by the controller in order to meet all the previously described requirements is presented below:

```
{                         *** Basic Notations ***
  t       : time slot - variable.
  Xi,j(t) : set of stations that have transmitted on wavelength
            λi a packet destined to node uj at time slot t -
            variable.
  Sj(t)   : set with the following wavelengths as elements:
            {wavelength λk: Xk,j(t) ≠ ∅} - variable.
  Di(t)   : set with the following destinations as elements
            {destination node uk: Xi,k(t) ≠ ∅} - variable.
  G       : set of unmatched wavelengths - variable.
  R       : set of unmatched destinations - variable.
* The algorithm selects one destination uj for each
  wavelength λi, *
* such that uj∈ Di(t) and allows only one source node
  uk ∈ Xi,j(t) to *
* access the star coupler. *}
```

```
PROCEDURE PACKET_SELECTION;
BEGIN
 FOR k  := 1 TO N DO F_k := OFF;
 G := {wavelength λ_i: D_i(t) ≠ Ø};
 R := {destination node u_j: S_j(t) ≠ Ø};
 WHILE G ≠ Ø DO
 BEGIN
       Select randomly a wavelength  λ_i ∈ G;
       G := G − {λ_i};
       IF (R ∩ D_i(t)) ≠ Ø THEN
       BEGIN
             Select randomly a destination node
             u_j ∈ (R ∩ D_i(t));
             R := R - {u_j};
             Select randomly a source node u_k ∈ X_i,j(t);
             Set F_k := ON;
       END;
 END;
END;
```

As is shown in the example of Figure 4.12, one of the output ports of the star coupler is connected to a fixed filter which is tuned to always pass the control wavelength. The output of the filter is connected to a photodetector, which performs the necessary optoelectronic conversion before the signal reaches the controller. In this way, the controller receives a control packet containing information about the source and destination nodes of the transmissions that were announced by all stations $D/2$ slots before. Note that one control slot is assumed to include N minislots that are accessed in a TDM manner by all stations and also is taken to be equal to one data slot in length (thus, a control packet includes N transmission announcements and is equal in length to a data packet). Since the controller is aware of the source nodes, it can straightforwardly derive the wavelength where each packet is transmitted, because allocation of wavelengths to source nodes is known. Hence, the controller has all the required information in order to run the packet selection algorithm previously presented.

It should be pointed out that the same algorithm is executed by all the stations when the control information arrives at them, i.e. after a round-trip propagation delay of D slots. By executing the algorithm and figuring out how the filters were tuned, the stations that had transmitted a packet should know whether they should reschedule their packets for transmission (in the event of a blocked transmission) or just delete them from the buffers that keep the transmitted but unacknowledged packets (in the event of a successful transmission). Furthermore, the destination nodes realize how they should tune their filters for reception.

Another observation that must be made is that the packet selection algorithm uses random numbers. Since it is assumed to run at all stations and at the controller of the hub as well, it is understood that in order to give the same results as needed, it is necessary to ensure that the same random numbers are used each time by every station and by the controller. For details on two satisfactory solutions of this problem, please refer to [19]. Finally, for a discussion on the run-time of the packet selection algorithm and a relevant numeric example the reader may refer to [19] as well.

4.4.2.3 Overview of Simulation Results for CPF

The CPF protocol is simulated and compared to the slotted ALOHA and TDM protocols in [19]. The former generally performs well under low load conditions due to the comparably low waiting time before transmission, while the latter performs better under high load conditions owing to the absence of channel collisions and the fact that there are few or no idle slots (which favours its operation). However, both suffer from receiver conflicts, and also the performance of slotted ALOHA is poor under high load conditions due to the channel collisions.

For the slotted ALOHA protocol the transmission probability p was taken to be equal to W/N [19] because this is the optimum value under medium or high load conditions, where the number of stations waiting to transmit on each wavelength is exactly N/W; recall that N/W was assumed to be an integer and also that in slotted ALOHA, the optimum value of p for a wavelength is equal to $1/M$, if we let M denote the number of waiting transmitters for this wavelength. However, it should be noted that this value of transmission probability results in high packet delays, when the network operates under low load conditions.

The performance of slotted ALOHA is generally found to be lower than the other two protocols [19]. Furthermore, the proposed CPF protocol outperforms TDM under any load conditions. Note that the performance of the latter two protocols and the next presented scheme (CWC) was simulated and compared in [30]. Representative results and diagrams from [30] are presented in Subsection 4.4.3 (about CWC). There, we will also see schematically the performance advantage of CPF over TDM for some example network configurations.

4.4.2.4 Concluding Remarks about CPF

The second protocol belonging to the family of centralized packet filtering protocols that we examined, i.e. CPF, is a new protocol for WDM star networks of the broadcast-and-select architecture. As we saw, its operation is based on an array of electro-optic tunable filters which are placed at the network hub and control the passing of the packets to the star coupler. CPF was shown to be both channel collision and receiver conflict-free, i.e. at most one packet per wavelength is allowed to pass to the star coupler by the filters and, also, the selected destination for each packet is always unique. It is applied on the same CC-FT^2-FRTR system assumed by DT-WDMA and RCALA, but CPF has the additional virtue of allowing the sharing of each wavelength by two or more stations.

It was shown that the performance of CPF is very high under any load conditions, mainly due to the intelligent filtering mechanism that prevents channel collisions and receiver conflicts from occurring. The only additional hardware needed is the centralized packet filtering mechanism that has to be placed at the network hub; therefore, a DT-WDMA network can be transformed to a CPF network quite easily. Filters may be arrayed in a common device. This limits the additional implementation cost, as discussed in the previous protocol. The main disadvantage of CPF is the common one for centralized schemes, i.e. the fact that a failure in a single point (the hub) would affect the operation of all stations.

4.4.3 CWC

Another recent MAC protocol for multiwavelength optical LANs is presented here. The proposed *Centralized Wavelength Conversion* (*CWC*) protocol [30–31, 35] achieves a high network performance owing to a wavelength conversion mechanism placed at the network hub, which allows at most one packet per wavelength to pass to the star coupler at each time slot. The basis of the conversion mechanism is an array of *tunable wavelength converters*. Moreover, according to CWC, the selection of the passing packets is performed in such a way that receiver conflicts are avoided (note that, as in the case of RCALA protocol, tunability is assumed at the receiving part of stations). Hence, the CWC protocol eliminates both channel collisions and receiver conflicts, which explains the important increase in network performance.

Another important feature of CWC is that, unlike RCALA and other important schemes proposed for an analogous network model (as DT-WDMA), CWC takes out the important restriction that the number of wavelengths should be equal to the number of stations, as the previous CPF protocol and therefore allows the sharing of a specific wavelength between two or more stations.

Besides the previous benefits acquired by the application of CWC to WDM star networks, it should be noted that the proposed scheme also takes into account the bursty nature of traffic in gigabit LANs and achieves a *balancing of the offered load* between the available wavelengths, so that the number of idle slots tends to be minimized. This contributes to the performance improvement as well. Next, let us present the network model considered by the CWC protocol.

4.4.3.1 Network Model of CWC

The network model considered by CWC can be described by the WDM passive star broadcast-and-select network shown in Figure 4.9, which is also used by the RCALA, DT-WDMA and CPF protocols. Let us define the set of stations as $U = \{u_1, u_2, \ldots, u_N\}$ and the set of available wavelengths for data as $\Lambda = \{\lambda_1, \lambda_2, \ldots, \lambda_W\}$, where N and W are the number of stations and wavelengths respectively. We notice that, as in the case of CPF, there is a small modification of the model considered in Figure 4.9 by DT-WDMA and RCALA, in that the number of available wavelengths may not necessarily be equal to the number of stations; specifically, a limited number of wavelengths may be shared by an arbitrary number of stations. This is an important advantage of the CWC protocol, as we mentioned in the introduction.

Since the protocol is pretransmission coordination-based, a control wavelength, denoted as λ_c in the figure, is also assumed to be used by all stations in order to preannounce their transmissions. Each station is provided with two fixed-tuned transmitters (lasers), one for data and one for control packets. These are connected to a 2×1 combiner before reaching the $N \times N$ passive star coupler via an optical fibre. The N outputs of the star coupler are connected via N separate fibres to the receiving parts of stations. These include 1×2 splitters that separate the data wavelength from the control wavelength. The former is led to a tunable receiver (filter) capable of tuning over the entire range of the W available data wavelengths, while the latter is connected to a receiver (filter) that is fixed-tuned to the control wavelength. Thus, the system considered in CWC is a CC-FT^2-TTFR system, according to the relevant discussion of Chapter 1. As in RCALA, DT-WDMA and CPF, the CWC protocol has the tell-and-go feature.

Obviously, each station has a home wavelength on which it always transmits its packets independently of destination, which, however, is not unique, but may be used by other stations as well. The allocation of home wavelengths to transmitters is done in a similar way to the CPF scheme: for each source node u_k $(k = 1, \ldots, N)$ data wavelength λ_{i_k} is allocated, where $i_k = \lceil kW/N \rceil$; thus $1 \leq i_k \leq W$. We shall assume that N/W is an integer in order to simplify the presentation of the protocol. Thus each data wavelength is shared by N/W stations.

The round-trip propagation delay from a station to the hub and back is assumed to be equal to D slots for all stations, where D is taken to be an even integer

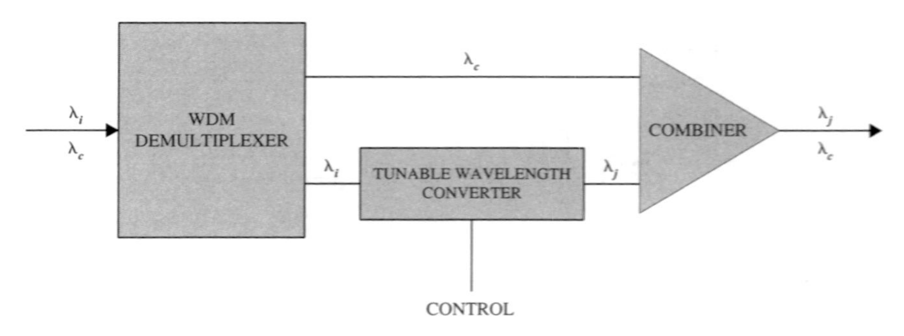

Figure 4.13 The internal architecture of the converters $C_k(k = 1, \ldots, N)$ used by CWC [30]. Each converter is capable of converting the data wavelength λ_i as required, but always preserves the control wavelength λ_c. (© 1999 Wiley)

number, so that the time required for a packet to reach the hub is exactly $D/2$ slots.

4.4.3.2 Detailed Description of CWC

The hub of a network operating under the CWC protocol is embellished with a wavelength conversion mechanism. Specifically, an array of N tunable wavelength converters C_k $(k = 1, \ldots, N)$, is placed at the network hub. The converters are capable of converting the wavelengths of the transmitted packets, while preserving the control wavelength λ_c, as shown in Figure 4.13. Each converter $C_k(k = 1, \ldots, N)$ corresponds to a specific station u_k and controls its transmitted packets. The output of the 2×1 combiner of u_k, which is an optical signal containing both the control and data wavelengths (λ_c and λ_{i_k} respectively), is connected via an optical fibre to converter C_k. An example of a network with four stations and two data wavelengths operating under the CWC protocol is depicted in Figure 4.14.

Generally, wavelength converter C_k may be tuned to convert the fixed incoming λ_{i_k} wavelength to any other of the available wavelengths. From here on, by $C_k = \lambda_{j_k}$, we imply that C_k converted wavelength λ_{i_k} to wavelength λ_{j_k}. An additional characteristic assumed for the converters of CWC, is that they may block the incoming data signal by converting it to an unused wavelength $\lambda_0 \notin \Lambda$. If we would not like to waste a wavelength for this purpose, we could place an optical switch for blocking the incoming signal when desired, after each wavelength converter. However, we will assume in our presentation of CWC that blocking is done by using the extra wavelength λ_0. The output port of each converter is connected to the star coupler, as illustrated in the example of Figure 4.14.

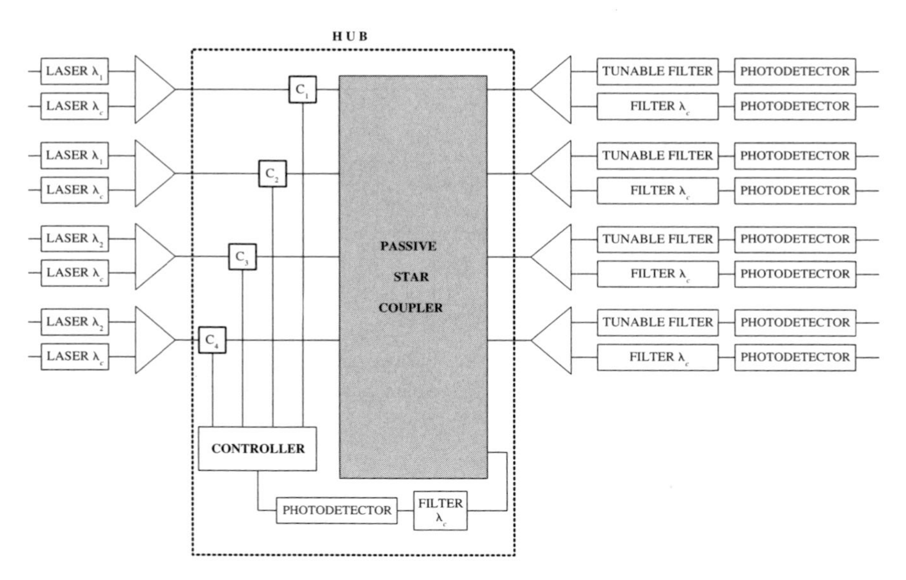

Figure 4.14 An example of a network with four stations and two data wavelengths operating under the CWC protocol [30]. (© 1999 Wiley)

A controller, also placed at the network hub, undertakes the task of determining the outputs of each converter at each time slot. Note that a unique controller is used for all the N available wavelength converters. The decisions made by the controller about the conversions should ensure the following:

1. At most one packet per data wavelength is allowed to pass to the star coupler, so that channel collisions are eliminated.
2. All the passing packets have different destination nodes, so that receiver conflicts are eliminated too.
3. The number of idle slots tends to be minimized.

In addition, the control of the converters must be implemented by a relatively simple algorithm so as to introduce the less possible computational overhead to the controller and not negatively affect the overall implementation and performance of CWC. The algorithm used by the controller in order to meet all the previously described requirements is the following:

```
{                          *** Basic Notations ***
  t     : time slot - variable.
  Xj(t) : set of source nodes that have transmitted a
          packet destined to node uj at time slot t-
          variable.
  λ0    : unused wavelength for blocking as described
          above - constant.
  Λ     : set of available wavelengths - constant.
  G     : set of unmatched source nodes - variable.
  R     : set of unmatched wavelengths - variable.
* The algorithm selects one source node ux ∈ Xj(t) for each *
* destination node uj with Xj(t) ≠ ∅. Then it distributes *
* the selected source nodes to the available wavelengths and *
* tunes each converter to the corresponding wavelength *
PROCEDURE WAVELENGTH_CONVERSION;
BEGIN
      G := 0;
      FOR i := 1 TO N DO Ci := λ0;
      FOR j  := 1 TO N DO IF Xj(t) ≠ ∅ THEN
      BEGIN
            Select randomly a source node ux ∈ Xj(t);
            G := G ∪ {ux};
      END;
      R := Λ;
      WHILE G ≠ ∅ AND R ≠ ∅ DO
      BEGIN
            Select randomly a source node uy ∈ G;
            Select randomly a wavelength λz ∈ R;
            G := G − {uy};
            R := R − {λz};
            Set Cy := λz;
      END;
END;
```

As is shown in Figure 4.14, one of the output ports of the star coupler is connected to a fixed filter that is tuned to always pass the control wavelength. The output of the filter is connected to a photodetector, which performs the necessary opto-electronic conversion before the signal reaches the controller. In this way, the controller receives a control packet containing information about the source and destination nodes of the transmissions that were announced by all stations $D/2$ slots before. Note that one control slot is assumed to include N minislots that are accessed in a TDM manner by all stations and also is taken to be equal to one data slot in length (thus, a control packet includes N transmission announcements and is equal in length to a data packet). Since the controller is aware of the source nodes, it can directly derive the wavelength where each packet is transmitted, because the allocation of wavelengths to source nodes is known. Hence, the controller has all the required information, in order to run the wavelength conversion algorithm presented above.

It should be pointed out that the same algorithm is executed by all stations when the control information arrives at them, i.e. after a round-trip propagation delay of D slots. By executing the algorithm and figuring out how the wavelength converters were tuned, the stations that had transmitted a packet should know whether they should reschedule their packets for transmission (in the event of a blocked transmission) or just delete them from the buffers that keep the transmitted but unacknowledged packets (in the event of a successful transmission). Furthermore, the destination nodes realize how they should tune their filters for reception.

Another observation that must be made is that the wavelength conversion algorithm uses random numbers. Since it is assumed to run at all stations and at the controller of the hub as well, it is understood that in order to give the same results as needed, it is necessary to ensure that the same random numbers are used each time by every station and the by the controller. This problem was present in the CPF protocol as well. For details on two satisfactory solutions of this problem, please refer to [19].

4.4.3.3 Overview of Analytical and Simulation Results for CWC

The performance of the proposed CWC protocol is analyzed and a noteworthy result is obtained [35]; to be exact, the CWC protocol achieves *higher throughput than any protocol which does not use wavelength conversion*. For details on how the previous statement can be proved, please refer to [35].

The CWC protocol was simulated and compared to the TDM and the CPF protocols. Both protocols are channel collision-free, while the CPF is also receiver conflict-free, as we described in section 4.4.2. A performance comparison with these protocols, especially with the CPF one which has been shown to lead to a high network performance, would be very helpful for realizing the improvement introduced by CWC.

Selectively, we shall present the simulation results for the following two network configurations from [30, 35]:

1. Network N_1: $N = 100$, $W = 20$, $D = 0$ and $Q = 5$.
2. Network N_2: $N = 160$, $W = 20$, $D = 5$ and $Q = 10$.

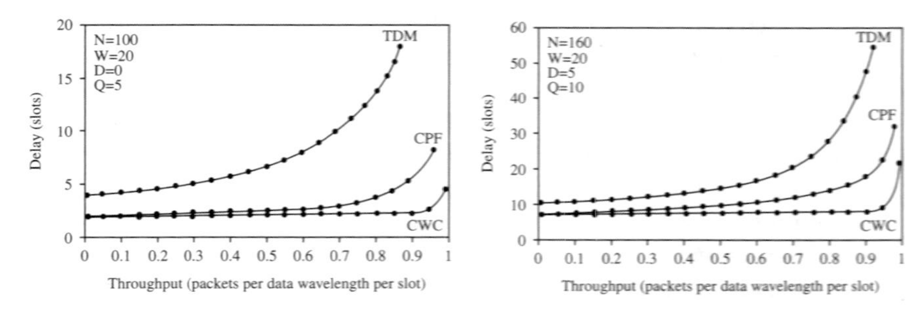

Figure 4.15 Delay versus throughput characteristic of TDM, CPF and CWC for two example network configurations [30]. (a) Delay versus throughput of the three protocols for network N_1. (b) Delay versus throughput of the protocols for network N_2. (© 1999 Wiley)

N and W are the number of stations and wavelengths respectively. The round-trip propagation delay is denoted by D, while the maximum length of each station's queue is denoted by Q.

The protocols are simulated under both bursty and non-bursty traffic. The offered traffic is modelled in a way similar to the one used in [14]. More details can be found in [35].

The performance metric used is the delay versus throughput characteristic. Parts (a) and (b) of Figure 4.15 illustrate the delay versus throughput characteristic of the three compared protocols for networks N_1 and N_2 respectively. The delay versus throughput characteristic was constructed by using 21 points in all graphs. Each point corresponds to a different value of the offered load, from 0.005 up to 1.00 packets per wavelength per slot. Additional simulation results along with the corresponding graphs for more network configurations may be found in [30, 35].

By looking at the graphs of Figure 4.15, we observe that the performance of CWC is very high and remains at about the same level. CPF outperforms TDM but not CWC, because it lacks the virtue of wavelength conversion. Note that some additional results from [30] for these two and other network configurations show that both the TDM and CPF protocols are negatively affected by the presence of bursty traffic. However, even when the offered traffic is assumed to be non-bursty, the CWC protocol achieves a considerably higher performance than the other two schemes. This is explained by the fact that CWC, apart from being channel collision and receiver conflict-free, also tends to minimize the number of idle slots by using the centralized wavelength conversion mechanism in a way that balances the offered load between available wavelengths.

4.4.3.4 Concluding Remarks about CWC

The CWC protocol is a new powerful MAC protocol for WDM star networks of the broadcast-and-select architecture. As we saw, its operation is based on an array of tunable wavelength converters, which are placed at the network hub. This centralized wavelength conversion mechanism balances the offered load between available wavelengths, so that the number of idle slots tends to be minimized. CWC also controls the passing of packets to the star coupler in a way that makes it both channel collision and receiver conflict-free, i.e. it always allows at most one packet per wavelength pass to the star coupler and, also, the selected destination for each packet is always unique. It is applied on the same CC-FT^2-FRTR system assumed by DT-WDMA and RCALA, but CWC has the additional virtue of allowing the sharing of each wavelength by two or more stations (like CPF).

It was shown that the performance of CWC is very high under both bursty and non-bursty traffic conditions. The only additional hardware needed is the centralized wavelength conversion mechanism that has to be placed at the network hub. Therefore, a DT-WDMA network can be transformed to a CWC network quite easily. The wavelength converters may be arrayed in a common device and this limits (to a certain extent) the additional implementation cost, as discussed in the section about HARP protocol. The main disadvantage of CPF is the common one for centralized schemes, i.e. the fact that a failure in a single point (the hub) would affect the operation of all stations.

4.4.4 OCON

We will close the discussion about centralized packet filtering protocols with a recent scheme for WDM star networks of the broadcast-and-select architecture, which is quite innovative and different from the others, as far as implementation technology is concerned. The proposed *Optically Control Optical Network (OCON)* protocol [27–28], as the name suggests, is a powerful scheme that is exclusively based on optical technology.

The centralized filtering protocols described so far indeed achieve a high network throughput as compared to other well-known contention-oriented (like slotted ALOHA) or round-robin (like R-TDMA) schemes, owing to the filtering mechanism applied in each case. Yet, there is still scope for additional improvements, if we consider for example the limits imposed on the performance by the (typically) quite slow tuning times of filters and the extensive use of electronics in a single point (hub), which implies an overall questionable reliability of the system.

The OCON architecture and protocol suggest some noteworthy improvements in both the performance and reliability of the network. It is based on optical technology, thus overall we have an all-optical centralized packet filtering scheme. The passing of packets to the star coupler is controlled by means of *optical logic circuits* (*OLCs*) described later, which allow at most one packet to pass to the star coupler at each time slot. The need for electronic processing of network feedback and for optoelectronic (O/E) translation, found in other schemes, is therefore eliminated. As a consequence, the proposed protocol accomplishes a notable improvement of network performance and, furthermore, increases the reliability of the system. Next, we will focus on the network model of OCON before describing it in more detail.

4.4.4.1 Network Model of OCON

The network model considered by OCON is generally described by the network of Figure 4.2. Of course as we mentioned in the introduction, the hub does not comprise only a passive star coupler, but it is embellished with an all-optical packet filtering mechanism.

The protocol assumes tunability in the part of the transmitters (lasers) of stations. The set of stations can be defined as $U = \{u_1, u_2, \ldots, u_N\}$, while the set of wavelengths is defined as $\Lambda = \{\lambda_1, \lambda_2, \ldots, \lambda_W\}$; thus N and W are the number of stations and wavelengths respectively. The lasers are assumed to be capable of tuning to any one of the available wavelengths, so that full connectivity is ensured for the network. Another difference in comparison with the network of Figure 4.2 is that each station is equipped with R fixed-tuned filters, which accept distinct wavelengths. Hence, each station can receive up to R packets carried on different wavelengths, at the same time slot, i.e. has R home wavelengths for reception. Note that R is generally assumed to be very small, e.g. up to three, in order to maintain a relative simplicity as far as the hardware implementation is concerned. Consequently, the system may be characterized as TT-FR^{R} according to the relevant discussion of Chapter 1.

It should also be noted that the round-trip propagation delay is assumed to be negligible. This implies that source nodes can be immediately informed about the success or blocking of their transmissions (using one of the methods in [36]) and, accordingly, choose to delete the transmitted packet from their buffers or reschedule it for transmission in the near future. At any time they get ready to transmit, they tune their lasers to one of the R home channels of the desired destination and transmit the packet (with probability one) in the next time slot. If they have two or more waiting packets they choose one in random. Thus, source nodes transmit in a slotted ALOHA fashion (with $p = 1$) with the random transmission strategy.

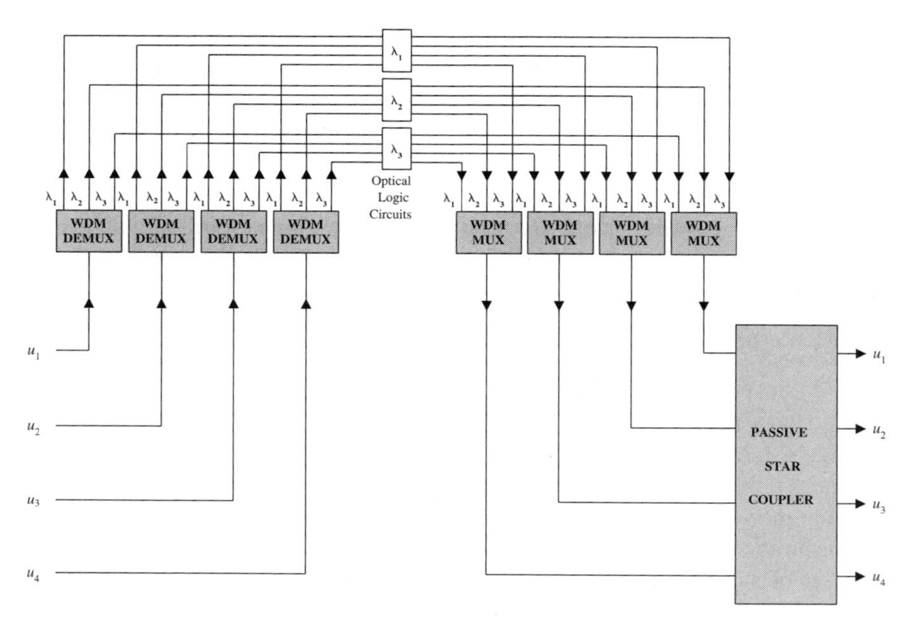

Figure 4.16 Example of the hub of a network with four stations and three wavelengths operating under the OCON protocol [28]. (© 1998 IEEE)

4.4.4.2 Detailed Description of OCON

In this section we focus on two key issues about OCON, i.e. on how the optical filtering mechanism operates and on how it is implemented. During the description of the operation of the packet filtering mechanism, the reader may refer to Figure 4.16, which illustrates the hub of an example network with four stations and three wavelengths operating under the OCON protocol.

The optical filtering mechanism comprises, first of all, an array of N WDM demultiplexers (each one corresponding to a transmitting station), which separate the different wavelengths from the source nodes. At each time slot, only one wavelength is used by a source node; thus, at most one output of each WDM demultiplexer contains an optical signal. The separated wavelengths are driven appropriately to an array of W OLCs, so that the outputs with the same wavelength of all the WDM demultiplexers are connected to a single OLC controlling this wavelength. At each time slot, each OLC allows just one packet, i.e. from only one of its inputs, to pass to the star coupler; therefore collisions on the corresponding wavelength controlled by the OLC are eliminated. All the packets which are possibly carried on other input ports are blocked by the OLC. Finally, an array of N WDM multiplexers combines the optical signals back as shown in Figure 4.16 and drives each combination to the appropriate input of the star coupler.

Next, let us see how an OLC is implemented, which is obviously of critical importance for the OCON protocol. We shall focus on the implementation of OLCs at a logic gate level, while the discussion on the actual implementation of the utilized logic gates with optical technology will be brief and deferred till later.

According to the description above and Figure 4.16 as well, each OLC must have exactly N input and N output ports. Moreover, at each time slot, it should allow only one of the input signals to pass to the corresponding output port. Let us describe the OLC structure that should be used in the example network of Figure 4.16, i.e. an OLC with four input and four output ports.

First, let us consider the simplified version depicted in Figure 4.17. We notice that the OLC has N AND-NOT gates H_k, $(k = 1, \ldots, N)$, where $N = 4$ in our example. These gates undertake the task of blocking or allowing the incoming packets to pass to the corresponding outputs. The OLC has also N control ports C_k, actually the 'NOT' input ports of the AND-NOT gates H_k, N input ports I_k and N output ports O_k $(k = 1, \ldots, N)$. Obviously $O_k = \overline{C_k} I_k$, therefore the AND-NOT gates allow an input to pass to the corresponding output if $\overline{C_k} = 0$. Otherwise, i.e. for $\overline{C_k} = 1$, the input port is blocked. Since only one of the outputs should contain an optical signal (or equivalently be equal to 1), it follows that the control ports must be defined in such a way that for only one passing signal we have output $O_p = \overline{C_p} I_p = 1$, while for all other outputs we have $O_m = \overline{C_m} I_m = 0$, for all $m \neq p$. As it is obvious in the example of the Figure 4.17, in the simplified version of the OLC the control ports C_k are in fact the outputs of an array of N OR gates G_k $(k = 1, \ldots, N)$ that are serially connected; i.e. C_k is the output of OR gate G_k and so on. In the simplified version of OLC it is assumed that $C_1 = 0$, while for $k = 2, \ldots, N$ we have:

$$C_k = C_{k-1} + I_{k-1} = (C_{k-2} + I_{k-2}) + I_{k-1} = \cdots = I_1 + I_2 + \cdots + I_{k-1}. \tag{4.3}$$

Thus:

$$O_k = \overline{C_k} I_k \overset{(4.3)}{\longrightarrow} O_k = \overline{I_1 + I_2 + \cdots + I_{k-1}} I_k = \overline{I_1}\,\overline{I_2} \cdots \overline{I_{k-1}} I_k. \tag{4.4}$$

According to this scheme we get the desirable $O_k = 1$, only when $I_k = 1$ and all the previous inputs are zero, i.e. $I_m = 0$ for all $m < k$. Note also that, since $I_k = 1$, the complement will be zero and according to Equation (4.4) it will give $O_n = 0$, for all $n > k$. Hence, indeed only this output port will have a signal. The scheme above suggests essentially a way to sense the input ports serially, starting always from

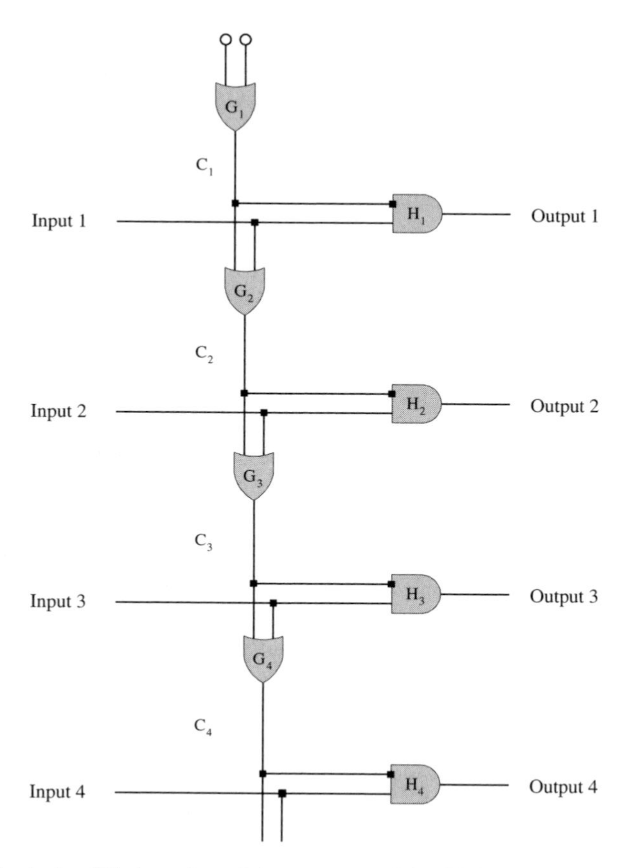

Figure 4.17 A simplified version of the structure of the OLC that should be used in the example network of Figure 4.16 [28]. (© 1998 IEEE)

input port I_1, until we reach an input port equal to one. The corresponding output becomes one, since all previous input ports where equal to zero and, therefore, from Equation (4.3) the corresponding control port is also set to zero, hence giving output a value equal to one, if we consider Equation (4.4). Moreover, according to Equation (4.3), since an input equal to one was encountered, all subsequent control ports will be set to one; thus, from Equation (4.4), all subsequent outputs become equal to zero, as desired.

This simplified version of the OLC has a serious drawback, namely it lacks fairness. Indeed, we mentioned that the serial search for an active input port always

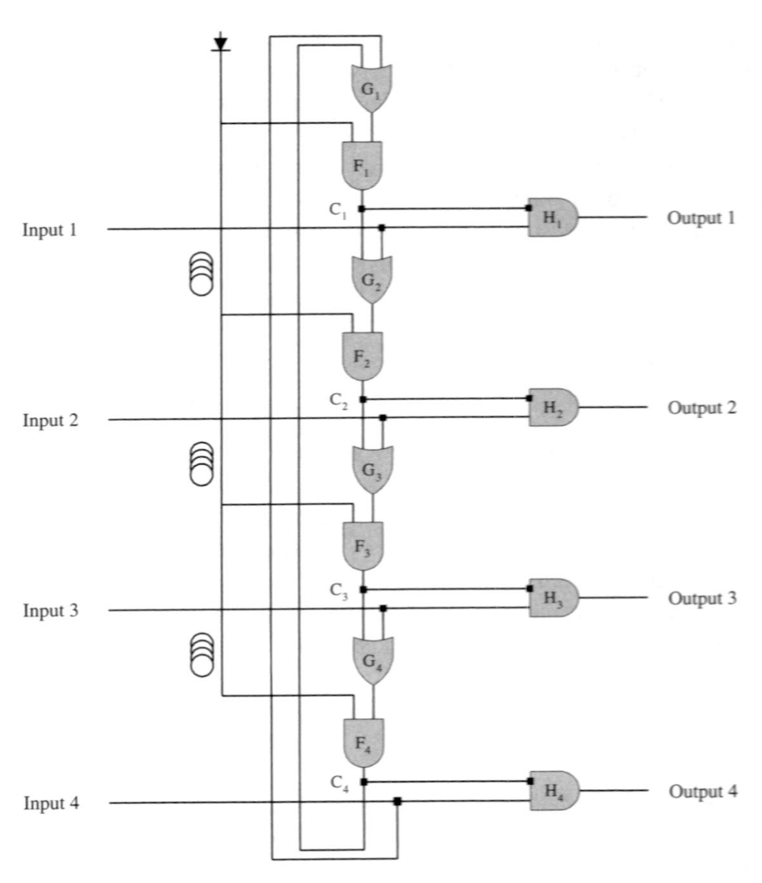

Figure 4.18 The full structure actually used for the OLCs of the example network of Figure 4.17 [28]. (© 1998 IEEE)

begins from port I_1, which obviously is not fair. This limitation can be overcome, if we cyclically connect the OR gates and place an AND-NOT gate $F_k (k = 1, \ldots, N)$ between each pair of consecutive OR gates. For a schematic representation of this please refer to Figure 4.18. In this figure, the laser diode which controls the AND-NOT gates transmits a light pulse every N time slots, while each pulse is delayed for a time slot before reaching the next gate, by means of a delay line (optical buffer). In this way, only one AND-NOT gate $F(t)$ is disabled at each time slot t. Also, the disabled gate is cyclically shifted at each time slot. Thus, if $F(t) = F_m$, then $F(t+1) = F_{(m \bmod N)+1}$. If $F(t) = F_m$, the serial search for an active input port begins from port I_m. We notice that, since F_m is cyclically shifted at each time slot, the OLC does not favour any input port (unlike the simplified version); hence, it becomes fair.

Details about the actual implementation of the described OLC (and thus the OCON protocol) by making use of directional couplers can be found in [27]. It should be noted that this implementation calls for electronic control of some optical devices, thus the overall scheme is all-optical in the sense that the signal remains on optical form throughout the network; however, due to the limits imposed by present-day technology, the control of the OCON centralized filtering mechanism is decided to be electronic.

4.4.4.3 Overview of Simulation Results for OCON

The OCON protocol was simulated and compared to protocols R-TDMA and HARP. The reasoning behind this choice is that both of these schemes are applied to the same TT-FR network model and are collision-free, like OCON.

We shall present simulation results regarding only one performance metric (namely delay versus throughput) of the protocols for no more than two (out of the six overall considered in [27–28]) network configurations:

1. Network N_1: $N = 40$, $W = 20$, $Q = 5$ and $R = 2$.
2. Network N_2: $N = 60$, $W = 20$, $Q = 10$ and $R = 3$.

Parameters N, W and Q are the number of stations, wavelengths and packets that can fit in each station's queue, respectively. Parameter R stands for the number of fixed-tuned receivers available at each station. Note also that the tuning time of the acousto-optic filters for HARP was given two values, i.e. $T = 0$ and $T = 0.2$ slots. Separate results were obtained for each value of the tuning time. The performance characteristic was constructed using 21 points, with each point corresponding to a value for the offered load ranging from 0.005 up to 1.00 packets per wavelength per slot.

The delay versus throughput characteristic of the simulated protocols for networks N_1 andN_2 are illustrated in parts (a) and (b) of Figure 4.19 respectively. From the graphs shown here (as well as the additional ones in [27]), it becomes obvious that the OCON protocol achieves a significantly higher performance than R-TDMA and HARP protocols under *any load conditions*. The performance improvement as compared to HARP becomes even higher as the tuning time for the acousto-optic filters of HARP increases (i.e. equal to 0.2 slots in the figure).

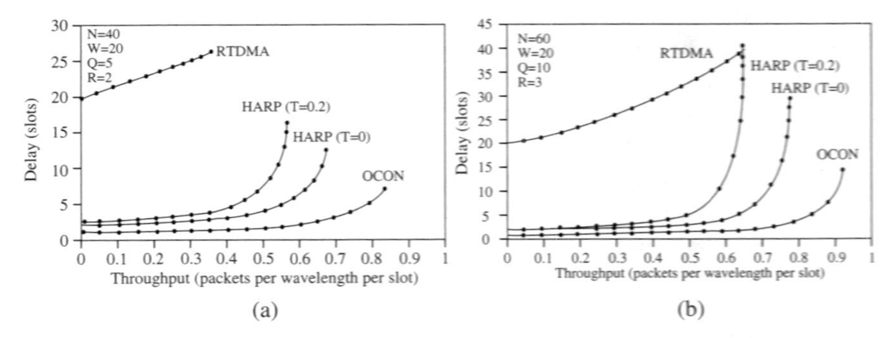

Figure 4.19 Delay versus throughput characteristic of R-TDMA, HARP and OCON for two example network configurations [28]. (a) Delay versus throughput characteristic of the protocols for network N_1. (b) Delay versus throughput characteristic of the protocols for network N_2. (© 1998 IEEE)

4.4.4.4 Concluding Remarks about OCON

The proposed OCON protocol was shown to be a quite innovative scheme with important differences as compared to the other centralized packet filtering protocols described earlier. As we saw, the most crucial innovation is the introduction of optical technology at the network hub for the implementation of the packet filtering mechanism. This mechanism was shown to be based on optical logic circuits, the design of which (at a logic gate level) was presented extensively. It was explained how the design of the filtering mechanism ensures that only one packet per wavelength is allowed to pass to the star coupler each time. The latter implies that OCON has the virtue of eliminating collisions. Furthermore, OCON protocol significantly improves network performance owing to the absence of slowly tunable optical filters (that were included in the previous schemes). Finally, we should point out that OCON has another very important advantage: it minimizes the use of electronics at the network hub and therefore its application results in a significant increase of the overall system's reliability, as compared to the previous centralized packet filtering schemes.

NOTES

1 We assume synchronous protocols where data (and any control, if present) channels are time-slotted.

2 The neighbour allocation, for which $i_k = \left\lfloor \frac{k-1}{\lceil N/W \rceil} \right\rfloor + 1$, is also proposed in [1].

3 That is why the same set of stations, u_1 through u_N, appear both in the left and right side of the hub in Figure 4.2.
4 The algorithm is quite similar to the one in [7]. The latter differs only in that it selects stations randomly and not by considering any choice probabilities, as is the case for SALP.
5 However, even under these queuing conditions, we shall shortly see that SALP still outperforms both of these schemes by far.
6 As we saw the results (decisions) of the learning automata are exactly the same in every station.
7 Refer to the relevant discussion in the introduction to SALP and also in [11] for example.
8 The absence of a collision is enough information for the learning automata to know that they should increase the transmission probability for the next time slot; i.e. the learning automata do not care to additionally know whether the transmission was actually successful or the wavelength was idle, as is obvious from the probability updating scheme.
9 However, it is noted in [24] that the LABRA protocol can be applied to networks with non-negligible propagation delays by making use of pipelining, see e.g. [3, 24].
10 Note that in [24] an additional network configuration is considered and thus the delay versus throughput and throughput versus offered load characteristics of the protocols are presented for four networks in total. The results, however, do not differ as they are always in favour of the LABRA protocol.
11 The learning algorithm that determines the passing probabilities for each wavelength will be presented later in this subsection.
12 Note that in [26] three additional network configurations are considered and thus the delay versus throughput and throughput versus offered load characteristics of the protocols are presented for four networks in total. The results, however, do not differ as they are always in favour of the LABON protocol.
13 The exact network model of the RCALA protocol will be presented in the following subsection. Note that tunability is assumed at the opposite side (i.e. receivers) unlike all the adaptive protocols examined so far, which considered a TT-FR network model.
14 Network N_1 will be described in more detail later, when we present an overview of the simulation results for RCALA from [23].
15 Actually, we have already described one protocol that uses centralized packet filtering, namely the LABON protocol, in Section 4.2. However, due to its purely random access nature and its many similarities with the adaptive random

access LABRA protocol, LABON was presented in Section 4.2 about adaptive random access schemes, right after LABRA.

16 We mentioned in the introduction that in this case the choice is made among all stations randomly.

17 For the three-port operation of the PIATOFs considered in the context of HARP, as shown for example in Figure 4.11, the reader may also refer to [9].

18 The network model of CPF will be presented in detail in the next subsection.

REFERENCES

[1] Bogineli K, Sivalingam KM and Dowd PW, "Low complexity multiple access protocols for wavelength division multiplexed photonic networks", *IEEE Journal on Selected Areas in Communications*, vol. 11, pp. 590–604, 1993.

[2] Cao XR, "The maximum throughput of a nonblocking space-division packet switch with correlated destinations", *IEEE Transactions on Communications*, vol. 43, no. 5, pp. 1898–1901, 1995.

[3] Chen MS, Dono NR and Ramaswami R, "A media access protocol for packet switched wavelength division multiaccess metropolitan networks", *IEEE Journal on Selected Areas in Communications*, vol. 8, no. 6, pp. 1048–1057, 1990.

[4] Cheung KW, "Acoustooptic tunable filters in narrowband WDM networks: system issues and network applications", *IEEE Journal on Selected Areas in Communications*, vol. 8, no. 6, pp. 1015–1025, August 1990.

[5] Chlamtac I and Fumagalli A, "QUADRO-Star: high performance optical WDM star networks", *IEEE Transactions on Communications*, vol. 42, no. 8, pp. 2582–2590, August 1994.

[6] Danielsen SL, Joergensen C, Mikkelsen B and Stubkjaer KE, "Analysis of a WDM packet switch with improved performance under bursty traffic conditions due to tunable wavelength converters", *IEEE Journal of Lightwave Technology*, vol. 16, pp. 729–735, May 1998.

[7] Ganz A and Koren Z, "Performance and design evaluation of WDM stars", *IEEE Journal of Lightwave Technology*, vol. 11, no. 2, pp. 358–366, 1993.

[8] Green PE, *Fiber Optic Networks*, Prentice Hall, Englewood Cliffs, NJ, 1993.

[9] Irshid MI and Kavehrad M, "A fully transparent fibre-optic ring architecture for WDM networks", *IEEE Journal of Lightwave Technology*, vol. 10, no. 1, pp. 101–108, 1992.

[10] Jain R and Routhier S, "Packet trains – measurement and a new model for computer network traffic", *IEEE Journal on Selected Areas in Communications*, vol. SAC-4, no. 6, pp. 986–995, 1986.

[11] Kung HT, "Gigabit local area networks: a system's perspective", *IEEE Communications Magazine*, vol. 30, no. 4, pp. 79–89, 1992.

[12] McKeown N, "The iSLIP scheduling algorithm for input-queued switches", *IEEE/ACM Transactions on Networking*, vol. 7, no. 7, pp. 188–201, 1999.

[13] McKeown N and Anderson TE, "A quantitative comparison of iterative scheduling algorithms for input-queued switches", *Elsevier Computer Networks and ISDN Systems*, vol. 30, no. 24, pp. 2309–2326, 1998.

[14] McKinnon MW, Rouskas GN and Perros HG, "Performance analysis of a photonic single-hop ATM switch architecture with tunable transmitters and fixed frequency receivers", *Performance Evaluation Journal*, vol. 33, no. 5, pp. 113–136, June 1998.

[15] Narendra KS and Thathachar MAL, "Learning automata - a survey", *IEEE Transactions in Systems, Man and Cybernetics*, vol. SMC-4, no. 8, pp. 323–334, July 1974.

[16] Narendra KS and Thathachar MAL, "On the behavior of a learning automaton in a changing environment with application to telephone traffic routing", *IEEE Transactions on Systems, Man and Cybernetics*, vol. SMC-10, no. 5, pp. 262–269, 1980.

[17] Oommen BJ, "Absorbing and ergosic discretized two-action learning automata", *IEEE Transactions on Systems, Man and Cybernetics*, vol. SMC-16, pp. 282–293, March/April 1986.

[18] Papadimitriou GI, "Centralized packet filtering protocols: a new class of high-performance protocols for single-hop lightwave WDM networks", in *Proceedings IEEE IPCCC'99*, Phoenix/Scotsdale, Arizona, pp. 419–425, February 1999.

[19] Papadimitriou GI, "Centralized packet filtering protocols: a new family of MAC protocols for WDM star networks", *Elsevier Computer Communications*, vol. 22, pp. 11–19, 1999.

[20] Papadimitriou GI and Maritsas DG, "HARP: a hybrid random access and reservation protocol for WDM passive star networks", in *Proceedings IEEE Globecom'94*, San Francisco, USA, pp. 1521–1527, November 1994.

[21] Papadimitriou GI and Maritsas DG, "LABON: a learning automata-based optical network", in *Proceedings IEEE ICC'94*, New Orleans, USA, pp. 1352–1358, May 1994.

[22] Papadimitriou GI and Maritsas DG, "Learning automata-based random access protocols for WDM passive star networks", in *Proceedings IEEE Globecom'93*, Houston, Texas, vol. 2, pp. 1169–1175, November/December 1993.

[23] Papadimitriou GI and Maritsas DG, "Learning automata-based receiver conflict avoidance algorithms for WDM broadcast-and-select star networks",

IEEE/ACM Transactions on Networking, vol. 4, no. 3, pp. 407–412, June 1996.

[24] Papadimitriou GI and Maristas DG, "Self-adaptive random-access protocols for WDM passive star networks", *IEEE Proceedings of Computers and Digital Technology*, vol. 142, no. 4, pp. 306–312, July 1995.

[25] Papadimitriou GI and Maritsas DG, "WDM star networks: hybrid random access and reservation protocols with high throughput and low delay", *Elsevier Computer Networks and ISDN Systems*, vol. 28, pp. 773–787, 1996.

[26] Papadimitriou GI and Maritsas DG, "WDM passive star networks: a learning automata-based architecture", *Elsevier Computer Communications*, vol. 19, pp. 580–589, 1996.

[27] Papadimitriou GI, Miliou AN and Pomportsis AS, "OCON: an optically controlled optical network", *Elsevier Computer Communications*, vol. 22, pp. 811–824, 1999.

[28] Papadimitriou GI, Miliou AN and Pomportsis AS, "Optical logic circuits: A new approach to the control of fibre optic LANs", in *Proceedings IEEE 23rd Annual Conference on Local Computer Networks* (*LCN'98*), Boston, Massachusetts, pp. 326–335, October 1998.

[29] Papadimitriou GI, Obaidat MS and Pomportsis AS, "Adaptive protocols for optical LANs with bursty and correlated traffic", *Wiley International Journal of Communication Systems*, vol. 15, pp. 115–125, December 2001.

[30] Papadimitriou GI and Pomportsis AS, "A class of centralized high-performance protocols for single-hop lightwave networks", *Wiley International Journal of Communication Systems*, vol. 12, pp. 363–374, 1999.

[31] Papadimitriou GI and Pomportsis AS, "Centralized wavelength conversion protocols for WDM broadcast-and-select star networks", in *Proceedings IEEE International Conference on Networks* (*ICON'99*), Brisbane, Australia, pp. 11–18, September/October 1999.

[32] Papadimitriou GI and Pomportsis AS, "Learning-automata-based MAC protocols for photonic LANs", in *Proceedings IEEE International Conference on Networks* (*ICON'00*), Singapore, pp. 481, September 2000.

[33] Papadimitriou GI and Pomportsis AS, "On the use of learning automata in medium access control of single-hop lightwave networks", *Elsevier Computer Communications*, vol. 23, pp. 783–792, 2000.

[34] Papadimitriou GI and Pomportsis AS, "Self-adaptive TDMA protocols for WDM star networks: A learning-automata-based approach", *IEEE Photonics Technology Letters*, vol. 11, no. 10, pp. 1322–1324, October 1999.

[35] Papadimitriou GI and Pomportsis AS, "Wavelength-conversion-based protocols for single-hop photonic networks with bursty traffic", *Kluwer Photonic Network Communications*, vol. 1, no. 4, pp. 263–271, 1999.

[36] Sivalingam KM, Bogineli K and Dowd PW, "Pre-allocation media access control protocols for multiple access WDM photonic networks", *ACM SIGCOMM*, Baltimore, MD, pp. 235–246, August 1992.

[37] Thathachar MAL and Harita BR, "Learning automata with changing number of actions", *IEEE Transactions on Systems, Man and Cybernetics*, vol. SMC-17, no. 6, pp. 1095–2000, November/December 1987.

[38] Unsal C, "Intelligent navigation of autonomous vehicles in an automated highway system: Learning methods and interacting vehicles approach", PhD dissertation Electrical Engineering Department, Virginia Polytechnic Institute and State University, January 1997.

INDEX